Electrical Engineering

Electrical Engineering

Edited by
John Fenmore

www.willfordpress.com

Published by Willford Press,
118-35 Queens Blvd., Suite 400,
Forest Hills, NY 11375, USA

ISBN: 978-1-68285-753-3

Cataloging-in-Publication Data

Electrical engineering / edited by John Fenmore.
p. cm.
Includes bibliographical references and index.
ISBN 978-1-68285-753-3
1. Electrical engineering. 2. Engineering. I. Fenmore, John.
TK145 .E44 2019
621.3--dc23

For information on all Willford Press publications
visit our website at www.willfordpress.com

Contents

Preface

This book has been a concerted effort by a group of academicians, researchers and scientists, who have contributed their research works for the realization of the book. This book has materialized in the wake of emerging advancements and innovations in this field. Therefore, the need of the hour was to compile all the required researches and disseminate the knowledge to a broad spectrum of people comprising of students, researchers and specialists of the field.

The branch of engineering that is concerned with the study and use of electronics, electromagnetism and electricity is known as electrical engineering. It is subdivided into a number of significant subfields, which include power engineering, control engineering, telecommunications, robotics, signal processing, microelectronics, etc. Electrical engineering has been instrumental in the development of various innovative technologies, such as electrical power generation, GPS technology, etc. The lighting and wiring of buildings, the operation of power stations, the electrical control of industrial machinery, the design of telecommunication systems and the design of household appliances are all within the scope of this field. Modern electrical engineering uses computer-aided design for the design of electrical systems. A wide range of instruments and tools are used in electrical engineering, such as multimeter, oscilloscope, network and spectrum analyzers, etc. This book covers in detail some existing theories and innovative concepts revolving around electrical engineering. There has been rapid progress in this field and its applications are finding their way across multiple industries. Students, researchers, experts and all associated with this field will benefit alike from this book.

At the end of the preface, I would like to thank the authors for their brilliant chapters and the publisher for guiding us all-through the making of the book till its final stage. Also, I would like to thank my family for providing the support and encouragement throughout my academic career and research projects.

Editor

1

Analysis and Design of Solar Photo Voltaic Grid Connected Inverter

M. Satyanarayana, P. Satish Kumar
Department of Electrical Engineering, University college of Engineering,
Osmania University (Autonomous), Hyderabad, Telangana, India
e-mail: mail2satyanarayana.m@gmail.com

Abstract

This paper presents common mode voltage analysis of single phase grid connected photovoltaic inverter. Many researchers proposed different grid tie inverters for applications like domestic powering, street lighting, water pumping, cooling and heating applications, however traditional grid tie PV inverter uses either a line frequency or a high frequency transformer between the inverter and grid but losses will increase in the network leading to reduced efficiency of the system. In order to increase the efficiency, with reduced size and cost of the system, the effective solution is to remove the isolation transformer. But common mode (CM) ground leakage current due to parasitic capacitance between the PV panels and the ground making the system unreliable. The common mode current reduces the efficiency of power conversion stage, affects the quality of grid current, deteriorate the electric magnetic compatibility and give rise to the safety threats. In order to eliminate the common mode leakage current in Transformerless PV systm two control algorithms of multi-carrier pwm are implemented and compared for performance analysis.The shoot-through issue that is encountered by traditional voltage source inverter is analyzed for enhanced system reliability. These control algorithms are compared for common mode voltage and THD comparisons. The proposed system is designed using MATLAB/SIMULINK software for analysis.

***Keywords**: common mode leakage current, transformerless grid connected PV inverter, SPWM, PD*

1. Introduction

Grid tie photovoltaic (PV) systems, particularly low-power single-phase systems up to 5 kW, are becoming more important worldwide. They are usually private systems where the owner tries to get the maximum system profitability. Issues such as reliability, high efficiency, small size and weight, and low price are of great importance to the conversion stage of the PV system [1]–[3]. Quite often, these grid-connected PV systems include a line transformer in the power-conversion stage, which guarantees galvanic isolation between the grid and the PV system, thus providing personal protection. Furthermore, it strongly reduces the leakage currents between the PV system and the ground, ensures that no continuous current is injected into the grid, and can be used to increase the inverter output voltage level [1],[2],[4]. The line transformer makes possible the use of a full-bridge inverter with unipolar pulse width modulation (PWM). This inverter is simple and it requires only four insulated gate bipolar transistors (IGBTs) and has a good trade-off between efficiency, complexity and price [5].

Due to its low frequency, the line transformer is large, heavy and expensive. Technological evolution has made possible the implementation, within the inverters, of both ground-fault detection systems and solutions to avoid injecting dc current into the grid. The transformer can then be eliminated without impacting system characteristics related to personal safety and grid integration [1],[4],[6]–[8]. In addition, the use of a string of PV modules allows maximum power point (MPP) voltages large enough to avoid boosting voltages in the conversion stage. This conversion stage can then consist of a simple buck inverter, with no need of a transformer or boost dc–dc converter, and it is simpler and more efficient. But if no boost dc–dc converter is used, the power fluctuation causes a voltage ripple in the PV side at double the line frequency. This in turn causes a small reduction in the average power generated by the PV arrays due to the variations around the MPP. In order to limit the reduction, a larger input capacitor must be used. Typical values of 2 mF for this capacitor limit the reduction in the MPPT efficiency to 1% in a 5-KW PV system [8]. However, when no transformer is used, a galvanic connection between the grid and the PV array exists. Dangerous leakage currents (common-mode currents) can flow through the large stray capacitance between the PV array and the ground if the inverter generates a varying common-mode voltage [1]-[4].

Recently, several transformerless inverter topologies have been presented that use super junction MOSFETs devices as main switches to avoid the fixed voltage-drop and the tail-current induced turn-off losses of IGBTs to achieve ultra high efficiency. However, this topology has high conduction losses due to the fact that the current must conduct through three switches in series during the active phase. Another disadvantage of the H5 is that the line-frequency switches S1 and S2 cannot utilize MOSFET devices because of the MOSFET body diode's slow reverse recovery. The slow reverse recovery of the MOSFET body diode can induce large turn-on losses, has a higher possibility of damage to the devices and leads to EMI problems. Shoot-through issues associated with traditional full bridge PWM inverters remain in the H5 topology due to the fact that the three active switches are series-connected to the dc bus Replacing the switch S5 of the H5 inverter with two split switches S5 and S6 into two phase legs and adding two freewheeling diodes D5 and D6 for freewheeling current flows, the H6 topology was proposed [9]-[10].

The H6 inverter can be implemented using MOSFETs for the line frequency switching devices, eliminating the use of less efficient IGBTs. The fixed voltage conduction losses of the IGBTs used in the H5 inverter are avoided in the H6 inverter topology improving efficiency; however, there are higher conduction losses due to the three series-connected switches in the current path during active phases. The shoot-through issues due to three active switches series connected to the dc-bus still remain in the H6 topology. Another disadvantage to the H6 inverter is that when the inverter output voltage and current has a phase shift the MOSFET body diodes may be activated. This can cause body diode reverse-recovery issues and decrease the reliability of the system. The adjustment to improve the system reliability comes at the cost of high zero-crossing distortion for the output grid current one key issue for a high efficiency and reliability transformerless PV inverter is that in order to achieve high efficiency over a wide load range it is necessary to utilize MOSFETs for all switching devices. Another key issue is that the inverter should not have any shoot-through issues for higher reliability [11]-[13]. The Figure 1 shows the improvements to traditional systems for common mode voltage analysis.

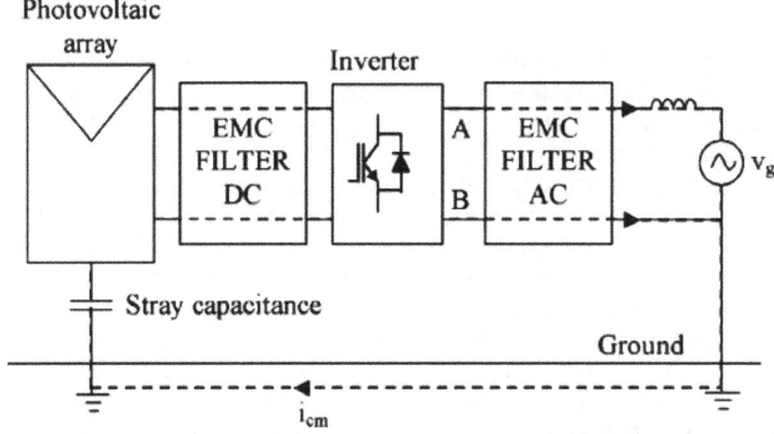

Figure 1. Common-mode currents in a transformerless conversion stage

In order to address these two key issues, a new inverter topology is proposed for single-phase transformerless PV grid-connected systems in this paper. The proposed converter utilizes two split ac-coupled inductors that operate separately for positive and negative half grid cycles. This eliminates the shoot-through issue that is encountered by traditional voltage source inverters, leading to enhanced system reliability. Dead time is not required at both the high-frequency pulse width modulation switching commutation and the grid zero crossing instants, improving the quality of the output ac-current and increasing the converter efficiency.

This paper is organized as section I is about the literature survey on transformerless PV inverter, sections II is presented about proposed topology with Sine PWM its principle of operation, section III is about common voltage analysis of proposed system, section IV matlab implementation of the proposed system with sine PWM and Phase Disposition technique. Comparison of two techniques for THD of output voltages with reduced leakage current is shown.

2. The Proposed Topology and Operational Analysis

The proposed topology is shown in Figure 2. Circuit diagram of the proposed transformerless PV inverter, which is composed of six MOSFETs switches (S1–S6), six diodes (D1–D6), and two split ac-coupled inductors *L*1 and *L*2. The diodesD1–D4 performs voltage clamping functions for active switches S1–S4. The ac-side switch pairs are composed of S5, D5 and S6, D6, respectively, which provide unidirectional current flow branches during the freewheeling phases decoupling the grid from the PV array and minimizing the CM leakage current.

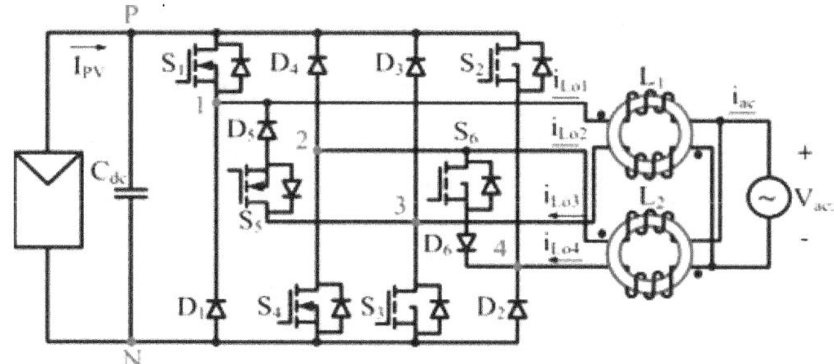

Figure 2. Proposed high efficiency and reliability PV transform less inverter topology.

Compared to the HERIC topology the proposed inverter topology divides the ac side into two independent units for positive and negative half cycle. In addition to the high efficiency and low leakage current features, the proposed transformerless inverter avoids shoot-through enhancing the reliability of the inverter. The inherent structure of the proposed inverter does not lead itself to the reverse recovery issues for the main power switches and as such super junction MOSFETs can be utilized without any reliability or efficiency Penalties.

Figure 3 illustrates the PWM scheme for the proposed inverter. When the reference signal *V*control is higher than zero, MOSFETs S1 and S3 are switched simultaneously in the PWM mode and S5 is kept on as a polarity selection switch in the half grid cycle; the gating signals G2, G4, and G6 are low and S2, S4, and S6 are inactive. Similarly, if the reference signal –*V*control is higher than zero, MOSFETs S2 and S4 are switched simultaneously in the PWM mode and S6 is on as a polarity selection switch in the grid cycle; the gating signals G1, G3, and G5 are low and S1, S3, and S5 are inactive.

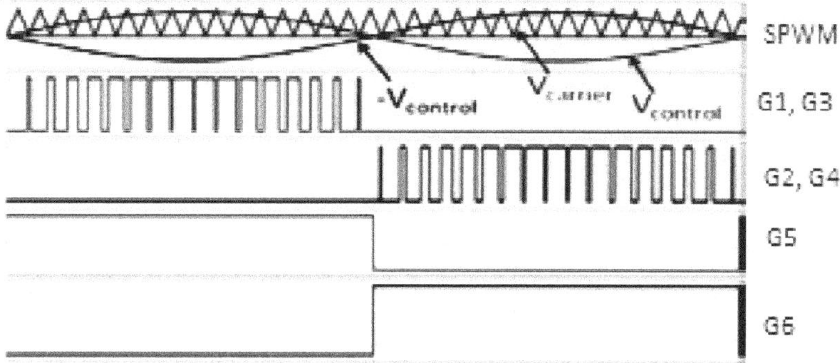

Figure 3. Phase disposition PWM signal used to control the system

Table 1. Switching states and respective common mode voltages

S_1	S_2	S_3	S_4	S_5	S_6	V_{cm}	Sequence
pwm	off	off	pwm	on	off	$U_{dc}/2$	positive
off	off	off	off	off	off	$U_{dc}/2$	
off	pwm	pwm	off	of	on	$U_{dc}/2$	negative
off	off	off	off	off	off	$U_{dc}/2$	

Figure 4 shows the four operation stages of the proposed inverter within one grid cycle. In the positive half-line grid cycle, the high-frequency switches S1 and S3 are modulated by the sinusoidal reference signal $V_{control}$ while S5 remains turned ON.

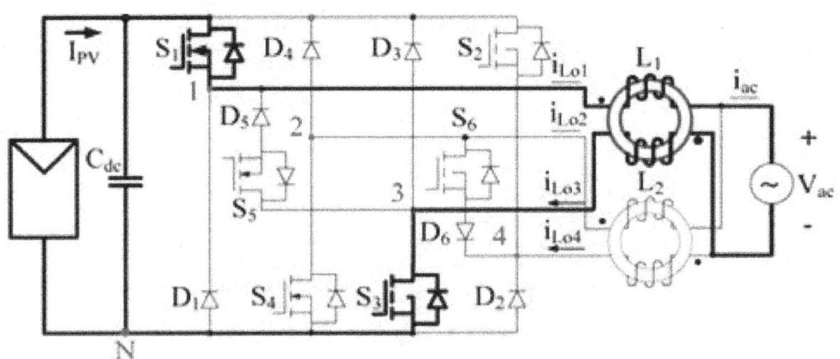

Figure 4. Active stage of positive half-line cycle

When S1 and S3 are ON, diode D5 in Figure 5 is reverse-biased, the inductor currents of $iLo1$ and $iLo3$ are equally charged, and energy is transferred from the dc source to the grid.

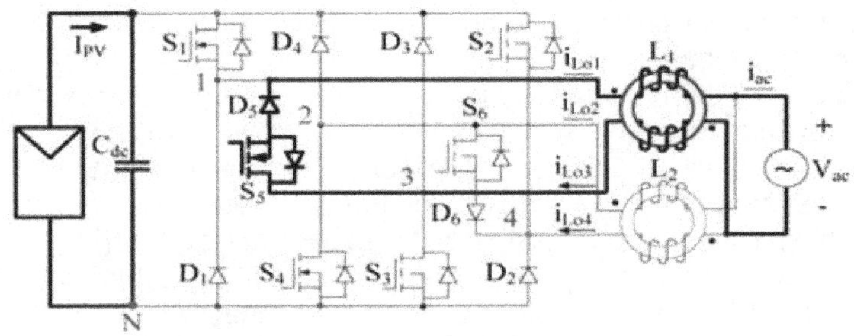

Figure 5. Freewheeling stage of positive half-line cycle

When S1 and S3 are deactivated, the switch S5 and diode D5 in Figure 6 provide the inductor current $iL1$ and $iL3$ a freewheeling path decoupling the PV panel from the grid to avoid the CM leakage current. Coupled-inductor $L2$ is inactive in the positive half-line grid cycle.

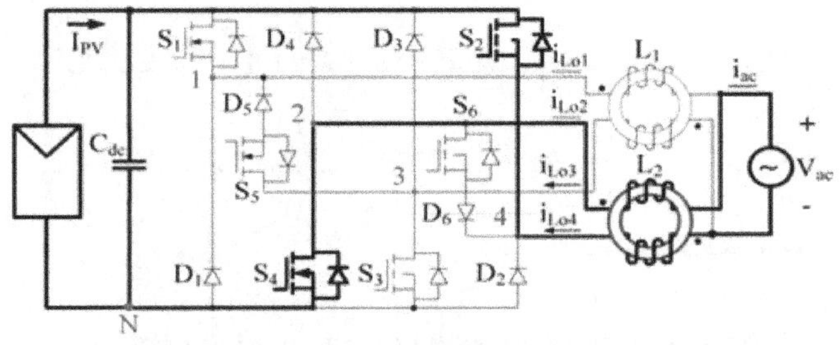

Figure 6. Active stage of negative half-line cycle

Similarly, in the negative half cycle, S2 and S4 in Figure 7 are switched at high frequency and S6 remains ON. Freewheeling occurs through S6 and D6. When S2 and S4 are ON, diode D6 is reverse-biased, the inductor currents of *i*Lo2 and *i*Lo4 are equally charged, and energy is transferred from the dc source to the grid; when S2 and S4 are deactivated, the switch S6 and diode D6 provide the inductor current *i*L2 and *i*L4 a freewheeling path decoupling the PV panel from the grid to avoid the CM leakage current.

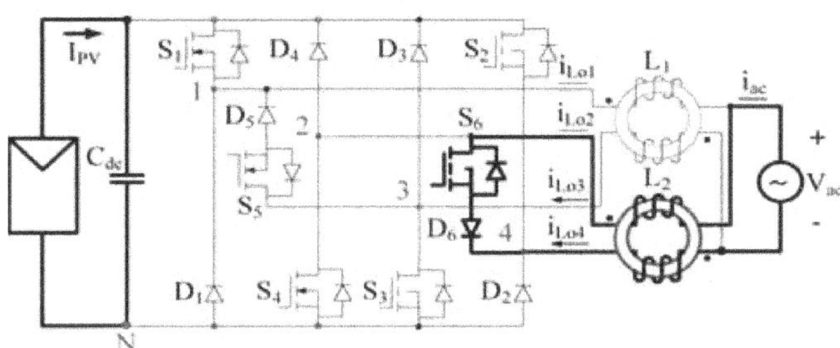

Figure 7. Freewheeling stage of negative half-line cycle

3. Ground Loop Leakage Current Analysis for the Proposed Transformerless Inverter

A galvanic connection between the ground of the grid and the PV array exists in transformerless grid-connected PV systems. Large ground leakage currents may appear due to the high stray capacitance between the PV array and the ground. In order to analyze the ground loop leakage current, Figure 8 shows a model with the phase output points 1, 2, 3, and 4 modeled as controlled voltage sources connected to the negative terminal of the dc bus (N point).

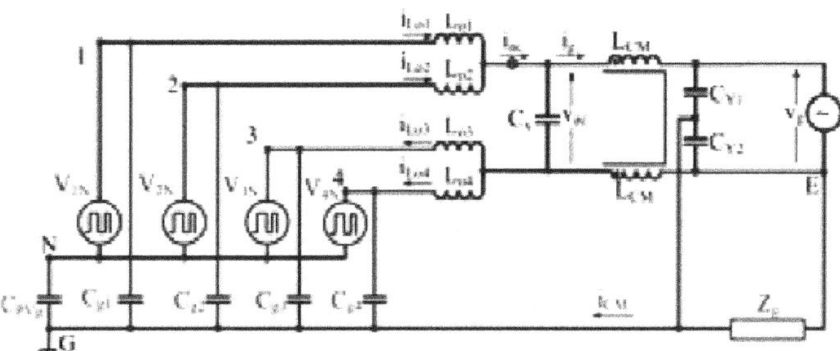

Figure 8. Simplified CM leakage current analysis model for positive half-line cycle

The value of the stray capacitances *Cg*1, *Cg*2, *Cg*3, and *Cg*4 of MOSFETs is very low compared with that of *C*PVg, therefore the influence of these capacitors on the leakage current can be neglected. It is also noticed that the DM capacitor *Cx* does not affect the CM leakage current. Moreover, during the positive half-line cycle, switches S_2, S_4, and S_6 are kept deactivated; hence the controlled voltage sources V_{2N} and V_{4N} are equal to zero and can be removed. Consequently, a simplified CM leakage current model for the positive half-line cycle is derived as shown in Figure 8.

4. Matlab Verification of the Proposed Circuit

The Figure 9 is the Matlab design of proposed system with unipolar PWM with the switching frequency of 20KHz. Sine PWM is used to generate the control signals to convert DC of supply into AC supply. The subsystem of Solar PV system is shown in Figure 10.

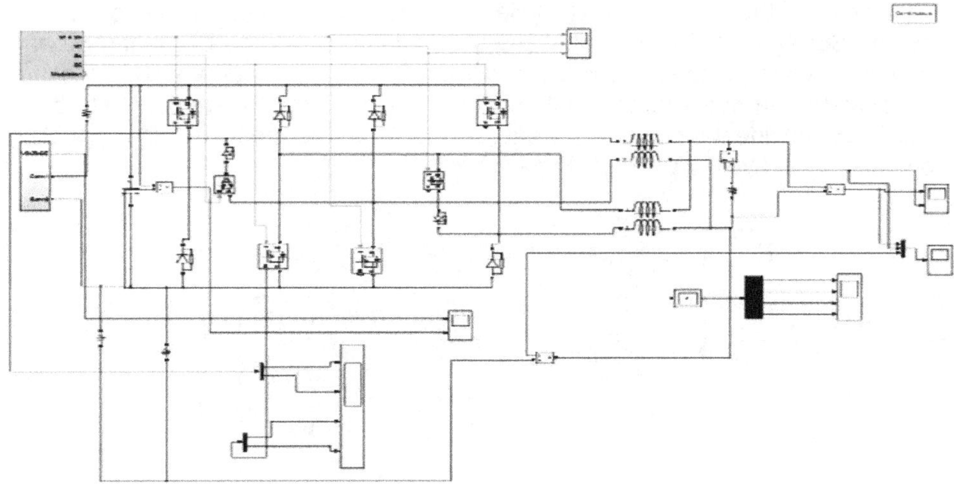

Figure 9. Proposed circuit in Matlab

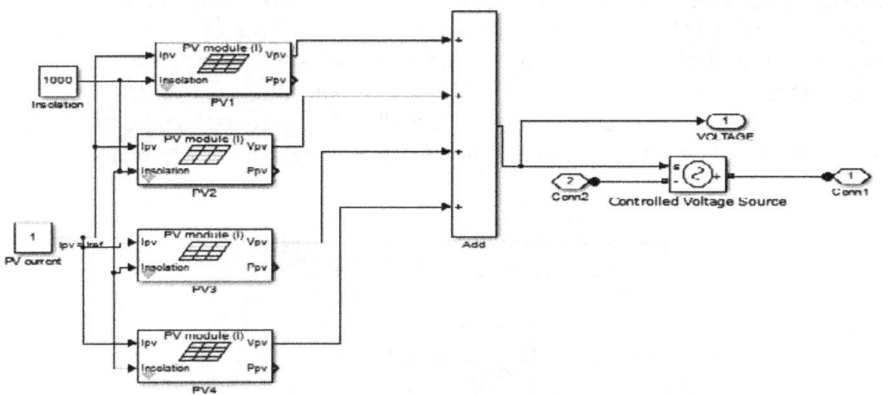

Figure 10. Solar pv system designed in Matlab

Phase of Disposition is the one multi-carrier pwm applied to the proposed system for common mode analysis. Multiple carriers were chosen based on the principle of POD pwm technique is shown in Figure 11.

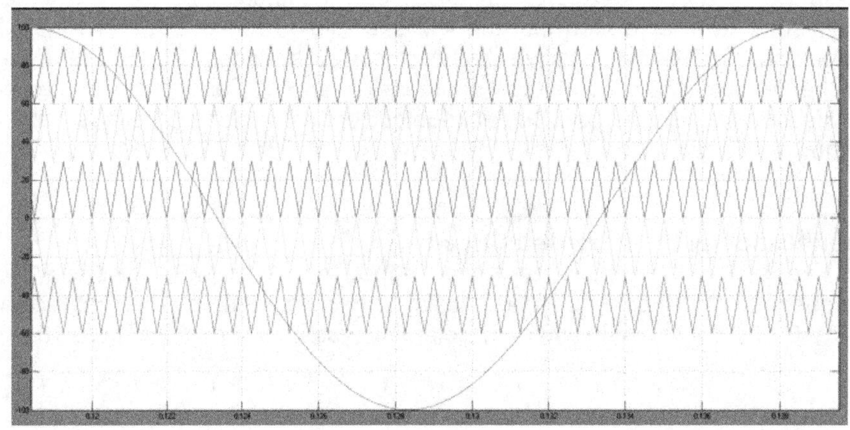

Figure 11. POD pwm applied to proposed system

The inverter output voltage and current waveform is shown in Figure 12 with output voltage of 230V, 50Hz and 4 amps of current is obtained as AC grid tie output. The green waveform is shown in figure represents the leakage currents due to common mode voltages.

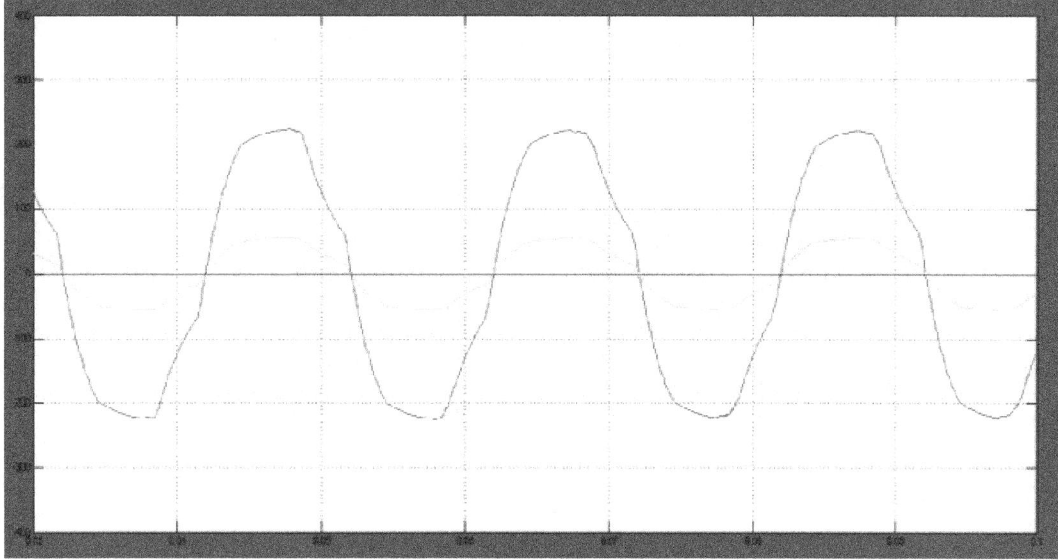

Figure 12. Grid voltage and current

As shown in the proposed circuit the output of inverter is not directly connected to grid, two inductive filters are employed for positive half and negative half cycle of the output independently. The waveforms represent the currents through the inductors for positive and negative half of full cycle. Figure 13 shows the closer image of the leakage current due to the common mode voltage. The Figure 14 shows the individual currents that flow through the filter inductor during both half cycles.

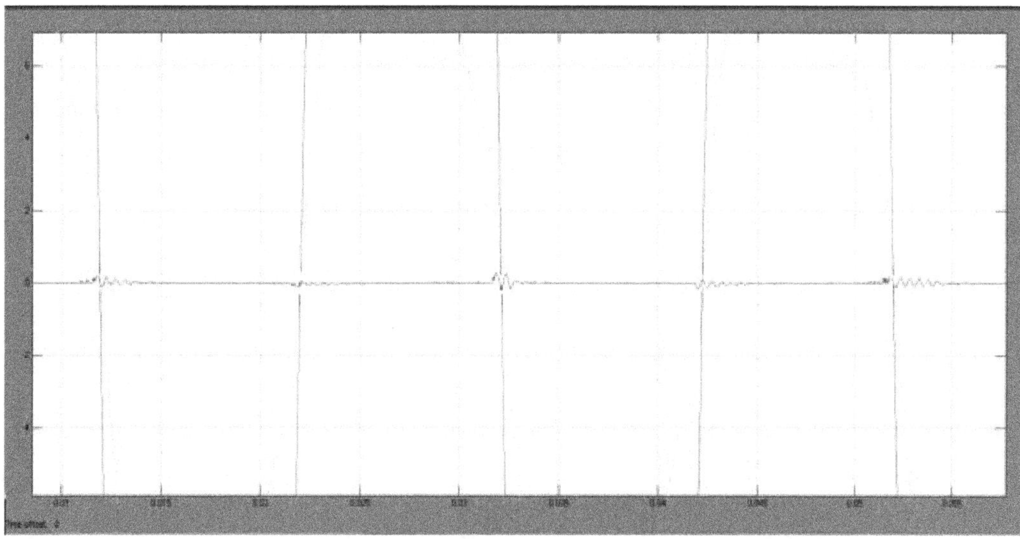

Figure 13. Common mode leakage current with POD PWM

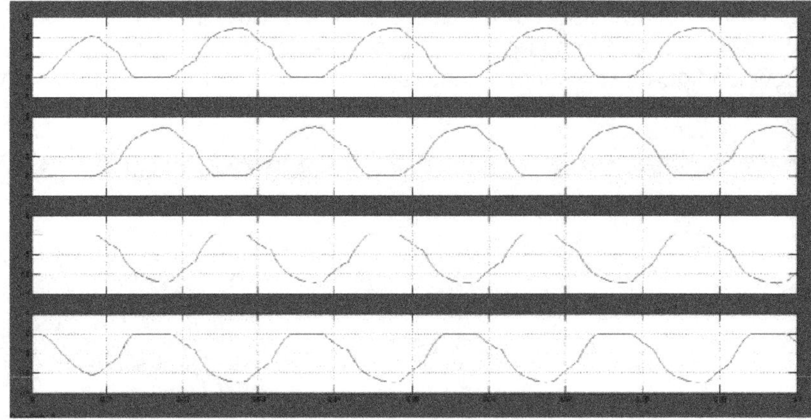

Figure 14. Inductor currents of *i*Lo1, *i*Lo2, *i*Lo3 and *i*Lo4

The Figure 15 shows the total Harmonic distortion of output voltage tied to grid while using sine PWM as the pulse generator, it is found that the THD is about 14.60%.

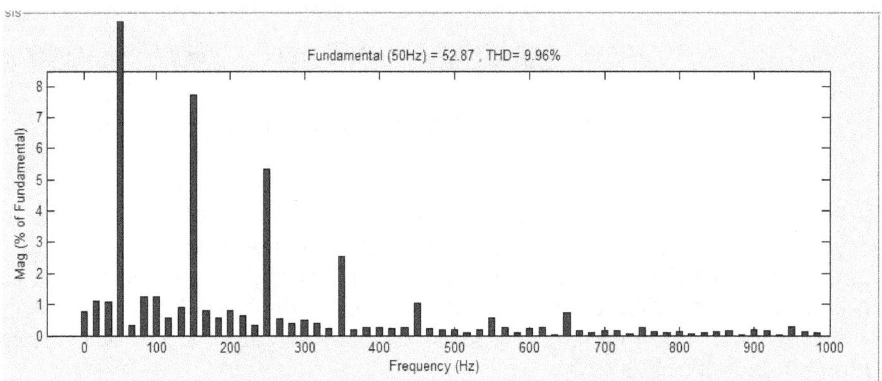

Figure 15. THD of output voltage using POD PWM technique

Phase opposition is the one of the efficient technique among the PD, POD, APOD, Figure 16 shows the PD technique implemented by using Matlab for generating gate signals it is evident from Figure 17 that the leakage current due to common mode voltage is became nearly to zero and the total harmonic distortion is reduced to 9.86%.

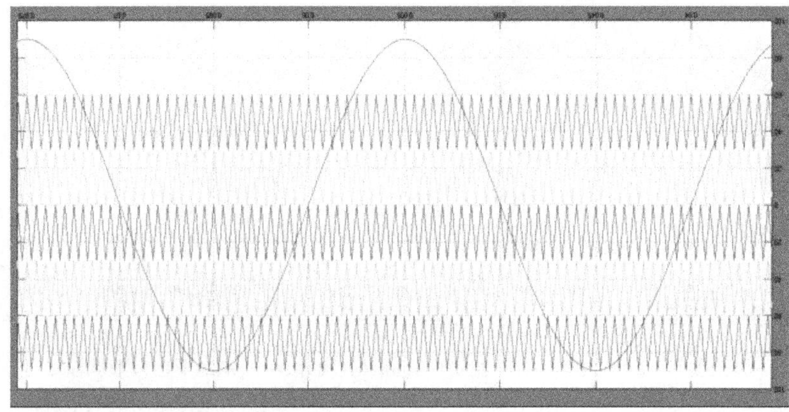

Figure 16. Phase Disposition modulation technique applied to proposed system

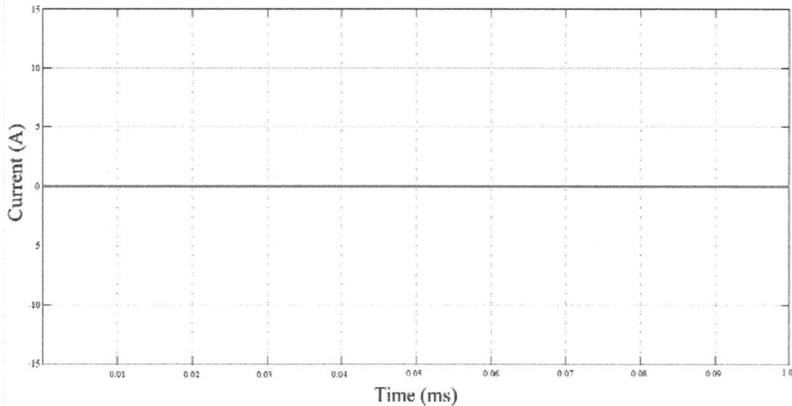

Figure 17. Reduced Leakage currents when applying PD technique

The Figure 6 is the gate pulse generation for the proposed converter for 20KHZ operating frequency of converter. The figure 11 is the grid voltages and current at pcc. The Figure 14 gives grid voltage, inductor currents of *i*Lo1 and *i*Lo2. The main of this project is reducing common mode currents is presented in Figure 17. The Figure 18 shows the THD of output voltage is about 2.22% shows that power quality is up to the mark. According to IEEE standard 5% of THD is acceptable limit.

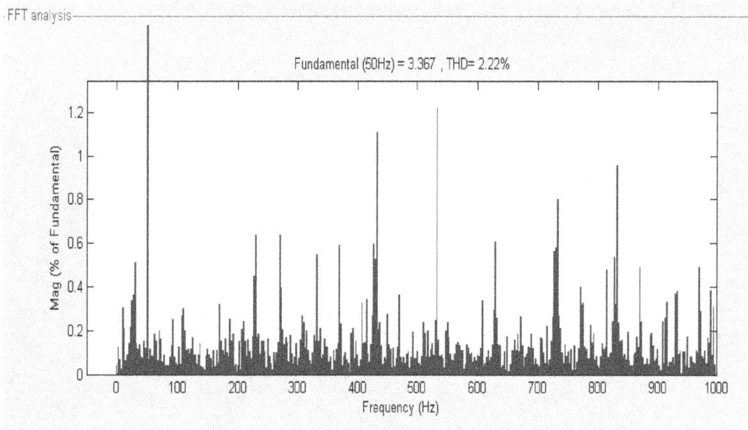

Figure 18. FFT analysis using PD technique

5. Conclusion

A high reliability and efficiency inverter for transformerless solar PV grid-connected systems is presented in this paper using Matlab/Simulink model design. The leakage current present due to the effect of common mode voltage with POD PWM and PD PWM are applied on the proposed converter performance analysis interms of THD. It is observed from FFT analysis of POD and PD pwm techniques that PD has very low thd and also the characteristics of the proposed transformerless inverter shows the reduced shoot-through issue leads to greatly enhanced reliability, low ac output current distortion is achieved because a high dead time is not needed at PWM switching commutation instants in the case of PD technique. It is shown that the proposed transformerless PV grid tie inverter is efficient when using PD as control technique for controlling the switching operation with overall improved efficiency.

The asymmetry of the switch arrangements in less usage of the number of high frequency switches in order to reduce the losses and increase the efficiency of proposed system will be a good option.

Appendix

Parameters	Specifications
Input voltage	440V DC
Grid voltage/ Frequency	230V/50Hz
Rated Power	1000W
Switching Frequency	20KHz
Dc bus capacitor	470μF
Filter capacitor	4.7μF
Filter Inductors	2mH
Parasitic capacitors	750nF

References

[1] SB. Kjaer, JK. Pedersen, F. Blaabjerg. A review of single-phase grid-connected inverters for photovoltaic modules. *IEEE Trans. Ind. Appl.* 2005; 41(5): 1292–1306.

[2] R. Gonzalez, E. Gubia, J. Lopez, L. Marroyo. Transformerless single phase multilevel-based photovoltaic inverter. *IEEE Trans. Ind. Electron.* 2008; 55(7): 2694–2702.

[3] SV. Araujo, P. Zacharias, R. Mallwitz. Highly efficient single-phase transformer-less inverters for grid-connected photovoltaic systems. *IEEE Trans. Power Electron.* 2010; 57(9): 3118–3128.

[4] W. Yu, JS. Lai, H. Qian, C. Hutchens. High-efficiency MOSFET inverter with H6-type configuration for photovoltaic non-isolated AC-module applications. *IEEE Trans. Power Electron.* 2011; 56(4): 1253–1260.

[5] T. Kerekes, R. Teodorescu, P. Rodriguez, G. Vazquez, E. Aldabas. A new high-efficiency single-phase transformerless PV inverter topology. *IEEE Trans. Ind. Electron.* 2011; 58(1): 184–191.

[6] B. Yang, W. Li, Y. Zhao, X. He. Design and analysis of a grid connected photovoltaic power system. *IEEE Trans. Power Electron.* 2010; 25(4): 992–1000.

[7] H. Xiao, S. Xie, Y. Chen, R. Huang. An optimized transformerless photovoltaic grid-connected inverter. *IEEE Trans. Ind. Electron.* 2011; 58(5): 1887–1895.

[8] B. Yang, W. Li, Y. Gu, W. Cui, X. He. Improved transformerless inverter with common-mode leakage current elimination for photovoltaic grid-connected power system. *IEEE Trans. Power Electron.* 2012; 27(2): 752–762.

[9] M. Calais, VG. Agelidis. Multilevel converters for single-phase grid connected photovoltaic systems: an overview. Proc. IEEE Int. Symp. Ind. Electron. 1998; 1: 224–229.

[10] M. Calais, JMA. Myrzik, VG. Agelidis. Inverters for single phase grid connected photovoltaic systems—Overview and prospects. Proc. 17th Eur. Photovoltaic Solar Energy Conf., Munich, Germany. 22–26 Oct. 2001: 437–440.

[11] IEEE Standard for Interconnecting Distributed Resources with Electric Power Systems. IEEE Std. 1547. 2003.

[12] V. Verband der Elektrotechnik. Elektronik und Informationstechnik (VDE). Std. V 0126-1-1, Deutsches Institut für Normung. 2006.

[13] WN. Mohan, T. Undeland, WP. Robbins. *Power Electronics: Converters, Applications, and Design.* New York. Wiley. 2003.

2

Flux based Sensorless Speed Sensing and Real and Reactive Power Flow Control with Look-up Table based Maximum Power Point Tracking Technique for Grid Connected Doubly Fed Induction Generator

DVN. Ananth[1], GV. Nagesh Kumar[2]
[1] Department of EEE, Viswanadha Institute of Technology and Management, Visakhapatnam, 531173, India
[2] Department of EEE, GITAM University, Visakhapatnam, 530045, Andhra Pradesh, India
email: drgvnk14@gmail.com

Abstract

This aim of this paper is to design controller for Doubly Fed Induction Generator (DFIG) converters and MPPT for turbine and a sensor-less rotor speed estimation to maintain equilibrium in rotor speed, generator torque, and stator and rotor voltages. It is also aimed to meet desired reference real and reactive power during the turbulences like sudden change in reactive power or voltage with concurrently changing wind speed. The turbine blade angle changes with variations in wind speed and direction of wind flow and improves the coefficient of power extracted from turbine using MPPT. Rotor side converter (RSC) helps to achieve optimal real and reactive power from generator, which keeps rotor to rotate at optimal speed and to vary current flow from rotor and stator terminals. Rotor speed is estimated using stator and rotor flux estimation algorithm. Parameters like tip speed ratio; coefficient of power, stator and rotor voltage, current, real, reactive power; rotor speed and electromagnetic torque are studied using MATLAB simulation. The performance of DFIG is compared when there is in wind speed change only; alter in reactive power and variation in grid voltage individually along with variation in wind speed.

Keywords: *doubly fed induction generator (DFIG), maximum power point tracking (MPPT), real and reactive power control, rotor & grid side converter (RSC & GSC), sensor-less speed estimation, wind energy conversion system (WECS)*

1. Introduction

Wind and solar electric power generation systems are popular renewable energy resources and are getting significance due to retreating of primary fuels and because of eco-friendly nature and is available from few kilo-watt power to megawatt rating [1]. The DFIG is getting importance compared to permanent magnet synchronous generator (PMSG) or asynchronous generator because of the operation under variable speed conditions [3]-[5], capability to extract more or maximum power point tracking theorm (MPPT) [2] and fast and accurate control of reactive power [6]-[13], better capability in low voltage and high voltage fault ride through situation, low cost of converters [14], effective performance during unbalanced and flickering loads. The efficiency enhancement and capability to meet desired reactive power demand from grid can be obtained by adopting robust rotor side control (RSC) for DFIG. But in general, RSC is rated from 25% to 35% of generator stator rating, which allows only ±25% variation in rotor speed. However, due to low power rating of converters, the cost incurred on controllers is low.

In general, the stator and rotor windings of DFIG can deliver both real and reactive power to the grid. The direction of real and reactive power flow from rotor can me varied to meet the desired reactive power requirements from grid with the help of sophisticated RSC controller scheme. The MPPT algorithm will be designed mostly for wind turbines to extract more mechanical power by adjusting the rotor blades and to make the generator shaft to rotate at optimal speed. This algorithm makes the blades to sweep maximum area to make wind sweeping turbine shaft with more mechanical force so that maximum mechanical power can be achieved at that particular wind speed. However more mechanical power can be obtained naturally at higher wind speed from the wind turbine.

The increase in wind generators connected to grid leads to penetration issues causes many problems to sensitive generators. If any generators among them are unable to convene desired grid codes, make them to trip, causing voltage at point of common coupling (PCC) decreases tending the other generators to oscillate if any small disturbance like change in wind speed occurs. It will tend WECS system to weaken or work at marginal stable situation for certain time. The solution to above penetration issue for making healthy system is effective control of reactive power.

To achieve desired reactive power requirement for grid, rotor side converter (RSC) plays a vital role in coordination with grid side controller (GSC). With change in grid voltage (due to reason like faults or so) stator voltage also needs to be adopted for not loosing synchronism with grid and to maintain stability, GSC controller is necessary. Maximum power extraction from DFIG using pitch angle controller and optimal power coefficient at low and high speed is analyzed in [15]. Direct and indirect control of reactive power control with an aim to meet active and reactive power equal to the reference values as achieved in [16]. MPPT based WECs design facilitates the wind turbine has to operate in variable speed as per the ideal cube law power curve [17]–[21]. The constant power mode of operation can be achieved by (i) including energy storage devices [21]–[25] and (ii) employing pitch control [17],[26],[27]. Introduction to tuning of PID controllers [28], pitch angle control with neural network for optimal tracking of real power is given in [29]. Advanced techniques like hybrid fuzzy sliding mode [29] and growing natural gas based MPPT algorithm is proposed in [30]. These advanced methods can improve overall DFIG system performance with robust control, faster in action and enhanced tracking of real and reactive power.

In this paper, performance of DFIG was compared and analyzed under situations like, with variation in wind speed alone, with reactive power variation and with grid voltage variation for same variation in wind speed. In these cases, variation in tip speed ratio and coefficient of turbine power, effect on real and reactive power flows, voltages and current from stator and rotor, rotor speed and electromagnetic torque are examined. The paper was organized with overview of WECS with wind turbine modeling and pitch angle controller in 2nd section, study of mathematical modeling of DFIG in 3rd section, the 4th section describe RSC architecture and design; section 5 analyses the performance of DFIG for three cases like effect of variation on wind speed variation, reactive power demand along with variation in wind speed and grid voltage variation with wind speed. Conclusion was given in Section 6. System parameters are given in appendix.

2. Wind Energy Conversion System (WECS)

The overview of wind energy conversion system (WECS) is shown in Figure 1. The construction has following dynamic models: Wind speed calculator is an anemometer sensor system with storage system to measure the actual wind speed in meters per second at that instant with air density and ambient temperature measurements also, and is given to WECS and turbine system. The WECS system consists of aerodynamics and wind turbine control model to extract maximum power during steady state and protection during transient or unstable state of operation. Mechanical and electrical model control system is to generate reference speed, power and voltage signals for controlling real and reactive power flow from generator to the grid and also contains protective system during abnormal situation. The mechanical model system gives command to turbine for extraction of maximum mechanical power extraction for a given wind speed and the electrical model system give command for the generator to produce respective real power and reactive power and to maintain synchronism under all operating condition.

The real and reactive power from DFIG is controlled by using two controllers namely rotor side (RSC) and grid side (GSC) controller using converter controller model as shown in Figure 2. The converter model is a bidirectional switches with IGBT (integrated bi-polar transistor), which controls the voltage, real and reactive power from stator and rotor to grid.

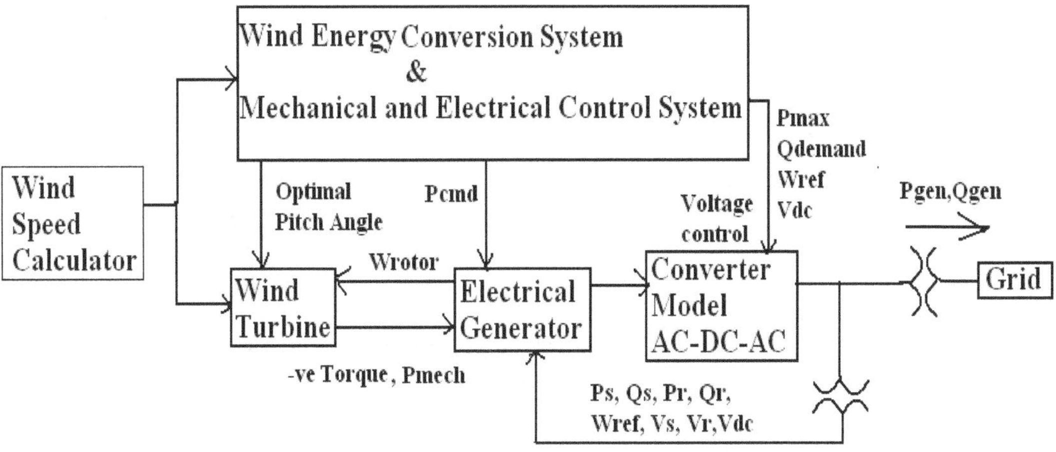

Figure 1. Block diagram of WECS for grid connected DFIG

The RSC controller aim is to maintain DFIG rotor to maintain optimal speed specified by MPPT and also to control the reactive power flow by varying rotor current direction. The GSC controller is maintain constant DC link voltage at back to back terminals across capacitor, so that this voltage can be maintained as per PCC point and also for rapid supply of leading or lagging reactive power without much deviation from generator real power.

The transformer near the grid is a step-up voltage and the other transformer is an isolation transformer.

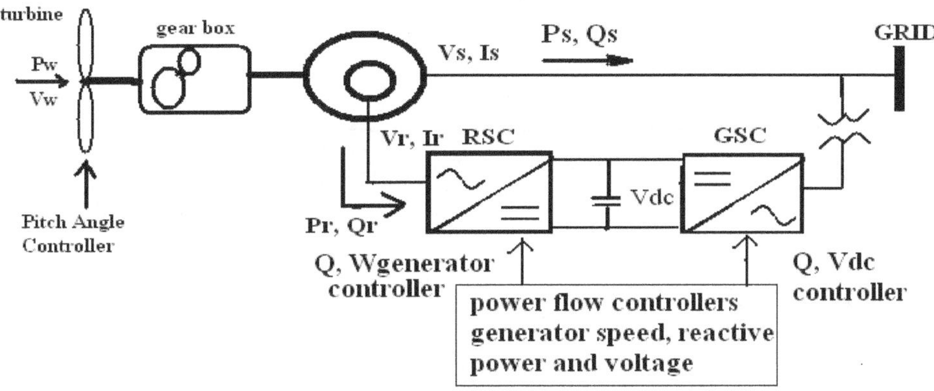

Figure 2. Schematic diagram of WECS for DFIG system connected to grid

2.1. The wind turbine modeling

The wind turbine is the prime mover which facilitates in converting kinetic energy of wind into mechanical energy which further converted into electrical energy. From basic theory of wind energy conversion, the output mechanical power from turbine is given by

$$P_{mech} = \frac{1}{2} Cp(\lambda, \beta) \rho \pi r^2 \upsilon_\omega^3 \quad (1)$$

Where P_{mech} is the mechanical power output from wind turbine, Cp is coefficient of wind power as a function of pitch angle (β) and tip speed ratio (λ), ρ is specific density of air, r is radius of wind turbine blade, υ_ω is wind speed.

$$Cp(\lambda, \beta) = 0.5176 \left(\frac{116}{\lambda_i} - 0.4\,\beta - 5\right) e^{-21/\lambda_i} + 0.0068\lambda \quad (2)$$

The tip sped ratio is a relation between turbine speed (ω_t), radius of turbine blades and wind speed and tip speed ratio at particular angle 'i' is given the relation as shown below

$$\lambda = \frac{\omega_t r}{\upsilon_\omega} \text{ and } \frac{1}{\lambda_i} = \frac{1}{\lambda + 0.08\beta} - \frac{0.035}{\beta^3 + 1} \quad (3)$$

the output power at nominal wind speed is given by the below equation

$$\upsilon_n = \sqrt[3]{\frac{2 P_{sh}}{\pi \rho r^3 C_{p\,max}}} \quad (4)$$

Where Psh is the turbine shaft power and Cpmax is maximum mechanical power coefficient. The maximum power P_{max} from wind turbine can be extracted by using the equation

$$P_{max} = \frac{1}{2\lambda_{opt}^3} \pi \rho C_{p\,max} r^5 \omega_{opt}^3 \quad (5)$$

2.2. Pitch angle controller

The wind turbine blade angles are controlled by using servo mechanism to maximize turbine output mechanical power during steady state and to protect the turbine during high wind speeds.

This control mechanism is known as pitch angle controller. When wind speed is at cut-in speed, the blade pitch angle is set to produce optimal power, at rated wind speed; it is set to produce rated output power from generator. At higher wind speeds, this angle increases and makes the turbine to protect from over-speeding. The pitch angle controller circuit is as shown in Figure 3.

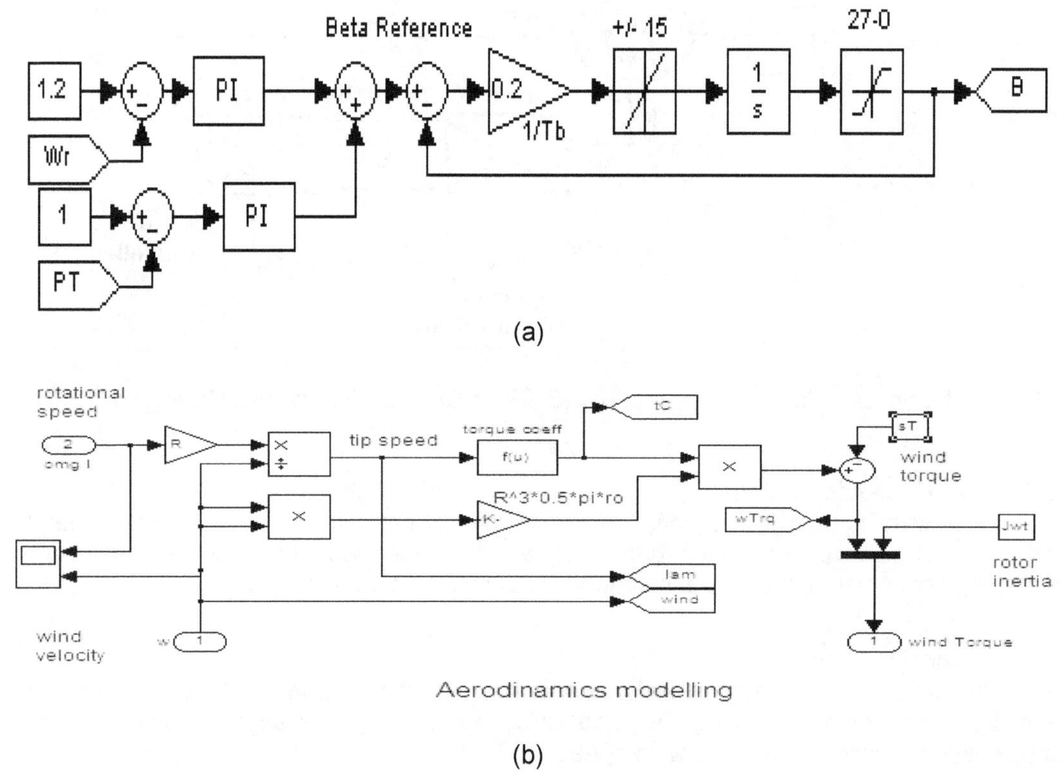

Figure 3. The pitch angle controller circuit, a) Pitch angle controller design for wind turbine, b) Wind turbine modeling- mechanical torque derivation from wind speed, tip speed ratio and other characteristic parameters

In this system reference generator speed is W_r^{ref} =1.2p.u or is obtained from MPPT algorithm and actual speed of the generator is Wr. The actual speed can be estimated using an encoder or using sensor-less estimation strategy. The error between reference and actual values is controlled using PI controller. In the similar way, the difference in reference (P*=1) and actual power outputs from turbine (PT) is controlled by PI controller. Both the outputs from PI controller are designed to get reference pitch angle controller (βref). The closed loop control of pitch angle is obtained as shown in Figure 3. The optimal pitch angle is written mathematically at this point as $\int \frac{1}{T_\beta}(\beta_{ref} - \beta)\, dt$. If pitch angle is set at zero degrees, maximum power can be extracted from the turbine. In general pitch angle is set between 0 to 7.5o/s to extract maximum power, also these values specifies the turbine is working in stable environment and if is set at 27o/s, it means either wind speed is high called cutout wind speeds or may be external fault in the external electric system. The reference real power (Pe) or actual mechanical output from turbine and mechanical output torque (Tm) is shown in Figure 4. Using Tip Speed Ratio (TSR) and Coefficient of Power (CP) are used to generate reference power and the control scheme is useful to extract maximum mechanical power, thereby more mechanical tore Tm by using the MPPT algorithm.

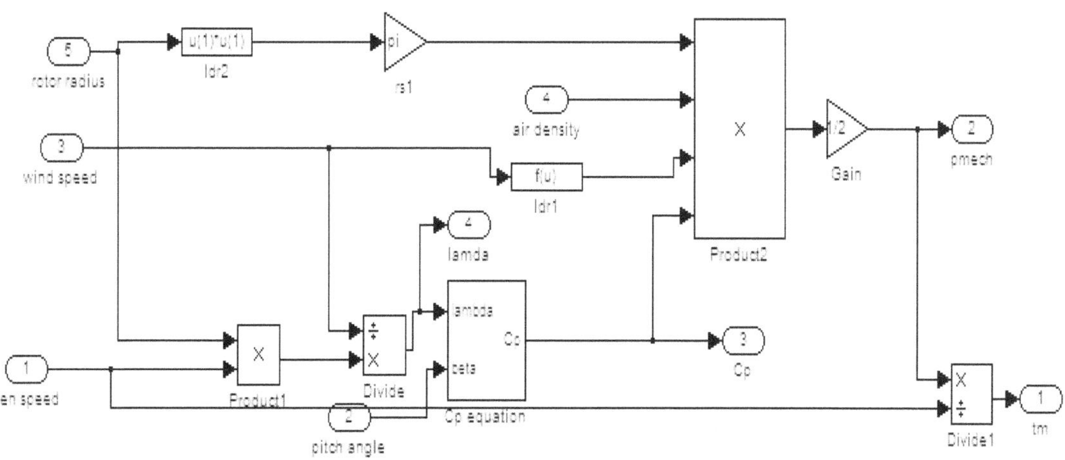

Figure 4. Reference electrical power generation control circuit and mechanical torque output from turbine with MPPT algorithm

The aim of MPPT algorithm as shown in Figure 4 is to generate optimal mechanical power output from turbine and mechanical input torque to be given to the doubly fed induction generator. The mechanical power is given as reference to grid side converter (GSC) to make the rotor to rotate at optimal speed. The optimal input torque to DFIG is so as to operate for extracting maximum power from the generator. The inputs to MPPT algorithm are radius of curvature 'R' of turbine wings, rotor speed (Wr), wind sped (Vw) and pitch angle (beta). Initially with R, Wr and Vw, tip speed ratio (g) is determined. Later using equation 2 and input parameter beta, coefficient of power (Cp) is calculated. Based on equation 5, optimal mechanical power (Pm_opt) is determined and dividing mechanical power by rotor speed, optimal torque (Tm_opt) is determined. The pitch angle (beta) is determined as shown in Figure 4. The application of Pm_opt and Tm_opt is shown in Figure 6 and 7.

When there is change in wind speed, turbine speed changes and thereby beta, Cp and mechanical power and torque changes [2]. If grid voltage varies due to fault or any conditions, the current flow in the rotor circuit varies. Huge requirement of reactive power for grid will come to picture to maintain low voltage ride through (LVRT) phenomenon. During this period, rotor speeds increases gradually and lose synchronism if not provided with LVRT capability [14]. In general, load varies continuously. So, active and reactive power demand from grid changes considerably. It has to be provided by DFIG. For this enhanced real and reactive power is necessary [12]. For this optimal real power generation is achieved by extracting optimal mechanical power output from turbine is derived and made to run at optimal loading and speed. For this MPPT algorithm proposed will be very helpful.

2.3. Mathematical Modeling of DFIG

There are many advantages of DFIG compared to squirrel cage induction generator or permanent magnet synchronous generator. Using DFIG independent control of active and reactive power, variable speed and constant frequency operation, over load capability, higher efficiency, higher ratings, are possible. The converters need to handle only 25% to 35% of generator capacity, thereby minimizing operating cost and switching losses, low hardware cost etc.

The equivalent circuit of DFIG in rotating reference frame at an arbitrary reference speed of ω is shown in Figure 5. The equations can be derived in dq reference frame were as follows:

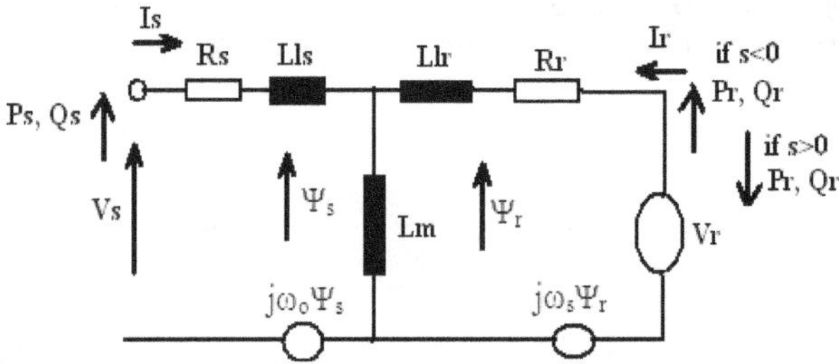

Figure 5. Equivalent circuit of DFIG in rotating reference frame at speed ω

The stator direct and quadrature axis (dq) voltages and flux of DFIG can be written as

$$V_{sd} = R_s I_{sd} - \omega_s \psi_{sq} + \frac{d\psi_{sd}}{dt} \tag{6a}$$

$$\psi_{sd} = \left((V_{sd} - R_s I_{sd})\frac{1}{s} - \sigma L_s I_{sd}\right)\frac{L_r}{L_m} \tag{6b}$$

$$V_{sq} = R_s I_{sq} + \omega_s \psi_{sd} + \frac{d\psi_{sq}}{dt} \tag{7a}$$

$$\psi_{sq} = \left((V_{sq} - R_s I_{sq})\frac{1}{s} - \sigma L_s I_{sq}\right)\frac{L_r}{L_m} \tag{7b}$$

The equations 6b and 7b are stator d and q axis flux written in terms of stator voltage, current and passive elements. The leakage factor σ can be stated as $1-\frac{L_m^2}{L_s L_r}$.

The rotor direct and quadrature axis are derives as

$$V_{rd} = R_r I_{rd} - (\omega_s - \omega_r)\psi_{rq} + \frac{d\psi_{rd}}{dt} \tag{8}$$

$$V_{rq} = R_r I_{rq} + (\omega_s - \omega_r)\psi_{rd} + \frac{d\psi_{rq}}{dt} \tag{9}$$

The difference between stator speed (ω_s) and rotor speed (ω_r) is known as slip speed $(s\omega_s)$. For motoring action, this difference is less than zero and for generating, the slip speed is negative. The stator and rotor flux linkages in axis frame are given below

$$\psi_{sd} = L_{ls} I_{sd} + L_m I_{rd} \tag{10}$$

$$\text{or } \psi_{sd} = L_m I_{sm} \tag{11}$$

$$\psi_{sq} = L_{ls} I_{sq} + L_m I_{rq} \tag{12}$$

$$\psi_{rd} = L_{lr} I_{rd} + L_m I_{sd} \tag{13}$$

$$\psi_{rq} = L_{lr} I_{rq} + L_m I_{sq} \tag{14}$$

The magnitude of rotor flux can be written as $\psi_r = \sqrt{\psi_{rd}^2 + \psi_{rq}^2}$.

The stator real power in terms of two axis voltage and current is

$$P_s = \frac{3}{2} (V_{sd} I_{sd} + V_{sq} I_{sq}) \tag{15}$$

The rotor real power in terms of two axis voltage and current is

$$P_r = \frac{3}{2} (V_{rd} I_{rd} + V_{rq} I_{rq}) \tag{16}$$

The stator reactive power in terms of two axis voltage and current is

$$Q_s = \frac{3}{2} (V_{sq} I_{sd} - V_{sd} I_{sq}) \tag{17}$$

The rotor reactive power in terms of two axis voltage and current is

$$Q_r = \frac{3}{2} (V_{rq} I_{rd} - V_{rd} I_{rq}) \tag{18}$$

The quadrature and direct axis rotor current in terms of stator parameters can be written as

$$I_{rq} = \frac{P_s}{|V_s|} = \frac{-L_{ls}}{L_m} I_{sq} \tag{19}$$

$$I_{rd} = \frac{Q_s}{|V_s|} + \frac{|V_s|}{\omega_s L_m} \tag{20}$$

The output electromagnetic torque is given by the equation

$$T_e = \frac{3}{2} \beta L_m (I_{sq} I_{rd} - I_{sd} I_{rq}) \tag{21}$$

The mechanical torque output from the turbine in terms of mechanical power and rotor speed is

$$T_m = \frac{P_{mech}}{\omega_r} \tag{22}$$

3. Rotor Side Controller (RSC) and Grid Side Controller (GSC) Architecture and Design

3.1. Operation of GSC and RSC controllers

The rotor side converter (RSC) is used to control the speed of rotor and also helps in maintaining desired grid voltage as demanded. The control circuit for grid side controller (GSC) is shown in 6a and rotor side controller (RSC) is shown in Figure 6b for the general network in Figure 8 with internal circuits for deriving RSC PLL for 2 phases to 3 phases inverse Parks transformation is shown in Figure 9. This Figure 9 helps to inject current in rotor winding at slip frequency. The GSC and RSC have four control loops each, later has one speed control loop, other is reactive power and last two are direct and quadrature axis current control loops. The speed and reactive power control loops are called outer control loop and direct and quadrature axis control loops are called inner control loops. The reference rotor speed is derived from the wind turbine optimal power output PmOpt as shown in Figure 4 and grid power demand. In total, the reference power input to the lookup table as shown in Figure 6b is Pm,gOpt. Based on the value of Pm,gOpt , the rotor is made to rotate at optimal speed so as to extract maximum power from DFIG set. The difference between reference speed of generator and actual speed of generator is said to be rotor speed error. Speed error is minimized and maintained nearly at zero value by using speed controller loop which is a PI controller with Kpn and Kin as proportional and integral gain parameters. The output from speed controller is multiplied with stator flux (Fs) and ratio of stator and rotor (Ls and Lr) inductances to get reference quadrature current (Iqr) for rotor. The error in reference and actual reactive power give reference direct axis current (Iqr). The difference between these reference and actual two axis currents is controlled by tuned PI controller to get respective direct and quadrature axis voltages. The output from each PI controller is manipulated with disturbance voltages to get reference voltage for pulse generation as shown in Figure 6b. It must be noted that the pulses are regulated at slip frequency for RSC rather than at fundamental frequency and slip frequency synchronizing for inverse Park's transformation can also be seen in the figure.

The MATLAB based block diagram of GSC is shown in Figure 6a. For a given wind speed, reference or control power from turbine is estimated using lookup table. From equation (15), stator real power (Pstator) is calculated and the error in powers is difference between these two powers (dP) which is to be maintained near zero by PI controller. The output from PI controller is multiplied with real power constant (Kp) gives actual controllable power after disturbance. The difference in square of reference voltage across capacitor dc link (Vdc*) and square of actual dc link voltage (Vdc) is controlled using PI controller to get reference controllable real power. The error in the reference and actual controllable power is divided by using 2/3Vsd to get direct axis (d-axis) reference current near grid terminal (Igdref). Difference in Igdref and actual d-axis grid current is controlled by PI controller to get d-axis voltage. But to achieve better response during transient conditions, decoupling d-axis voltage is added as in case of separately excited DC motor. This decoupling term helps in controlling steady state error and fastens transient response of DFIG during low voltage ride through (LVRT) or during sudden changes in real or reactive powers from/ to the system.

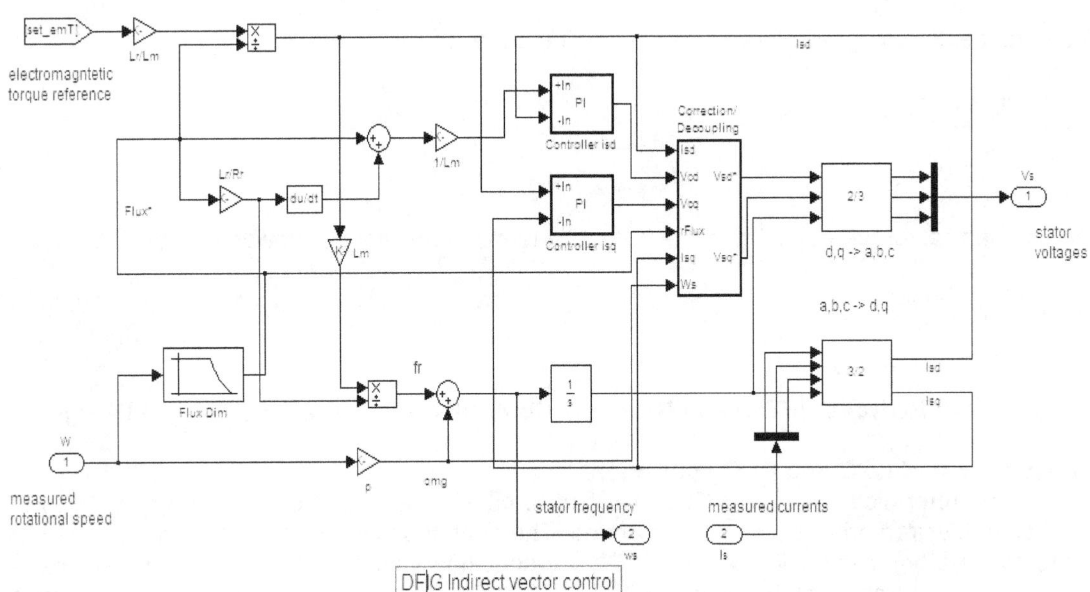

Figure 6a. Grid side controller with indirect vector control technique for DFIG

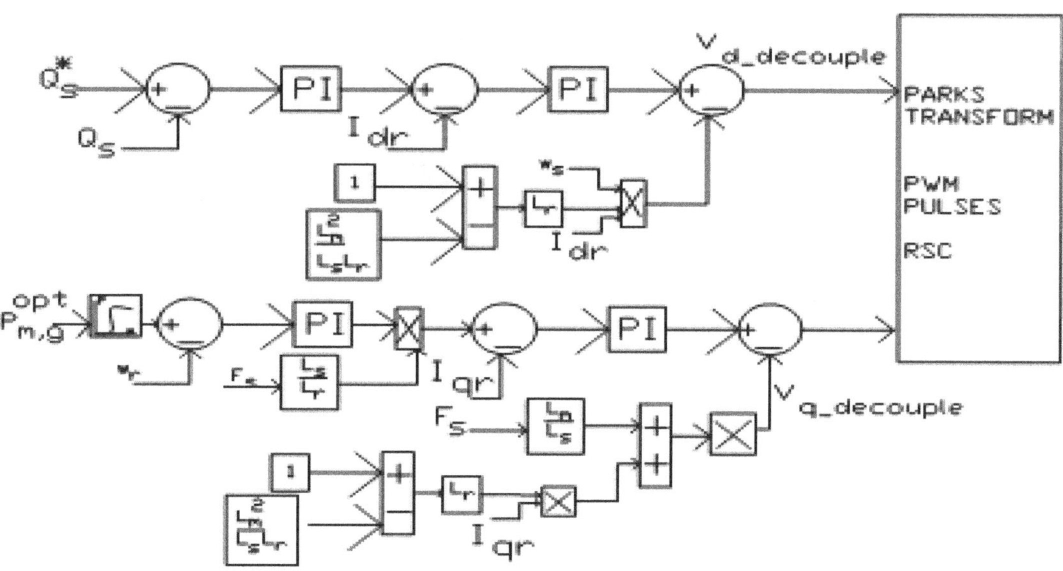

Figure 6b. Rotor side converter using power to speed conversion with improved vector control for DFIG

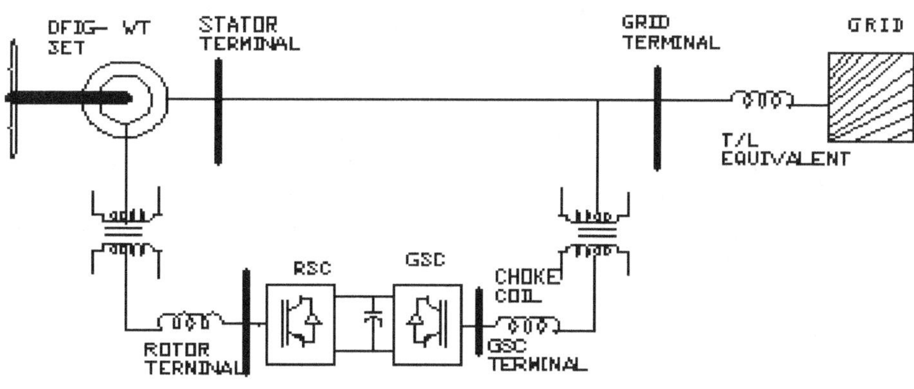

Figure 7. Design of overall DFIG based system with RSC and GSC

Similarly from stator RMS voltage (Vs) or reference reactive power, actual stator voltage or reactive power is subtracted by PI controller and multiplied with appropriate reactive power constant (Kq) to get actual reference reactive power compensating parameter. From equation (17), actual reactive power is calculated and the difference in this and actual compensating reactive power and when divided by 2/3Vsq, we get quadrature axis (q-axis) reference current (Iqref). When the difference in Iqref and stator actual q-axis current (Iq) is controlled by PI controller, reference q-axis voltage is obtained. As said earlier, to improve transient response and to control steady state error, decoupled q-axis voltage has to be added as shown in Figure 6a. Both d and q axis voltage parameters so obtained are converted to three axis abc parameters by using inverse Park's transformation and reference voltage is given to scalar PWM controller to get pulses for grid side controller.

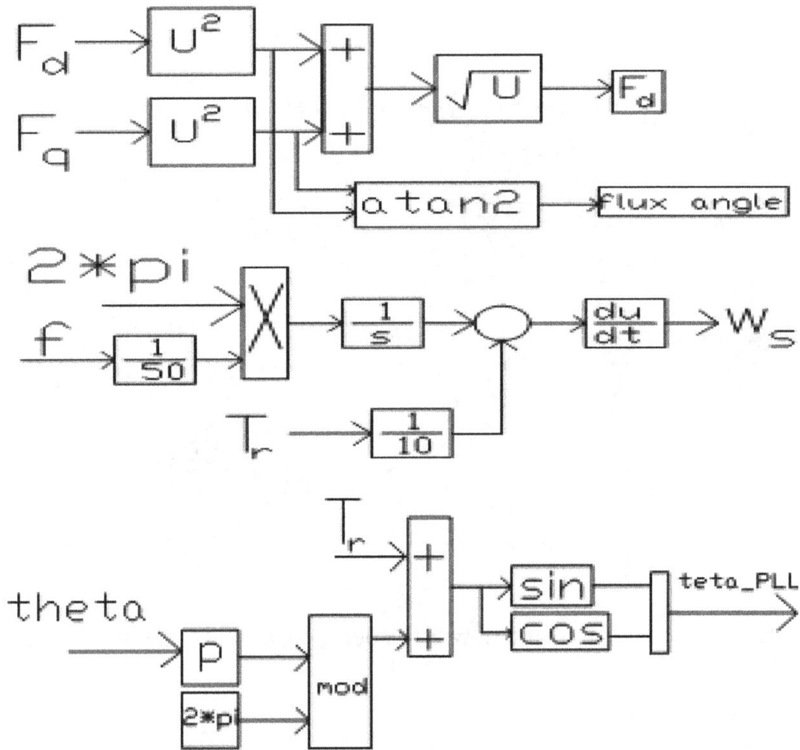

Figure 8. Internal circuits design of RSC for extracting rotor Parks transform PLL block for dq to abc conversion

The main purpose of rotor side controller (RSC) is to maintain desired generator speed and reactive power flow from rotor circuit, while grid side controller (GSC) is used to achieve nearly constant DC link voltage and to control bidirectional reactive power flow from GSC converter, stator and grid. This GSC is also capable in controlling real power from stator to achieve desired real power from generator stator.

The general form of speed regulation is given by

$$T_e = J\frac{d\omega_r}{dt} + B\,\omega_r + T_l \tag{23a}$$

$$= (Js+B)\,\omega_r + T_l \tag{23b}$$

Where T_e is electromagnetic torque, J is moment of inertia and B is friction coefficient, T_l is considered to be disturbance.

Multiplying both sides with ω_{error}, we get the equation as

$$T_e\,\omega_{error} = (Js + B)\,\omega_r\,\omega_{error} + T_l\,\omega_{error} \tag{24}$$

Considering ω_r constant and change in speed error is ω_{error} is control variable, the above equation becomes.

$$P_s^* = (K_{in}s + K_{pn})\,\omega_{error} + P_l \tag{25}$$

As product of torque and speed is power, we will be getting stator reference power and disturbance power as shown below.

$$P_s^* - P_l = (K_{in}s + K_{pn})\,\omega_{error} \tag{26}$$

Where, $K_{in} = J^*\omega_r$ and $K_{pn} = B^*\omega_r$

Finally direct axis reference voltage can be written by using equation (26) and is incorporated for Figure 6a. The equations for voltage and current control loops are

$$V_{rd}^{*}= -(\omega_{error})(K_{pn}+\frac{K_{in}}{s}) + (P_s)(K_{pt}+\frac{K_{it}}{s}) \tag{27}$$

$$V_{rq}^{*}= Q_{error}(K_{pQ}+\frac{K_{iQ}}{s}) \tag{28}$$

$$V_{gd}^{*} = K_{gp}(i_{gd}^{*} - i_{gd}) + k_{gi}\int(i_{gd}^{*} - i_{gd})dt - \omega_o L_g i_{gd} + +k_1 V_{sd} \tag{29}$$

$$V_{gq}^{*} = K_{gp}(i_{gq}^{*} - i_{gq}) + k_{gi}\int(i_{gq}^{*} - i_{gq})dt + \omega_o L_g i_{gd} + k_2 V_{sq} \tag{30}$$

$$i_{gq}^{*} = K_q sqrt(V_{dc}^{2*} - V_{dc}^{2}) + k_{qi}\int(V_{dc}^{*} - V_{dc})dt + R_{dc}V_{dc} \tag{31}$$

$$i_{gd}^{*} = K_d sqrt(V_s^{2*} - V_s^{2}) + k_{di}\int(V_s^{*} - V_s)dt \tag{32}$$

The rotating direct and quadrature reference voltages of rotor are converted into stationary abc frame parameters by using inverse parks transformation. Slip frequency is used to generate sinusoidal and cosine parameters for inverse parks transformation.

3.2. Rotor speed sensing by using sensor-less control technique

The sensor-less speed control for DFIG system with stator and rotor flux observers are shown in Figure 9a. The three phase stator voltage and currents are converted into two phase dq voltages and current by using Park's transformation. The dq axis stator voltage and current are transformed into dq axis stator flux based on equations 6b and 7b. The internal structure for dq axis flux derivations are shown in Figure 9b and 9c. The derived rotor and stator flux are compared and is controlled to estimate rotor speed by using PI controller. The blocks G1 and G2 are PI controller functional blocks. The speed is estimated and is termed as Wr and is integrated to get rotor angle. The angle is multiplied with trigonometric SIN and COS terms and is given to mux to get sin_cos parameters and the total setup can be used as phase locked loop (PLL). This sin_cos helps in estimating exact phase sequence and for locking the new system to reference grid.

The estimated speed Wr is given as input for RSC controller as shown in Figure 6b. From the lookup table, reference rotor speed is estimated from optimal power block, which is obtained from MPPT algorithm explained earlier.

3.3. Behaviour of mechanical and electrical system with the variation in wind sped and reactive power

The mechanical to electrical relationship is explained as follows. The rotor speed can be expressed as

$$\omega_r = (1 - s)\omega_s = p\eta\omega_{\omega t} \tag{33}$$

Where s is slip of DFIG, p is pair of poles of DFIG; η is gear box ratio and $\omega_{\omega t}$ is wind turbine speed. With the change in wind speed and depending on gears ratio and number of field poles, the rotor speed varies is shown in equation 33. When rotor speed varies, reference quadrature axis current changes, thereby current flow in the rotor circuit varies. The stator output also varies with variation in wind turbine speed and DFIG output power. When slip varies, the voltage in rotor circuit also varies which can be explained as per equations 8 and 9. Further change in rotor voltage leads to change in rotor current, there by rotor power flow also varies.

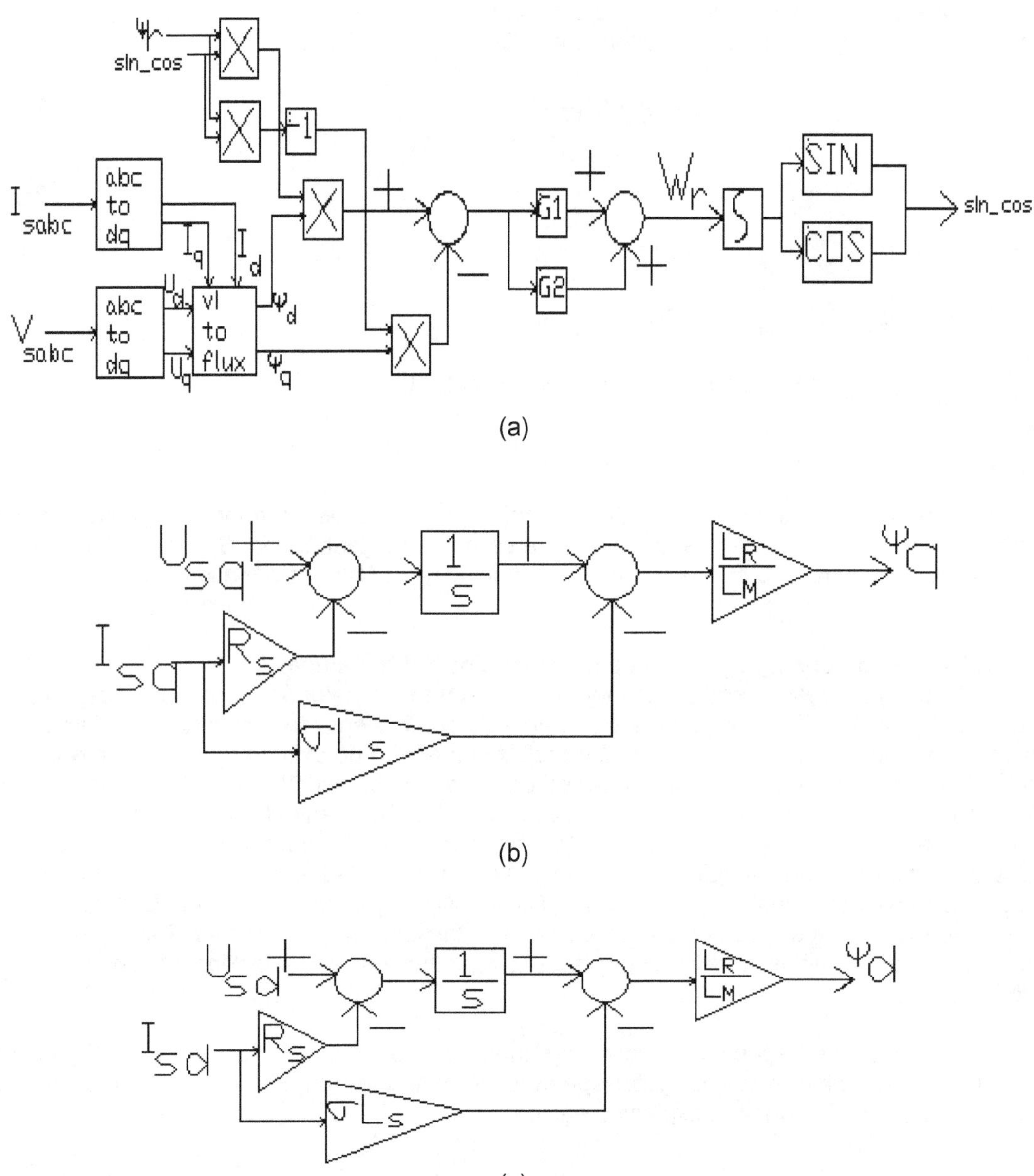

Figure 9. The sensor-less speed control for DFIG system, a). Estimation of rotor speed with stator voltage and current and rotor flux as inputs, b) Derivation of stator q-axis flux from q- axis stator voltage and current equation 7b, c). Derivation of stator d-axis flux from d- axis stator voltage and current from equation 6b

The mechanical turbine tip speed ratio (TSR) can be written in terms of radius of turbine wings (R), angular stator speed (ωs), pole pairs and gear box ratio as

$$\lambda = \frac{R\omega_s}{p\eta v_w}(1 - s) \tag{34}$$

Increase in stator or grid frequency, TSR increases and vice versa. Similarly with increase in rotor speed or wind speed, TSR decreases and vice versa. Hence when an electrical system gets disturbed, mechanical system also will get some turbulence and electrical to mechanical system is tightly interlinked. The steady state behavior of overall system must satisfy the relation below.

$$\Delta P = \frac{-P_{\omega t}}{(1-s)} - P_{em} = 0 \quad (35)$$

Under normal conditions, the change in turbine output has to be compensated by electrical power output from DFIG. Otherwise slip gets changed and thereby rotor speed changes. Hence imbalance in mechanical to electrical power output ratios, the slip changes. With the change in coefficient of power Cp, the mechanical power varies. The mechanical power changes mostly when wind speed or air density around the turbine wings changes. The electrical power from DFIG changes when mechanical power changes or rotor speed changes or load demand from grid varies.

4. Results and Discussion

The dynamic performance of the DFIG system is shown in Figure 9 is investigated under three different cases and the rating specifications for DFIG and wind turbine parameters are given in appendix. The wind speed change in all cases in meters per seconds as 8, 15, 20 and 10 at 15, 25 and 35 seconds. The reactive power and voltage value change in individual two cases with change in time is from -0.6pu at 12 seconds to 0pu change at 20 seconds. It was further changed from 0pu to +0.6pu magnitude at 30 seconds. In general wind speed will change with time which is a natural phenomenon, demand in lagging or leading reactive power requirement will come into picture because of change in load. Due to addition of large furnace or induction motor or non linear type load, leading reactive power greater than 0pu is required, while for light load lagging reactive power is required (<0pu). Hence DFIG will become better generator source if immediately it can supply any desired reactive power effectively. The change in grid terminal voltage takes place when suddenly switching on or off large loads or due to small faults near point of common coupling (PCC). The effect of change in wind speed, change in wind speed with reactive power and change in wind speed with grid terminal voltages on generator and turbine parameters are studied.

5. Case Studies

5.1. Case A: Change in TSR and Cp with wind speed, reactive power and grid voltage

The changes in tip- speed ration and power coefficient Cp with change in wind speed alone is shown in Figure 10 (a), with both reactive power and wind speed variation in Figure 11 (b) and variation with grid terminal voltage and wind speed both is shown in Figure 10 (c).

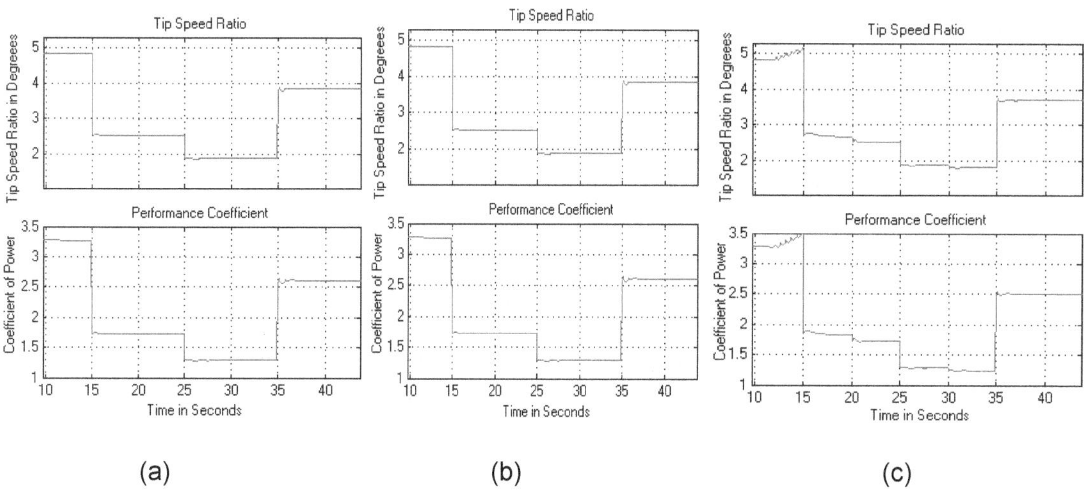

Figure 10. Tip speed ratio and Coefficient of power Cp for (a) change in wind speed alone, (b) reactive power change & wind speed variation, (c) both grid voltage & wind speed changes

It can be observed that when wind speed is at 8m/s, tip speed ratio (TSR) is high near 4.8 degrees and slowly decreases to 2.6 degrees at 15s when speed increases to 15m/s, further increased to 1.9o at 25s when speed of wind is 20m/s and decreased to 3.9o when wind speed decreased to 10m/s at 35s. In the similar way, Cp is also changing from 3.25 to 1.7 at 15s, and further decreased to 1.25 at 25s, and then increased to 2.55 at 35 seconds with wind speed variation from 8 to 15 and then to 20, and 10 m/s.

The variation in TSR and Cp with change in reactive power is independent and has no effect as shown in Figure 10 (a and b). However, with change in grid terminal voltage, a very small change in TSR and Cp can be observed. It is due to the fact that the TSR and Cp depends on parameters as described by equations 1 to 5 and is independent on voltage and reactive power. The TSR and Cp are blade size and shape with change in ambient temperature and wind speed dependant natural parameters.

5.2. Case B: Change in electromagnetic torque and rotor speed with wind speed, reactive power and grid voltage

The reference mechanical turbine torque and generator torque with magnitudes overlapping and variation of rotor speed for all three cases comparison is shown in Figure 11. In this the reference and actual torque waveform with blue color is turbine reference torque and pink color lines are for generator torque. It can be observed in Figure 11 (a, b and c), with increase in wind speed, torque is increasing and vice-versa. Till time up to 15 seconds, wind speed is at low value of 8m/s, so torque is at -0.2pu and increased to -0.5pu at 15s with increase in wind speed to 15m/s. The torque further increased to -0.9pu when wind speed is 20m/s and decreased to -0.28pu when speed decreased to 10m/s. there are small surges in torque waveform because of sudden change in wind speed. These surges can be minimized if a flywheel is used between turbine and generator, but has disadvantage of increase in weight, cost and maintenance. With the change in wind speed, rotor speed is also varying but is maintained nearly at constant value of 1.3pu RPM. In the first case, reactive power was at 0pu and grid terminal voltage is 1pu.

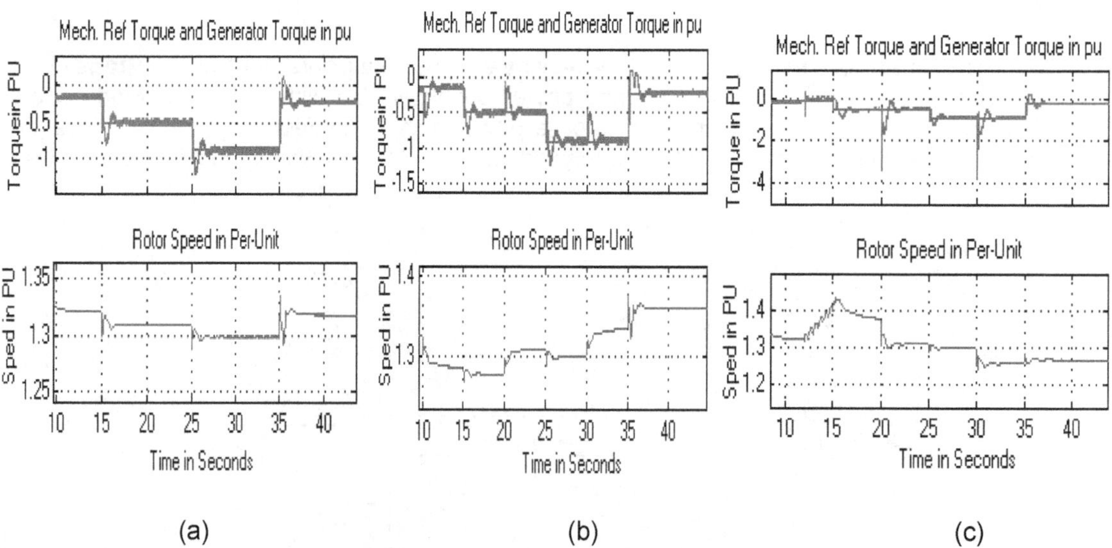

Figure 11. Reference and actual generator torque and rotor speed variation with time for: (a) change in wind speed alone, (b reactive power change & wind speed variation, (c) both grid voltage & wind speed changes

The changes in torque has effect with change in reactive power as in Figure 11 (b) and further more surges been observed when grid voltage disturbance occurred as in Figure 11 (c) is taking place. When reactive power is lagging at -0.6pu, there is a small surge in torque at 20 seconds. Generator speed is also low at 1.27pu at -0.6pu reactive power, while at 0pu reactive power, it is 1.32pu speed. But rotor speed increased to 1.4pu speed at low terminal grid voltage of 0.8pu. When, reactive power changes to 0pu from -0.6pu, rotor speed increased to 1.3pu from 1.27pu and grid terminal voltage changes to 1pu from 0.8pu between 20 to 30 seconds. Speed further increased to 1.35pu with leading reactive power of +0.6pu and decreased when grid voltage increased from 1pu to 1.2pu. Therefore rotor speed increases if reactive power changes from lagging (-ve) to leading (+ve) and rotor speed decreases with increase in grid terminal voltage beyond 1pu value in rms. The surges in torque will be observed very high when terminal grid voltage changes is due to the fact of change in mechanical power is not that faster than in comparison with electrical power change, which can be understandable using equal area criterion for SMIB system.

5.3. Case C: Change in stator voltage and current with wind speed, reactive power and grid voltage

The change in stator voltage and current with all three cases is shown in Figure 12 and zoomed voltage and current is shown in Figure 13. It can be observed that the stator terminal voltage is constant with change in wind speed as in Figure 12 (a) or with change in reactive power as in Figure 12 (b). There is an increase in current from 0.18pu to 0.5pu at 15 seconds with increase in wind speed from 8 to 15m/s and further increased to 0.9pu amps when speed increased to 20m/s and decreased to 0.3pu amps when speed of wind is 10m/s as shown in Figure 12 (c). But with change in reactive power, terminal voltage is nearly constant but there is a large change in current and voltage angle, hence large magnitude and angle change in current. When reactive power (Q) change from 0pu to -0.6pu, current increased from 0.15pu 0.5pu amps and decreased to 0.5pu amps when Q changes from -0.6pu to 0pu and increased to 1pu amps when speed of wind is 20m/s and reactive power is 0.6pu.

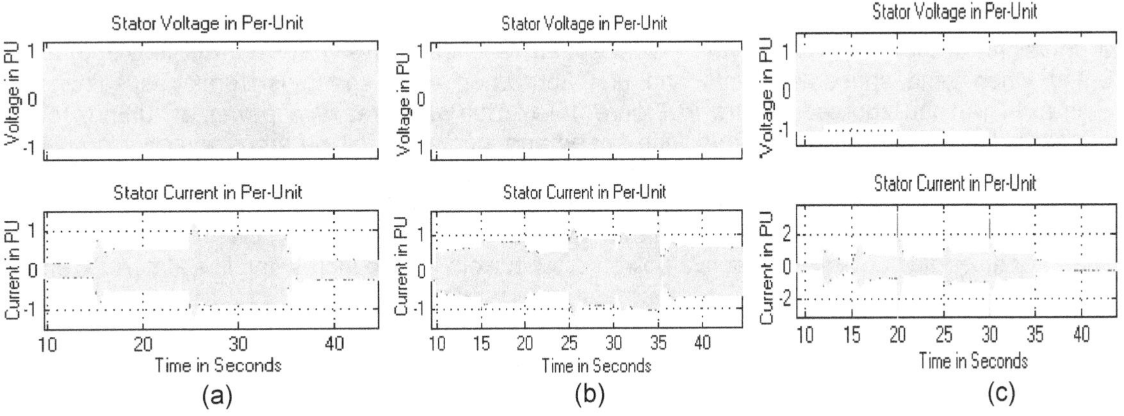

Figure 12. Stator voltage and current for (a) change in wind speed alone, (b) reactive power change & wind speed variation, (c) both grid voltage & wind speed changes

With sudden decrease in grid terminal voltage from 1pu to 0.8pu volts at 12 seconds, slowly stator current decreased exponentially when wind speed is very low of 0.1pu amps at 8m/s and this current was improved to 1pu when wind speed increased to 15m/s. But when terminal voltage changed to 1pu from 0.8pu at 20s, current again reached to normal value of 0.5pu amps as in case 1 and the current increased to again 1pu when wind speed reaches 20m/s. when the grid terminal voltage increased to 1.2pu from 1pu, the stator current again decreased to 0.8pu amps and when wind speed finally reaches 10m/s with voltage at 1.2pu, the current is 0.2pu Amps as in Figure 12 (c) and 12 (a). Hence with increase in voltage at constant wind speed, current decreases and with increase in wind speed at same voltage current will increase and vice-versa. The zoomed stator voltage and current waveform for a particular time period of nearly 1 second for all three cases is shown in Figure 13. In the similar way as does in stator voltage and current, rotor voltage and current will also vary, but rotor current is bi-directional unlike stator current does.

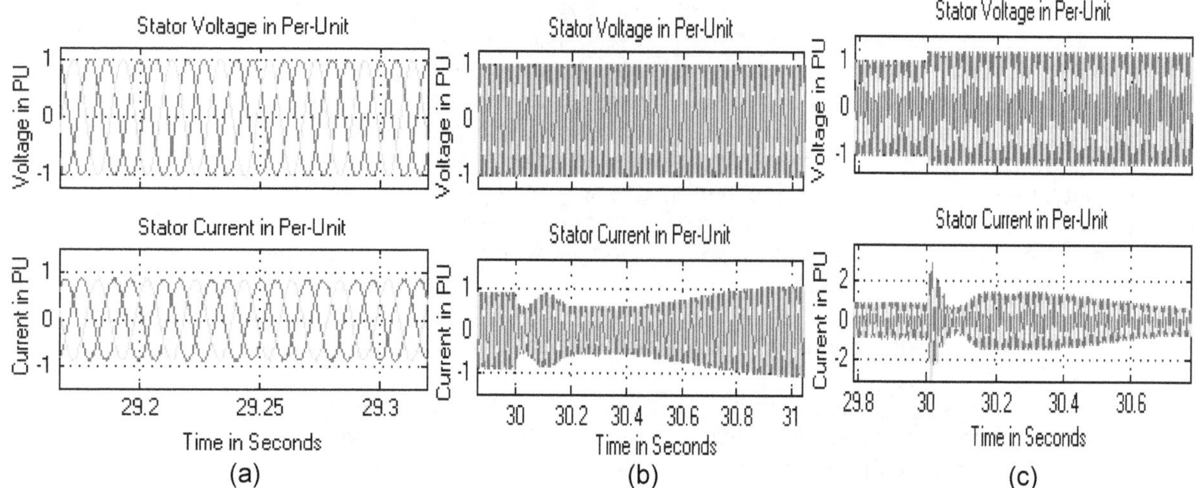

Figure 13. Stator voltage and current for (a) change in wind speed alone, (b) reactive power change & wind speed variation, (c) both grid voltage & wind speed changes

5.4. Case D: Change in rotor voltage and current with wind speed, reactive power and grid voltage

In all the three cases, rotor voltage is nearly constant at 0.32pu, but current is varying with variation in wind speed alone Figure 14 (a), with both wind speed and reactive power change in Figure 14 (b) and for voltage and wind speed variation as in Figure 14 (c). For the first case, with increase in rotor speed, rotor current increases and vice versa. When wind speed is at 8m/s, rotor current is 0.2pu, when wind speed reaches 15m/s, rotor current is 0.5pu, it is 0.9pu when wind speed is 20m/s and is 0.3pu when wind speed is 10m/s as shown in Figure 14 (a) and zoomed picture in Figure 15(a). But when reactive power at -0.6pu, rotor current is 0.8pu is even low at 8m/s wind speed and increased to 1pu amps when wind speed reaches 15m/s as in Figure 14 (b). When reactive power reaches 1pu, rotor current is 0.5pu amps at wind speed of 15m/s and for leading reactive power of +0.6pu, the rotor current is again 1pu at wind speed of 20m/s and 0.5pu amps at 10m/s wind speed. With increase in wind speed or at leading or lagging reactive power, rotor current is also increasing like stator current.

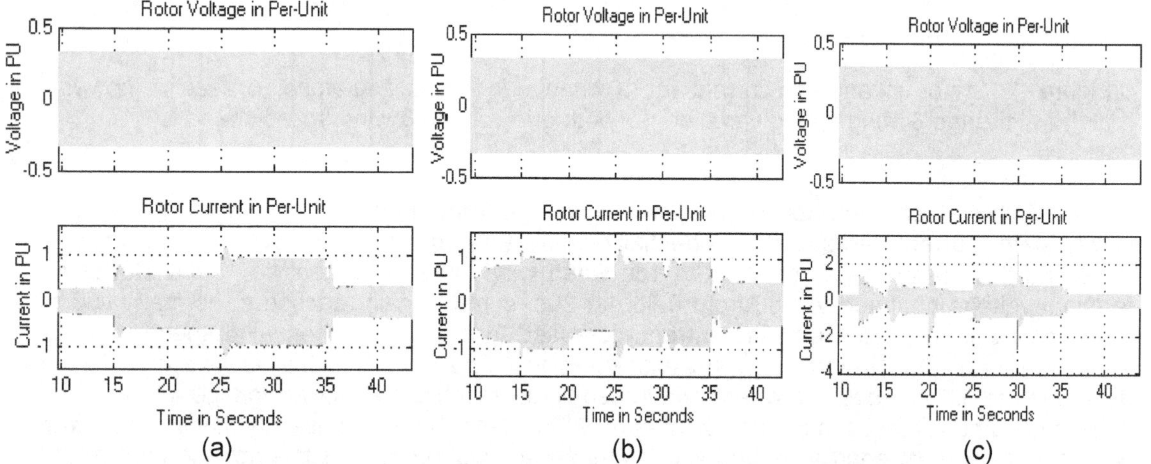

Figure 14. Rotor voltage and current (a) change in wind speed alone, (b) reactive power change & wind speed variation, (c) both grid voltage & wind speed changes

In the same scenario, rotor current is decreasing with increase in grid terminal voltage and vice-versa but without any appreciable change in rotor voltage. With sudden changes in voltage at 12, 20 and 30 seconds as in Figure 14 (c), there are few spikes in rotor current due to sudden reversal of current magnitude and angle with respect to terminal voltages respectively. The zoom in rotor voltage and current in the time period between 29.8 to 30.6 seconds for change in voltage and wind speed is shown in Figure 15 (c).

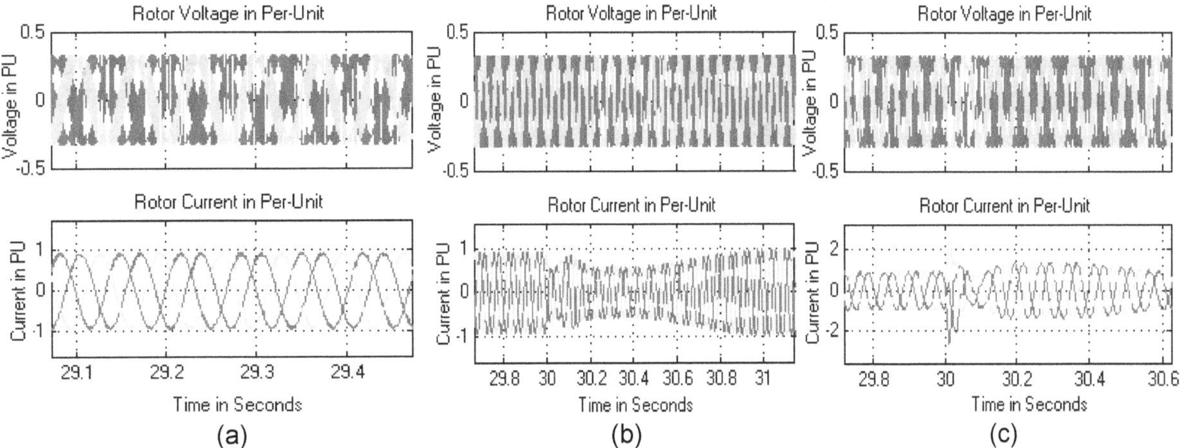

Figure 15. Rotor voltage and current for (a) change in wind speed alone, (b) reactive power change & wind speed variation, (c) both grid voltage & wind speed changes

5.5. Case E: Change in stator real and reactive power with wind speed, reactive power and grid voltage

The stator real and reactive power flow for all three cases is shown in Figure 16. The reference power which is the mechanical power output from turbine and actual generator real power change is shown in Figure 17. In the first case with change in wind speed, with very low wind speed of 8m/s, output stator real power is 0.1pu watts till 15 seconds. When wind speed reaches 15m/s, stator real power increased to 0.5pu and further increased to 0.8pu for 20m/s wind speed at 25s and decreased to 0.2pu power at 35s for 10m/s speed as shown in Figure 16 (a). During the change in wind speed, real power alone is changing and reactive power is constant at reference of 0pu. There are few surges in the reactive power due to change in voltage angle with respect to grid and also mainly due to change in stator and rotor current flows and rotor voltage change. With the change in reactive power demand from grid from 0pu to -0.6pu and +0.6pu at 12 and 30 seconds are shown in Figure 16 (b). It can be observed that with change in reactive power from 0pu to -0.6pu, reactive power from generator is changing with a small time lag of 0.8s and real power maintained nearly constant value of 0.1pu at 8m/s wind speed. Similarly with reactive power changing to 0pu and +0.6pu, the reactive power is changing within 1 second and real power is almost constant with small surges in real stator power during this transient.

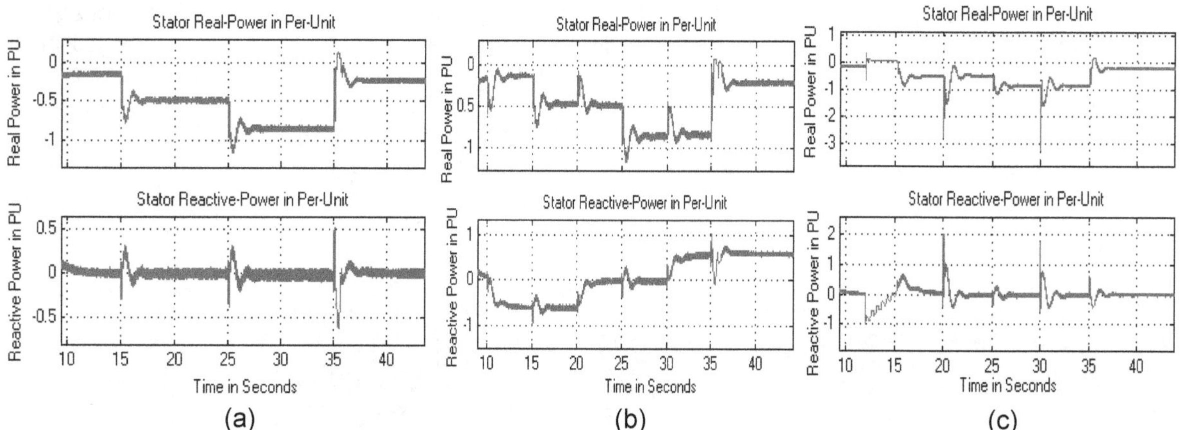

Figure 16. Stator real and reactive power waveform with time for (a) change in wind speed alone, (b) reactive power change & wind speed variation, (c) both grid voltage & wind speed changes

In third case with both voltage and wind speed changing, with the grid voltage variation from 1 to 0.8pu at 12th second, real power which is at 0.1pu changed to 0.05pu and reactive power which is at 0pu reached 1pu at this 12th second instant and slowly decaying to reach to reference 0pu value. This change in reactive power is to make voltage of stator to get adjusted to grid voltage without losing synchronism.

5.6. Case F: Comparison of reference and actual stator real and reactive power with change in wind speed, reactive power and grid voltage

The reference mechanical power output is shown with pink line and generator power is with blue line for the first case is shown in Figure 17 (a). It can be observed that, nearly generator actual power is matching with reference power and the mismatch is because of looses in turbine, gear wheels and generator and this mismatch is inevitable. With increase in wind speed, reference power is increasing and vice versa. When wind speed is 8m.s, output electrical real power is 0.1pu till 15s and reaches 0.4 and 0.8pu at 15 and 25 seconds with wind speed changing from 15 to 25m/s and then decreases to 0.2pu due to decrease in wind speed to 10m/s respectively. With the change in voltage at grid, stator terminal real power is maintained at constant value but with surges at instant of transient but reactive power is adjusting till stator voltage reaches the grid voltage for maintaining synchronism as shown in Figure 16 (c) and 17 (c). At the instant of 20 and 30 seconds, there is surge in real and reactive powers but were maintaining constant stator output real powers of 0.5 and 0.8pu watts and 0pu Var as in Figure 16 (c).

The Figure 6h is the output real and reactive powers from stator and rotor adding vector ally as shown in blue color lines and reference real power from turbine and reactive power from grid terminal in pink. To meet the desired grid reactive power, both stator and rotor has to supply for faster dynamics with an aid to RSC and GSC control schemes and is achieving as shown in Figure 17 (b). With the change in wind speed and reactive power, real power from generator is matching its reference value for case 2, but small deviation can be observed from time 30 to 35 seconds is due to sudden change in reactive power demand from grid and the deviation in real power is from 0.8pu to 0.7pu which is small. However reactive power is following its trajectory within 1 second.

In the case 3, both voltage and reactive power changing with time, total output real and reactive powers from stator and rotor is delivering to grid to meet the desired grid power demand. Unlike with change in reactive power, change in voltage is not affecting any deviation in real power and is following the trajectory nearly accurate with maximum deviation of 5% in real power. The reactive power change with grid voltage is high when voltage decreased from 1pu to 0.8pu volts.

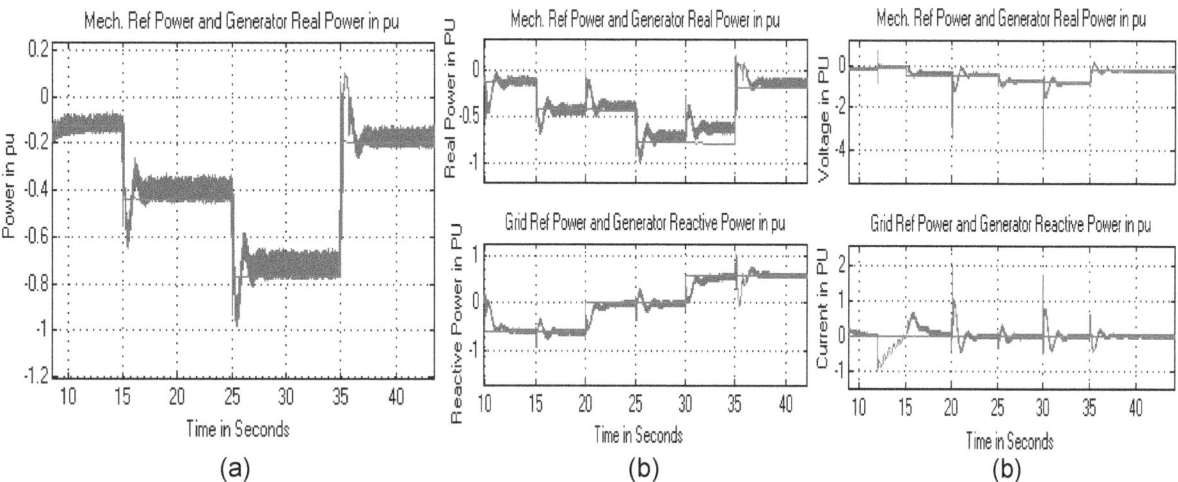

Figure 17. generator reference and actual real power waveform with time for (a) change in wind speed alone, generator real and reactive power flows for actual and reference for change in (b) reactive power change & wind speed variation, (c) both grid voltage & wind speed changes

When wind speed is increasing, mechanical and electrical torques are increasing without any change in stator reactive power. Variation in grid reactive power causes quadrature currents on both stator and rotor to change but torque, speed or real powers from stator or rotor remains unaltered. The variation in grid terminal voltage, a very small change in TSR and Cp can be observed. It is due to the fact that the TSR and Cp depends on parameters and is independent on voltage and reactive power. The TSR and Cp are blade size and shape with change in ambient temperature and wind speed dependant natural parameters. With increase in voltage at constant wind speed, current decreases but, with increase in wind speed at same voltage, current will increase and the decrease in wind speed caused current from stator to rotor decreases with stator voltage as constant as depicted by grid. In the similar way as does in stator voltage and current, rotor voltage and current will also vary, but rotor current is bi-directional unlike stator current does. The change in voltage at grid, stator terminal real power is maintained at constant value but with surges at instant of transient but reactive power is adjusting till stator voltage reaches the grid voltage for maintaining synchronism. To meet the desired grid reactive power, both stator and rotor has to supply for faster dynamics and it depends on faster action of RSC and GSC control schemes. The change in three cases is tabulated below. The variation in reactive power and grid voltage variations during the respective time period is shown in Table 2 and Table 3. In the Table 1, due to change in wind speed input to turbine alone, generator and wind turbine parameters change are summarized. There are surges produced in electromagnetic torque (EMT) due to variations in reactive power and grid voltage. Large spikes in stator current and rotor current are produced due to sudden increase or decrease in grid voltage. Certain deviations in rotor speed can be observed due to change in reactive power or grid voltage. It is due to variation in current flow in the rotor circuit, thereby variation in rotor flux and hence rotor speed.

Table 1. Change in Reactive power during the time period along with change in wind speed

Time range (s)	0.-20	20-30	30-50
Reactive power (pu)	-0.6	0	0.6

Table 2. Change in grid voltage during the time period along with change in wind speed

Time range (s)	0.-12	12-20	20-30	30-50
Reactive power (pu)	1	0.8	1	1.2

Table 3. Variation of turbine and generator parameters with change in wind speed input

Wind speed (m.s)	8	15	20	10
Time (s)	10	15	25	35
TSR (degree)	4.9	2.5	1.95	3.95
Cp	3.25	1.75	1.25	2.60
EMT (pu)	-0.2	-0.5	-0.8	-0.3
Rotor speed (pu)	1.32	1.31	1.30	1.33
Stator current (pu)	0.2	0.5	0.8	0.3
Rotor current (pu)	0.25	0.55	0.90	0.35
Stator power (pu)	-0.2	-0.5	-0.8	-0.3
Rotor power (pu)	0	0	0	0

6. Conclusion

From the proposed control scheme, the torque, speed and reactive power control of DFIG is very specific. With change in wind speed, electromagnetic torque surges are low and the variation in wind speed is not getting the generator rotor speed variation is due to better transition in gear wheel mechanism. Reactive power demand from grid is accurate which can be met by proper control action of RSC and GSC. The proposed methodology is accurate and following all basic mathematical equations explained in previous section. Distinct from reactive power variation, change in voltage is not affecting any deviation in real power and is following the trajectory nearly accurate with maximum deviation of 5% in real power. The reactive power change with grid voltage is high when voltage decreased from 1pu to 0.8pu volts. Hence the proposed control scheme can be applied with ever changing transients like large variation in wind speed, reactive power and grid voltage. The system is very stable without losing synchronism when grid voltage is increasing or decreasing to a ±0.2pu change from nominal voltage value.

Appendix

The parameters of DFIG used in simulation are,
Rated Power = 1.5MW, Rated Voltage = 690V, Stator Resistance Rs = 0.0049pu, rotor Resistance Rrl = 0.0049pu, Stator Leakage Inductance Lls = 0.093pu, Rotor Leakage inductance Llr1 = 0.1pu, Inertia constant = 4.54pu, Number of poles = 4, Mutual Inductance Lm = 3.39 pu, DC link Voltage = 415V, Dc link capacitance = 0.2F, Wind speed = 14 m/sec.
Grid Voltage = 25 KV, Grid frequency = 60 Hz.
Grid side Filter: Rfg = 0.3Ω, Lfg = 0.6nH
Rotor side filter: Rfr = 0.3mΩ, Lfr = 0.6nH
Wind speed variations: 8, 15, 20 and 10 at 15, 25 and 35 seconds.
Reactive power change: -0.6 to 0 and +0.6pu at 20 and 30 seconds.
Grid voltage change: 0.8 to 1 and to 1.2pu at 20 and 30 seconds.

Nomenclature

L_{ls}, L_{lr}, L_m Stator or Rotor leakage reactance and magnetizing reactance
R_s, R_r Stator or Rotor resistance
$V_{sd}, V_{rd}, V_{sq}, V_{rq}$ two axis stator or rotor voltage
$I_{sd}, I_{rd}, I_{sq}, I_{rq}$ two axis stator or rotor current
$\Psi_{sd}, \Psi_{rd}, \Psi_{sq}, \Psi_{rq}$ two axis stator or rotor flux linkage
P_s, P_r Stator or Rotor real power
Q_s, Q_r Stator or Rotor reactive power
ω_s, ω_r Stator or Rotor speed
T_e Electromagnetic torque
p Pole pairs
s Slip
p.u. per unit

References

[1] Aghanoori, N, Mohseni, M, Masoum, MAS. Fuzzy approach for reactive power control of DFIG-based wind turbines. IEEE PES Innovative Smart Grid Technologies Asia (ISGT). 2011: 1-6.

[2] Syed Muhammad Raza Kazmi, Hiroki Goto, Hai-Jiao Guo, Osamu Ichinokura. A Novel Algorithm for Fast and Efficient Speed-Sensor less Maximum Power Point Tracking in Wind Energy Conversion Systems. *IEEE Transactions On Industrial Electronics*. 2011; 58(1): 29-36.

[3] Iwanski, G, Koczara, W. DFIG-Based Power Generation System With UPS Function for Variable-Speed Applications. *IEEE Transactions on Industrial Electronics*. 2008; 55(8): 3047–3054.

[4] Aktarujjaman, M, Haque, ME, Muttaqi, KM, Negnevitsky, M, Ledwich, G. Control Dynamics of a doubly fed induction generator under sub- and super-synchronous modes of operation. IEEE Power and Energy Society General Meeting - Conversion and Delivery of Electrical Energy in the 21st Century. 2008: 1–9.

[5] Iwanski, G, Koczara, W. DFIG-Based Power Generation System With UPS Function for Variable-Speed Applications. *IEEE Transactions on Industrial Electronics*. 2008; 55(8): 3047-3054.

[6] Dawei Zhi, Lie Xu. Direct Power Control of DFIG With Constant Switching Frequency and Improved Transient Performance. *IEEE Transactions On Energy Conversion*. 2007; 22(1): 110–118.

[7] Lie Xu, Cartwright, P. Direct active and reactive power control of DFIG for wind energy generation. *IEEE Transactions on Energy Conversion*. 2010; 25(4): 1028-1039.

[8] Mustafa Kayıkc, Jovica V. Milanovi. Reactive Power Control Strategies for DFIG-Based Plants. *IEEE Transactions On Energy Conversion*. 2007; 22(2): 389-396.

[9] Stephan Engelhardt, Istvan Erlich, Christian Feltes, Jorg Kretschmann, Fekadu Shewarega. Reactive Power Capability of Wind Turbines Based on Doubly Fed Induction Generators. *IEEE Transactions On Energy Conversion*. 2011; 26(1): 364- 372.

[10] Gerardo Tapia, Arantxa Tapia, and J. Xabier Ostolaza. Proportional–Integral Regulator-Based Approach to Wind Farm Reactive Power Management for Secondary Voltage Control. *IEEE Transactions On Energy Conversion*. 2007; 22(2): 488- 498.

[11] Dongkyoung Chwa, Kyo-Beum Lee. Variable Structure Control of the Active and Reactive Powers for a DFIG in Wind Turbines. *IEEE Transactions On Industry Applications*. 2010; 46(6): 2545-2555.

[12] Lingling Fan, Haiping Yin, Zhixin Miao. On Active/Reactive Power Modulation of DFIG-Based Wind Generation for Interarea Oscillation Damping. *IEEE Transactions On Energy Conversion*. 2011; 26(2): 513-521.

[13] Hua Geng, Cong Liu, Geng Yang. LVRT Capability of DFIG-Based WECS Under Asymmetrical Grid Fault Condition. *IEEE Transactions On Industrial Electronics*. 2013; 60(6): 2495-2509.

[14] Shukla, RD, Tripathi, RK. Low voltage ride through (LVRT) ability of DFIG based wind energy conversion system II. Students Conference on Engineering and Systems (SCES). 2012: 1-6.

[15] Guzmán Díaz. Optimal primary reserve in DFIGs for frequency support. *International journal of Electrical Power and Energy Systems*. 2012; 43: 1193–1195.

[16] F. Poitiers, T. Bouaouiche, M. Machmoum. Advanced control of a doubly-fed induction generator for wind energy conversion. *Electric Power Systems Research*. 2009; 79: 1085–1096.

[17] Cardenas, R, Pena, R, Perez, M, Clare, J, Asher, G, Wheeler, P. Power smoothing using a flywheel driven by a switched reluctance machine. *IEEE Trans. Ind. Electron*. 2006; 53(4): 1086–1093.

[18] Muljadi, E, Butterfield, CP. Pitch-controlled variable-speed wind turbine generation. *IEEE Trans. Ind. Appl*. 2001; 37(1): 240–246.

[19] Wei, Q, Wei, Z, Aller, JM, Harley, RG. Wind speed estimation based sensorless output maximization control for a wind turbine driving a DFIG. *IEEE Trans. Power Electron*. 2008; 23(3): 1156–1169.

[20] Sharma, S, Singh, B. Control of permanent magnet synchronous generator-based stand-alone wind energy conversion system. *IET Power Electron*. 2012; 5(8): 1519–1526.

[21] Kazmi, SMR, Goto, H, Hai-Jiao, G, Ichinokura, O. A Novel algorithm for fast and efficient speed-sensorless maximum power point tracking in wind energy conversion systems. *IEEE Trans. Ind. Electron*. 2011; 58(1): 29–36.

[22] Mathiesen, BV, Lund, H. Comparative analyses of seven technologies to facilitate the integration of fluctuating renewable energy sources. *IET Renew. Power Gener*. 2009; 3(2): 190–204.

[23] Takahashi, R, Kinoshita, H, Murata, T, et al. Output power smoothing and hydrogen production by using variable speed wind generators. *IEEE Trans. Ind. Electron*. 2010; 57(2): 485–493.

[24] Bragard, M, Soltau, N, Thomas, S, De Doncker, RW. The balance of renewable sources and user demands in grids: power electronics for modular battery energy storage systems. *IEEE Trans. Power Electron*. 2010; 25(12): 3049–3056.

[25] Bhuiyan, FA, Yazdani, A. Reliability assessment of a wind-power system with integrated energy storage. *IET Renew. Power Gener*. 2010; 4(3): 211–220.

[26] Sen, PC, Ma, KHJ. Constant torque operation of induction motor using chopper in rotor circuit. *IEEE Trans. Ind. Appl*. 1978; 14(5): 1226–1229.

[27] Geng, H, Yang, G. Robust pitch controller for output power leveling of variable-speed variable-pitch wind turbine generator systems. *IET Renew. Power Gener*. 2009; 3(2): 168–179.

[28] Astrom, KJ, Hagglund, T. PID controllers: theory, design and tuning. Instrument Society of America, Research Triangle Park, NC. 1995.
[29] Lin, W, Hong, C. A new elman neural network-based control algorithm for adjustable-pitch variable-speed wind-energy conversion systems. *IEEE Trans. Power Electron.* 2011; 26(2): 473–481.
[30] Belkacem, Belabbas, et al. Hybrid fuzzy sliding mode control of a DFIG integrated into the network. *International Journal of Power Electronics and Drive Systems (IJPEDS).* 2013; 3(4): 351-364.

3

Cost Optimal Design of a Single-Phase Dry Power Transformer

Raju Basak[1], Arabinda Das[2], Amarnath Sanyal[3]
[1,2]Electrical Engineering Department, Jadavpur University, Kolkata -700032, West Bengal, India
[3]Calcutta Institute of Engineering and Management, Kolkata -700040, West Bengal, India
email: basakraju149@gmail.com

Abstract

The Dry type transformers are preferred to their oil-immersed counterparts for various reasons, particularly because their operation is hazardless. The application of dry transformers was limited to small ratings in the earlier days. But now these are being used for considerably higher ratings. Therefore, their cost-optimal design has gained importance. This paper deals with the design procedure for achieving cost optimal design of a dry type single-phase power transformer of small rating, subject to usual design constraints on efficiency and voltage regulation. The selling cost for the transformer has been taken as the objective function. Only two key variables have been chosen, the turns/volt and the height: width ratio of window, which affects the cost function to high degrees. Other variables have been chosen on the basis of designers' experience. Copper has been used as conductor material and CRGOS as core material to achieve higher efficiency, lower running cost and compact design. The electrical and magnetic loadings have been kept at their maximum values without violating the design constraints. The optimal solution has been obtained by the method of exhaustive search using nested loops.

Keywords: *transformer, optimal design, design variables, design constraints, exhaustive search*

1. Introduction

Dry type transformers have recently become much popular [1]. In the earlier days these were used for small rating power transformers only. But now-a-days these are also finding application as distribution transformers and they are being used for relatively much higher rating. In oil-filled transformers, the maintenance requirement is relatively more. A conservator is to be provided to take care of expansion and contraction of oil with load cycles. The ingress of moisture is to be counteracted by providing breather connected to the conservator. It has to be filled up with fused silica gel which need be checked at regular intervals. If oil-filled transformers are not properly maintained, there are chances of dielectric failure due to contamination of oil. In extreme cases the tank may burst and hot oil thrown off around the tank thus endangering near-by people. Moreover, the presence of oil is always associated with some amount of leakage which makes the near-by places dirty. For all these reasons, dry transformers are getting preference over the oil-filled type [2]-[4].

Oil is used as filler for two reasons- better cooling and enhancing the dielectric strength of oil. Cooling conditions are inferior in case of a dry transformer as the cooling medium is air. Therefore dry transformers are to be designed for lower current density. The cost of conductor, and as such the over-all cost, for the same rating, becomes a little higher in comparison to oil-filled transformers. But in consideration of compactness of the design, absence of oil-hazards and neatness of the environment makes dry transformers quite attractive. Dry type resin-cast transformers are still more expensive but their performance is highly satisfactory [5],[6].

2. Cost Optimality and Quality Design

As dry transformers are gradually replacing their oil-filled counterparts, their cost-optimal design is of great importance in the present context. In this paper, the cost of production (selling cost) has been taken as the objective function. However, the annual running losses or life-time losses may also be included in the cost function for dual optimization from the point of view of the customer and the manufacturer [7].

The core construction has been used as it is more economic than the shell type. Copper has been used as conductor for compactness of design. High price nomex-insulated conductors have been recommended as they can withstand higher temperature. Cold rolled grain-oriented silicon steel has been recommended as core material. The choice of materials has been made with a view to achieve higher efficiency and reduce the running losses [6],[7].

3. The Optimizing Techniques

A design problem does not have a unique solution. Generally there are many feasible solutions. Optimization is the way to find out the best possible solution out of them. Several techniques have evolved through centuries to deal with optimal solution. Some of them are traditional based on mathematical formulation and some of them are non-traditional based on soft-computing techniques. The chief contribution is from the mathematicians. The engineers and the technicians have merely applied them to real world industrial problems and designs. Searching for the optimal solution requires much iteration. It is difficult to expedite the same through long-hand calculations using a calculator. For reaching optimal solution recourse must be made to a digital computer as it can perform calculations at an extremely high rate and it has got large amount of memory. De facto, it is the advent of computer which has revolutionized the field of optimization [8]-[10].

The process of optimization starts by choosing the design variables for a particular problem and listing the design constraints. The key variables are to be identified and their bounds defined. Then the objective function (or the cost function) is to be formulated. Then we are to search for optimality (maximality or minimality as the case may be) using an appropriate mathematical technique.

In this section, it is explained the results of research and at the same time is given the comprehensive discussion. Results can be presented in figures, graphs, tables and others that make the reader understand easily [11],[12]. The discussion can be made in several sub-chapters.

3.1. Exhaustive search method

Using nested loops is the simplest of all search methods. The optimum of a function is bracketed by calculating the function values at a number of equally spaced points.Usually the search begin from a lower bound on the variable and the consecutive function values are compared at a time based on the assumption of unimodality of the function. Based on the outcome of comparison, the search is either terminated or continued by replacing one from the above points with a new point. Figure 1 shows the method graphically for a univaraiate objective function.

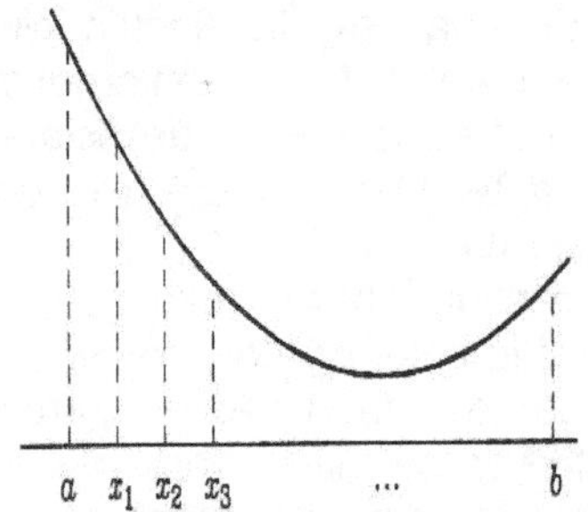

Figure 1. Exhaustive search plane

3.2. Multivariable search

Single variable optimization is simple. But unfortunately, the design optimization problems are multi-variable and constrained. Obviously, the problems are complicated by nature. These are addressed by methods like:

i) Repeated unidirectional search (in each direction defined by the variables)
ii) Direct search methods or
iii) Gradient-based methods.

The Direct search or gradient methods are mathematically sophisticated. On the contrary, the method of exhaustive search is much simpler. It is a bracketing method, normally used for single variable but it can be extended to multiple variables. Though the no of iterations is more in this method, the solution does not get stuck to local minima. If the step lengths are not too large, then it is almost sure to reach the optimal solution. Therefore, we have chosen the method of exhaustive search [12]-[14].

There are also powerful methods for design optimization using soft-computing techniques [15],[16] e.g., genetic algorithm, simulated annealing etc.

4. Design Variables and Constraints

There are several design variables for the design of a power transformer. Some of them affect the objective function to a large extent. These are called key variables. Other variables have rather less influence on the objective function. The optimal solution is sought by varying the key variables only. Other variables are kept constant at their usual values.

The variables may be continuous or integer. For example, the no of turns of the primary and the secondary windings in each limb must be an integer. However if the number is large, they can be treated as continuous variables. The following key variables have been identified:

K = emf constant (in $E_t = K\sqrt{S}$, where E_t = emf/turn, S = rating in KVA)

$R_w = H_w / W_w$ = Height: Width ratio of the transformer;

δ = current density in the conductor in A/mm^2; B_m= Maximum flux-density, Tesla

It has been found that the cost of production continuously reduces with increasing B_m and δ. So they are not candidate variables for optimization. They are fixed at their maximum values for which the constraints are not violated. Appropriate values for other variables have been chosen by consulting design data-books and text-books.

In addition, there are some decision-variables e.g.

(a) Core or shell construction: Core construction has been used for greater economy

(b) Conductor material- copper or aluminium: Higher efficiency is achievable with copper and the design is compact. Therefore, copper has been used as conductor materials.

(c) Core material- CRNOS or CRGOS: CRGOS has been used to reduce the iron loss to minimum and achieve greater efficiency.

(d) Cooling AN or ANAF: Air natural cooling has been used. It is enough for transformers of small ratings. To keep the temperature rise within statutory limits.

Constraints have been imposed on: Maximum % copper loss ≤ 2.5%; Maximum % iron loss ≤ 1.0%, with a view to limit the temperature rise; Maximum efficiency at CMR ≥ 96%. Maximum temperature rise of the coil ≤ 75 °C; Voltage regulation ≤ 3%

4. Algorithm

The following algorithm has been used to solve the design problem

Step 1: Input specifications of the power transformer: power and voltage ratings.

Step 2: Choose copper as conductor material, CRGOS as core material. Initialize: min ← a large number

Step 3: Input user-specified data for design variables: flux density (B_m), current density (δ), number of core steps (N_{st}) etc.

Step 4: For i = 0 to 30 do: K ←0.5 + i / 100 'Bound: 0.5 to 0.8

Step 5: For j = 0 to 20 do: R_w ← 2.5 + j / 20 'Bound 2.5 to 3.5

Step 6: Go to transformer design sub-routine, calculate the objective function, C (i, j) and the performance variables.

Step 7: If efficiency < 96% Go to step 12

Step 8: If voltage regulation > 3% Go to step 12

Step 9: If temp. Rise of coil > 90°C go to step 12

Step 10: if C (i, j) > min go to step 12

Step 11: K_{min} ← K(i); $R_{w\,min}$ ← R_w(j); C_{min} ← C (I, j)

Step 12: end for

Step 13: end for

Step 14: Put K ← K_{min}; R_w ← $R_{w\,min}$

Step 15: Go to transformer design sub-routine,

Calculate the objective function, C(i) and the performance variables.
Step 16: Print out results
Step 17: Stop
Step 18: End

5. The Design Problem

Optimal design of a small dry type 1-phase transformer has been taken. Core construction has been preferred as it is more economic.

5.1. Transformer rating

KVA =5; frequency =50 Hz. Power factor = 0.8 lagging; Primary/Secondary voltage: 230 V/115V

5.2. Specifications

Efficiency: efficiency $\geq$0.96; voltage regulation: $\leq$3%; temp. Rise $\leq$ 75 °C. CRGOS has been used as core material & Copper as conductor material. Cruciform core has been used. Cylindrical type coils have been used for both primary and secondary. Flux-density/ current density have been kept at maximum values withoutviolating the constraints. Chosen value of magnetic and electric loading:
The flux-density= 1.6 Tesla and current density= 2.5 A/ mm^2.
The resistivity of copper at operating temperature has been taken as 0.021 Ω/m/mm^2
Other constants: stacking factor= 0.92; window space factor= 0.45
With reference to the table of convergence given below the objective function reduces toits minimum: Rs. 12100/- for: K = 0.65; R_w = 3.1

With these values of the design variables, we get the following results:
EMF/turn= 1.5541 Volt; No of turns of primary/secondary: 148 / 74
Core flux= 7.0002E-03 Waber; Net area of iron = 4.3752E-03 m2
Diameter of circumscribing cycle of the core= 8.7439E-02 m
Length of the sides in m: 0.074 / 0.046; Distance between core centers= 0.12787 m
Total width= 0.20187 m; Window height/width, m: 0.1415; 4.043E-02
Height of yoke/ total height in m: 6.4265E-02; 0.2700

5.3. Coil design

Primary and secondary coils are split into two halves in two limbs.
The primary/secondary current, a: 21.739; 43.478
Cylindrical windings are being used for both primary and secondary.
Required cross-section of the primary/secondary windings, mm^2: 8.6957; 17.391
Cross-sections are large for both primary and secondary. So strip conductors are being used.
Chosen dimension of the strip for secondary: 3.4 mm x 5.1 mm
Chosen dimension of the strip for primary: 1.6 mm x 5.4 mm
Actual conductor cross-section of primary and secondary, mm^2: 8.64; 17.34
Thickness of insulation on the core = 1.5 mm
Gap between LV and HV windings= 5 mm
Inside/Outside radius of secondary, mm: 90.439 ; 100.64
Inside/Outside radius of primary, mm: 110.66; 121.44
Clearance between adjacent HV windings= 6.4256 mm
The clearance is adequate for L.V. windings.

5.4. Performance evaluation

Resistance of the primary/secondary winding, Ω: 0.13031; 2.6823E-02
Copper loss at rated current= 112.29 W; % copper loss= 2.2458
The % leakage reactance= 0.85873, % No Load current = 1.558.
Voltage regulation at full load and 0.8 lagging p.f. = 2.312
Iron loss/kg= 1.9632 W
Iron loss at rated voltage= 45.12 W; % iron loss= 0.9024
Efficiency at rated load and nominal pf= 0.96214
Cooling coefficient has been taken as 0.05 (on the higher side as there is no forced cooling.)

Exposed area of the coil = 0.18568 m^2; Average temperature rise of coils=42.48 °C
The constraints on voltage regulation, efficiency and temperature rise have not been violated.

5.5. Cost analysis

Weight of iron= 22.984 Kg.; Cost of iron= Rs. 3677/-
Weight of copper = 7.614343 Kg.; Cost of copper= Rs. 4378/-
Total cost of iron and copper = Rs. 8056/-
Manufacturing cost allowing 50% on cost of materials= Rs. 12148/-
(The slight difference in cost arises out of integer variables replacing real e.g. no. of turns etc.)
Annual loss on the basis of average 80% loading = 842.31 kW-h.
Price of a BOT unit= Rs. 5/-
Annual cost of lost energy = Rs. 4211/-

Table 1 shows the improvement parameters over conventional design. Comparison with a manufacturer's data for the same machine is given in Table 2.

Table 1. The improvement parameters over conventional design

K	R_w	Cost	K	R_w	Cost	K	R_w	Cost
0.60	2.50	12184	0.60	2.60	12171	0.60	2.70	12161
0.60	2.80	12152	0.60	2.90	12145	0.60	3.00	12139
0.60	3.10	12135	0.60	3.20	12132	0.60	3.30	12130
0.60	3.40	12128	0.60	3.50	12128	0.61	2.50	12166
0.61	2.60	12154	0.61	2.70	12145	0.61	2.80	12137
0.61	2.90	12131	0.61	3.00	12126	0.61	3.10	12122
0.61	3.20	12120	0.61	3.30	12118	0.61	3.40	12118
0.61	3.50	12119	0.62	2.50	12151	0.62	2.60	12140
0.62	2.70	12132	0.62	2.80	12125	0.62	2.90	12119
0.62	3.00	12115	0.62	3.10	12112	0.62	3.20	12111
0.62	3.30	12110	0.62	3.40	12111	0.62	3.50	12112
0.63	3.50	12139	0.63	2.60	12130	0.63	2.70	12122
0.63	2.80	12116	0.63	2.90	12111	0.63	3.00	12108
0.63	3.10	12106	0.63	3.20	12105	0.63	3.30	12105
0.63	3.40	12106	0.63	3.50	12108	0.64	2.50	12130
0.64	2.60	12121	0.64	2.70	12114	0.64	2.80	12109
0.64	2.90	12105	0.64	3.00	12103	0.64	3.10	12102
0.64	3.50	12107	0.65	2.50	12124	0.65	2.60	12116
0.65	2.70	12110	0.65	2.80	12105	0.65	2.90	12102
0.65	3.00	12101	**0.65**	**3.10**	**12100**	0.65	3.20	12101
0.65	3.30	12102	0.65	3.40	12105	0.65	3.50	12108
0.66	2.50	12120	0.66	2.60	12113	0.66	2.70	12108
0.66	2.80	12104	0.66	2.90	12102	0.66	3.00	12101
0.66	3.10	12101	0.66	3.20	12102	0.66	3.30	12105
0.66	3.40	12100	0.66	3.50	12111	0.67	2.50	12119
0.67	2.60	12113	0.67	2.70	12109	0.67	2.80	12106
0.67	2.90	12104	0.67	3.00	12104	0.67	3.10	12105
0.67	3.20	12106	0.67	3.30	12109	0.67	3.40	12113
0.67	3.50	12117	0.68	2.50	12121	0.68	2.60	12115
0.68	2.70	12112	0.68	2.80	12109	0.68	2.90	12108
0.68	3.00	12109	0.68	3.10	12110	0.68	3.20	12113
0.68	3.30	12116	0.68	3.40	12120	0.68	3.50	12125
0.69	2.50	12124	0.69	2.60	12120	0.69	2.70	12117
0.69	2.80	12115	0.69	2.90	12115	0.69	3.00	12116
0.69	3.10	12118	0.69	3.20	12121	0.69	3.30	12125
0.69	3.40	12130	0.69	3.50	12136	0.70	2.50	12130
0.70	2.60	12126	0.70	2.70	12124	0.70	2.80	12123
0.70	2.90	12124	0.70	3.00	12126	0.70	3.10	12128
0.70	3.20	12132	0.70	3.30	12137	0.70	3.40	12142
0.70	3.50	12148						

Table 2. Comparison with a manufacturer's data for the same machine

Item	Optimal Design	Manufacturers' data
Total cost of materials	Rs. 8056/-	Rs. 9285/-
% iron loss	0.9024	0.5824
% copper loss	2.2458	3.2431
Efficiency at full load, 0.8 lagging p.f.	0.96214	0. 9724
% no load current	1.558	1.3
% Leakage reactance	0.85873	0.7534
% Voltage regulation at full load, 0.8 lagging p.f.	2.312	3.61

8. Conclusion

The paper has dealt with design optimization of a single phase dry type power transformer. The dry transformers are used where we like to have a compact design and trouble-free operation at low cost of maintenance. These transformers now find application as small and medium-sized transformers where cost is not an impediment. Core constructions are preferred as they are more economic [17],[18]. Best grade materials e.g. CRGOS as core material and refined copper as conductor material are used to achieve higher efficiency and to reduce the running losses. The key variables which affect the cost of production have been identified as the emf constant, K and the window height: width ratio R_w. The bounds on the design variables have been found out from the text-books on design. The optimal solution has been found out by the method of exhaustive search using nested loops. The step lengths have been chosen judiciously so as not to skip the optima. A case study has been made on a 5 KVA, 230/115 V, 50 Hz, 1-phase power transformer. The design details of the optimal machine have been documented. All the specifications have been fulfilled- no constraints have been violated.

If the design is not methodically made, we cannot aspire to reach an optimal solution. A comparison of performance and cost is given in tabular form (Table 2), where manufacturer's data has been compared with the obtained optimal design data. It shows that the cost of the optimal machine is much less and most of the performance indices are better.

References

[1] R. Basak, A. Das, A. Sensarma, AN. Sanyal. Discrete design Optimization of small open type dry transformers. *Bulletin Teknik Elektrodan Informatika.* 2012; 1(1): 37-42. ISSN: 2089-3191.

[2] AK. Shawney. A course in electrical machine design. DhanpatRai& Sons. 2003.

[3] MG. Say. Performance and design of A.C. machines. CBS Publishers & Distributors. 2005.

[4] I. Dasgupta. Design of Transformers. TMH, New Delhi. 2002.

[5] A. Still. Principles of Transformer Design. John Wiley and Sons Inc. 2007.

[6] MJ. Heathcote. The J & P Transformer Book 12th. Edition- a practical technology of the power transformer. 1998.

[7] A. Khatri, OP. Rahi. Optimal design of transformer: a compressive bibliographical survey. 2012; 1(2): 159-167. ISSN: 2277-1581.

[8] SB. Williams, PA. Abetti, EF. Mangnusson. Application of digital computers to transformer design. *AIEE Trans.* 1956; 75(III): 728-35.

[9] WA. Shapley, JB. Oldfield. The digital computer applied to the design of large power transformers. Proc. IEE. 1958; 105: 112-25.

[10] SB. Williams, PA. Abetti, HJ. Mason. Complete design of power transformer with a large size digital transformer. *AIEE trans.* 1959; 77: 1282-91.

[11] OW. Anderson. Optimum design of electrical machines. *IEEE Trans. (PAS).* 1967; 86: 707-11.

[12] M. Ramamoorty. Computer-aided design of electrical equipment. Affiliated East-West press. 1987.

[13] K. Deb. Optimization for engineering design. PHI. 2010. ISBN 978-81-203-0943-2.

[14] SS. Rao. Engineering optimization- theory and practice. New Age Int. 2013. ISBN978-81-224-2723-3.

[15] AK. Yadav, et al. Optimization of Power Transformer Design using Simulated Annealing Technique. *International Journal of Electrical Engineering.* 2011; 4(2): 191-198, ISSN 0974-2158.

[16] RA. Jabr. Application of geometric programming to transformer design. *IEEE Trans Magnetics.* 2005; 41: 4261-4269.

[17] A. Shanmugasundaram, G. Gangadharan, R. Palani. Electrical machine design data book. Wiley eastern Ltd. ISBN 0 85226 813 0.

[18] R. Basak, A. Das, AN. Sanyal. Optimal Design of a 3-phase core type Distribution Transformer using Modified Hooke and Jeeves Method. *TELKOMNIKA Indonesian Journal of Electrical Engineering.* 2014; 12(10): 7114-7122.

4

Performance Evaluation of GA optimized Shunt Active Power Filter for Constant Frequency Aircraft Power System

Saifullah Khalid
Department of Electrical Engineering,
IET Lucknow, India
e-mail: Saifullahkhalid@Outlook.com

Abstract

Sinusoidal Current Control strategy for extracting reference currents for shunt active power filters have been modified using Genetic Algorithm and its performances have been compared. The acute analysis of Comparison of the compensation capability based on THD and speedwell be done, and recommendations will be given for the choice of technique to be used. The simulated results using MATLAB model are shown, and they will undoubtedly prove the importance of the proposed control technique of aircraft shunt APF.

***Keywords**: Aircraft electrical system, Shunt Active Filter (APF), Sinusoidal Current Control strategy, GA, THD*

1. Introduction

More advanced aircraft power systems [1]-[3] have been needed due to increased use of electrical power on behalf of other alternate sources of energy. The subsystems like flight control, flight surface actuators, passenger entertainment, are driven by electric power, which flowingly increased the demand for creating aircraft power system more intelligent and advanced. These subsystems have extensive increased electrical loads i.e. power electronic devices, increased feeding of electric power, additional demand for power, and above to all of that great stability problems.

In peculiarity to standard supply system, the source frequency is of 50 Hz, whereas, aircraft AC power system works on the source frequency of 400 Hz [1]-[3]. Aircraft power utility works on source voltage of 115/200V. The loads applicable to the plane a system differs from the loads used in 50 Hz system [1]. When we deliberate the generation portion; aircraft power utility will remain AC driven from the engine for the plane primary power. Novel fuel cell technology can be used to produce a DC output for ground power, and its silence process would match up to suitably with the Auxiliary Power Unit (APU). Though when considering the dissemination of primary power, whether AC or DC; each approach has its merits. In DC distribution, HVDC power distribution systems permit the most resourceful employ of generated power by antithetical loss from skin effect. This allows paralleling and loads sharing amongst the generators. In AC distribution, AC Flogging is very clear-cut at high levels too. Due to its high dependence on HVDC system, a wide range of Contactors, Relays can be exploited.

While talking regarding craft Power Systems we tend to conjointly ought to contemplate increased power electronics application in craft that creates harmonics, massive neutral currents, wave form distortion of each supply voltage and current, poor power issue, and excessive current demand. Besides if some non-linear loads is affected upon a supply, their effects are additive. Due to these troubles, there could also be nuisance tripping of circuit breakers or inflated loss and thermal heating effects which will provoke early element failure. This is a prodigious problem to every motor loads on the system. Hence, decent power quality of the generation system is of scrupulous attention to the Aircraft manufacturer. We discern that aircraft systems work on high frequency so even on the higher frequencies in the range of 360 to 900Hz; these components would remain very significant.

Today, advanced soft computing techniques are used widely in the involuntary control system, and optimization of the system applied. Several of them are such as fuzzy logic [4]-[8], optimization of active power filter using GA [9]-[12], power loss reduction using particle swarm optimization [13], Artificial neural network control [14]-[18] applied in together machinery and filter devices.

In this paper, GA has been used to mend the complete performance of active filter for the lessening of harmonics and other delinquents created into the aircraft electrical system because of the non-linear loads [1]. The simulation results clearly show their effectiveness. The simulation results acquired with the new model are much improved than those of traditional method.

The paper has been modified in a sequential manner. The APF outline and the load under contemplation are discussed in Section II. The control algorithm for APF converses in Section III. MATLAB/ Simulink based simulation results are presented in Section IV, and finally Section V concludes the paper.

2. System Depiction

The craft power grid may be a three-phase power grid with the frequency of four hundred cycle per second. As exposed in Figure 1, Shunt Active Power Filter improves the power quality and compensates the harmonic currents within the system [22], [24]-[25], [27]-[28]. The shunt APF is understood by using one voltage supply inverters (VSIs) connected at the PCC to a typical DC link voltage [20]-23].

The set of loads for aircraft system consist of three loads. The first load is a three-phase rectifier in parallel with an inductive load and an unbalanced load connected in a phase with the midpoint (Load 1). The second one is a three phase rectifier connects a pure resistance directly (Load 2). The third one is a three-phase inductive load linked with the ground point (Load 3). Finally, a combination of all three loads connected with system together at a different time interval to study the effectiveness of the control schemes has been used to verify the functionality of the active filter in its ability to compensate for current harmonics. For the case of all three load connected, Load 1 is always connected, Load 2 is initially connected and is disconnected after every 2.5 cycles, Load 3 is connected and disconnected after every half cycle. All the simulations have been done for 15 cycles. The circuit parameters are given in Appendix.

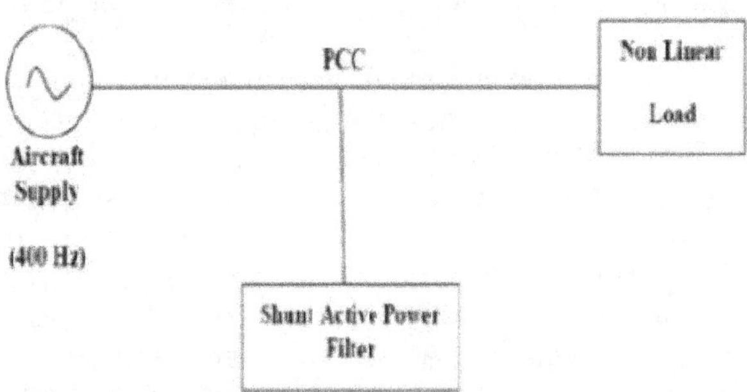

Figure 1. Aircraft system using Shunt Active Power Filter

3. Control Theory

The projected control of APF depends on Sinusoidal Current Control strategy, and it has been optimized for artificial intelligent technique e.g. Genetic Algorithm. Sinusoidal Current control strategy has been mentioned in short during this section.The subsequent section conjointly deals with the basic application of GA in the control schemes [19], [20], [29].

3.1. Sinusoidal Current Control Strategy (S.C.C.)

The sinusoidal current control strategy could be a changed version of constant instantaneous power control strategy, which might compensate load currents underneath unbalanced states too. The modification includes a positive sequence detector that replaces the 6.4 KHz cutoff frequency low-pass filters. It specifically matches the frequency and phase angle of the fundamental component. Thus, APF compensates the load reactive power. The extreme concern should be taken, while designing this detector, so that shunt active filter produces ac currents orthogonal to the voltage component. Otherwise, it will provide active power. i_α, i_β , p' and q' are attained after the calculation from α-β-0 transformation block and send to the α-β reference voltage block, which calculates $v_{\alpha'}$ and $v_{\beta'}$. Lastly, α-β-0 inverse transformation block calculates the $V_{'sa}$, $V_{'sb}$, and $V_{'sc}$. Instead of the filtered voltages used previously, $V_{'sa}$, $V_{'sb}$, and $V_{'sc}$ are reflected as input to the basic control circuit of figure 2. Now fundamental negative sequence power, harmonic power, and the fundamental reactive power, are also incorporated with the compensating powers.

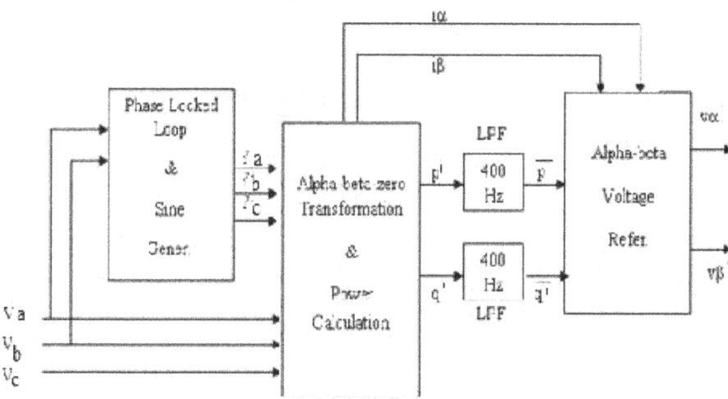

Figure 2. Block diagram of the fundamental positive-sequence voltage detector for sinusoidal current control strategy

3.2. Application of Genetic Algorithm

GA could also be a search technique that's used from generation to generation for optimizing performs. In fact, GA works on the rule of survival of the fittest. For the selecting the parameters used in controller using GA, the analysis methodology want a check, performed on-line on the particular plant or off-line with simulations on computer. Every on-line and offline methodology are having advantages and disadvantages each. If we've got a bent to means on-line approach, the foremost advantage is that the consistency of the final word answer, as a results of it's chosen on the idea of its real performances, whereas if we have a tendency to predict concerning its disadvantage, it always involve thousands of tests to attain an even result i.e. this optimization methodology will take long run for experiments to run on the real system. Simply just in case of the off-line approach, GA improvement relies on a so much plenty of precise model of the system in conjunction with all elements, all non-linearties and limits of the controllers. It has to be compelled to however be well-known that a negotiation must be met in terms of simulation accuracy and optimization time. Offline, computer simulation using MATLAB Simulink has been applied to hunt out the optimum value.

In In this paper, the GA is applied to figure out the appropriate APF parameters i.e. device filter (Lf). The boundary and limits of parameters inside the filter has been outlined and a program using genetic algorithm has been written to return up with the foremost effective value of the filter device.

For the program, the limits, inequality and bounds need to be defined. This paper has attempted to develop a single GA code program for optimizing objective function.

x0 = [V_{dc}; V_s; I_c; t; L_f];
lb = [V_{dcmin}; V_{smin}; I_{cmin}; t_{min}; L_{fmin}];
ub = [V_{dcmax}; V_{smax}; I_{cmax}; t_{max}; L_{fmax}];
Aeq = [];
beq = [];
A = [1 -1 1 -1 1; 1 1 -1 1 -1; 0 0 1 1 1; 1 1 -0 0 -1; 1 1 0 1 0];
b = [Values of V_{dc}; V_s; I_c; t; L_f depending upon the equations];
[x,fval,exitflag]=fmincon(@myobj,x0,A,b,Aeq,beq,lb,ub)

The boundary and limits of parameters in the filter has been defined using the data of ANN model. The data has been collected using MATLAB/Simulink. Finally, a program using genetic algorithm has been written to generate the best value of the filter inductor. After the calculation, GA generates the value of 0.187mH. After using this inductor value, total harmonic distortion of source current and voltage have been reduced so we can say that inductor value calculated is optimum.

4. Simulation Results & Discussions

The proposed scheme of APF is simulated in MATLAB environment to estimate its performance. Three loads have been applied together at a different time interval to check the affectivity of the control schemes for the reduction of harmonics. A small quantity of inductance is additionally connected to the terminals of the load to urge the foremost effective compensation. The simulation results clearly reveal that the scheme will with success cut back the numerous quantity of THD in supply current and voltage among limits.

4.1. Uncompensated System

Figure 3 shows the waveforms obtained after the simulation of an uncompensated system. It has been observed that the THD of source current calculated when loads connected with the system is 9.5% and THD of source Voltage were 1.55%. By observing these data, we can easily recognize supply has been polluted when loads have been connected and is obviously not within the limit of the international standard.

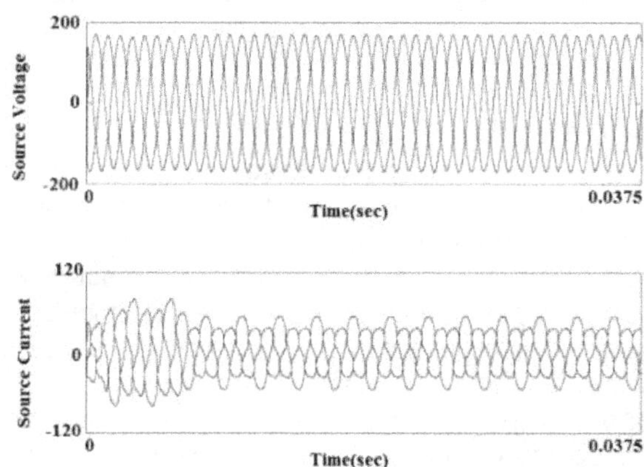

Figure 3. Source Voltage and source current waveforms of uncompensated system

4.2. Compensated System

The performance of APF under different loads connected, when utilizing GA has been discussed below for the control strategy given below.

4.2.1. For Sinusoidal Current Control Strategy

From Figure 4it has been empiric that that the THDs of source current and source voltage were 2.82% and 1.65% respectively. The compensation time was 0.01 sec. At t=0.01 sec, it is apparent that the waveforms for source voltage and source current have become sinusoidal. Figure 5 shows the waveforms of compensation current, DC capacitor voltage, and load current.

The aberration in dc voltage can be acutely apparent in the waveforms. As per claim for accretion the compensation current for accomplishing the load current demand, it releases the energy, and after that it accuses and tries to achieve its set value. If we carefully observe, we can acquisition out that the compensation current is, in fact, accomplishing the appeal of load current, and afterward the active filtering the source current and voltage is affected to be sinusoidal.

4.2.2. For Sinusoidal Current Control Strategy using GA

THDs of source current & source voltage have been found 1.92% and 1.60% respectively after making observations from the simulation results shown in figure 5. The waveforms for source voltage and source current have become sinusoidal at t=0.0066 sec. Compensation time is 0.0066 sec. The waveforms of compensation current, dc capacitor Voltage, and load current have been shown in figure 5. The aberration in dc voltage can be acutely apparent in the waveforms. As per claim for accretion the compensation current for accomplishing the load current demand, it releases the energy, and after that it accuses and tries to achieve its set value. If we carefully observe, we can acquisition out that the compensation current is, in fact, accomplishing the appeal of load current, and afterward the active filtering the source current and voltage is affected to be sinusoidal.

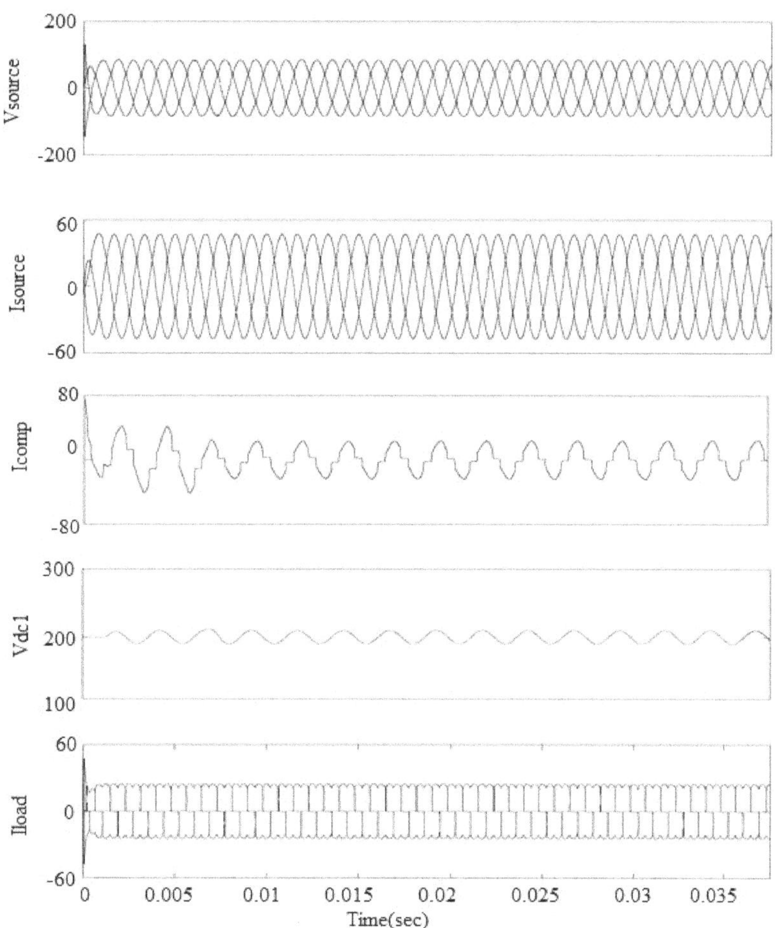

Figure 4. Source Voltage, source current, compensation current (phase b), DC link Voltage and load current waveforms of Active power filter using Sinusoidal Current Control strategy

4.3. Comparative Analysis of the Simulation Results

From the Table 1 we can easily say that Sinusoidal Current Control (SCC-GA) has been found best for current and voltage harmonic reduction. When these results have been compared based on compensation time, it has been also found that SCC-GA strategy is also fastest one.

Table 1. Summary of simulation results

Strategy	THD-I (%)	THD-V (%)	Compensation Time (sec)
SCC	2.83	1.65	0.0100
SCC-GA	1.92	1.60	0.0066

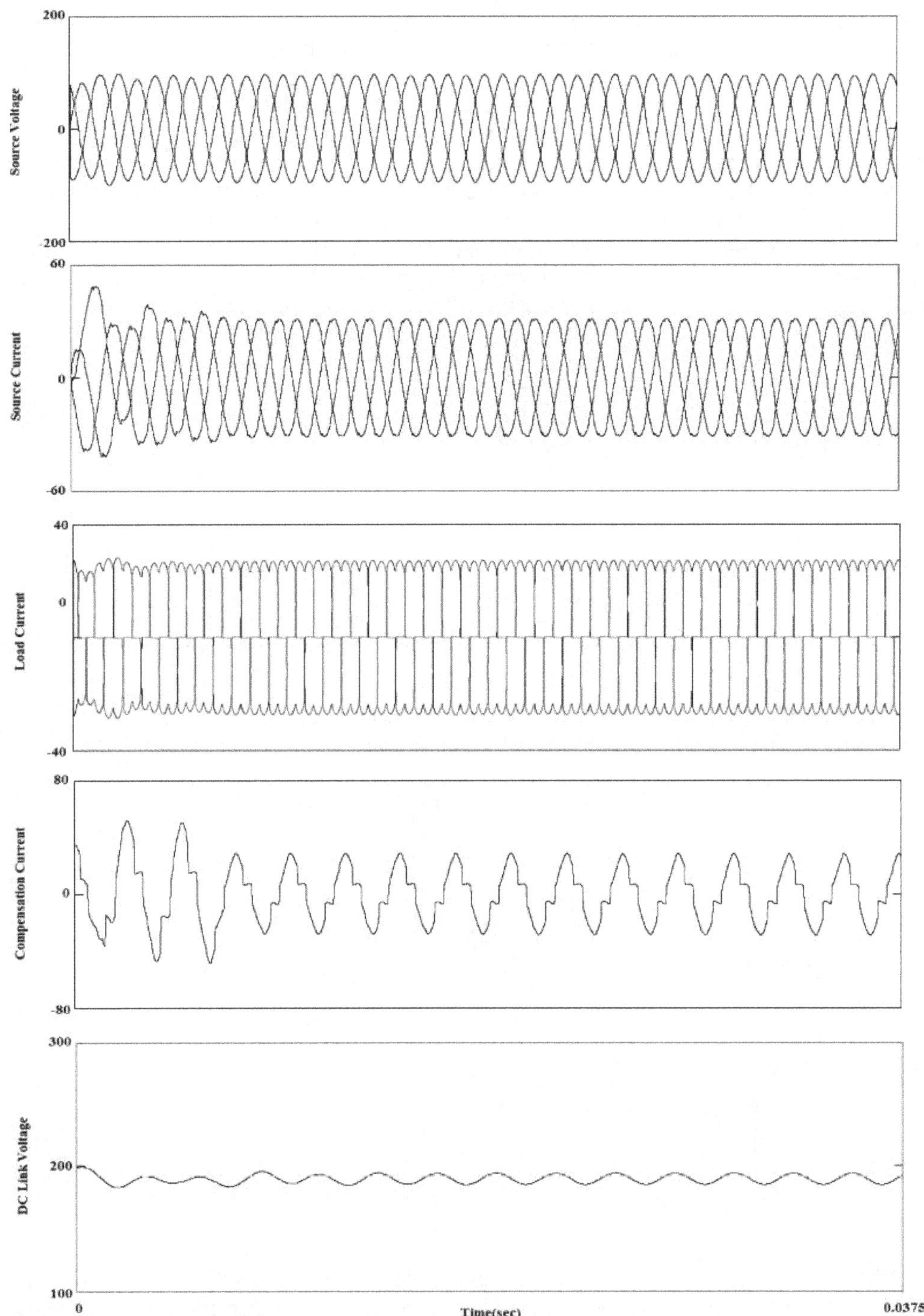

Figure 5. Source Voltage, Source Current, Load Current, Compensation Current (Phase b) and DC Link Voltage Waveforms of Active Power Filter based on *Sinusoidal* Current Control Strategy using Genetic Algorithm with All Three Loads Connected for Aircraft System

5. Conclusion

This paper has done an acute analysis of traditional control and GA applied system for shunt APF in aircraft power utility of 400 HZ. Optimum selection of control strategy based on compensation time and THD has been suggested. Overall Sinusoidal Current Control strategy (SCC-GA) has been observed as an optimum choice. Sinusoidal Current Control Strategy's performance has been improved, which itself an achievement for the case of optimization in traditional strategies.

Appendix

The aircraft system parameters are [1]:
Three-phase source voltage: 115V/400 Hz
Filter capacitor: 5 μF,
Filter inductor=0.25m H
Dc capacitor: 4700μF
Dc voltage reference: 400 V

References

[1] Chen Donghua, Tao Guo, Shaojun Xie, Bo Zhou. "Shunt Active Power Filters Applied in the Aircraft Power Utility". *36th Power Electronics Specialists Conference, PESC '05, IEEE*. 2005: 59-63.

[2] Khalid Saifullah, Dwivedi Bharti. Comparative Evaluation of Various Control Strategies forShunt Active Power Filters in Aircraft Power Utility of 400 Hz. *Majlesi Journal of Mechatronic Systems*. 2014; 3(2): 1-5.

[3] Khalid Saifullah, Dwivedi Bharti. *Application of AI techniques in implementing Shunt APF in Aircraft Supply System*. Proceeding ofSPRINGER- SOCROPROS Conference, IIT-Roorkee. 2013; 1: 333-341.

[4] Guillermin P. Fuzzy logic Applied to Motor Control. *IEEE Transactions on Industrial Application*. 1996; 32(1): 51-56.

[5] Abdul Hasib A Hew Wooi P, Hamzah A and Mowed HAF. Fuzzy Logic Control of a three phase Induction Motor using Field Oriented Control Method. *Society of Instrument and Control Engineers, SICE GAual Conference*. 2002: 264-267.

[6] S Jain, P Agrawal and H Gupta. "Fuzzy logic controlled shunt active power filter for power quality improvement". in *IEE Proceedings of the Electric Power Applications*. 2002; 149: 317-28.

[7] Mariun Norman, Bahari Samsul, Noor Mohd, Jasni Jasronita, Omar SB. *A Fuzzy logic Controller for an Indirect vector Controlled Three Phase Induction Motor*. Proceedings Analog and Digital Techniques In Electrical Engineering, TENCON 2004, Chiang Mai, Thailand. 2004; 4: 1-4.

[8] Afonso JL, Fonseca J, Martins JS and Couto CA. *Fuzzy Logic Techniques Applied to the Control of a Three-Phase Induction Motor*. Proceedings of the UKMechatronics Forum International Conference, Portugal. 142-146.

[9] Chiewchitboon P, Tipsuwanpom P, Soonthomphisaj N and Piyarat W. Speed Control of Three-phase Induction Motor Online Tuning by Genetic Algorithm. *Fifth International Conference on Power Electronics and Drive Systems, PEDS 2003*. 2003; 1: 184-188.

[10] Parmod Kumar and Alka Mahajan. "Soft Computing Techniques for the Control of an Active Power Filter". *IEEE Transactions on Power Delivery*. 2009; 24(1): 452-461.

[11] Bouserhane Ismail K., Hazzab Abdeldjebar, Boucheta Abdelkrim, Benyounes Mazari and Mostefa Rahli. Optimal Fuzzy Self-Tuning of PI Controller Using Genetic Algorithm for Induction Motor Speed Control. *Int. J. of Automation Technology*. 2008; 2(2): 85-95.

[12] Wang Guicheng, Zhang Min, Xinhe Xu and Jiang Changhong. Optimization of Controller Parameters based on the Improved Genetic Algorithms. *IEEE Proceedings of the 6th World Congress on Intelligent Control and Automation, Dalian, China*. 2006: 3695-3698.

[13] Thangaraj Radha, Thanga Raj Chelliah, Millie Pant, Abraham Ajit and CrinaGrosan. Optimal gain tuning of PI speed controller in induction motor drives using particle swarm optimization. *Logic Journal of IGPL Advance Access*. 2010: 1-14.

[14] Pinto Joao OP, Bose Bimal K and Borges da Silva Luiz Eduardo. A Stator-Flux-Oriented Vector-Controlled Induction Motor Drive with Space-Vector PWM and Flux-Vector Synthesis by Neural Networks. *IEEE Transaction on Industry Applications*. 2001; 37(5): 1308-1318.

[15] Rajasekaran S and Pai GA Vijayalakshmi. *Neural Networks, Fuzzy Logic and Genetic Algorithm: Synthesis and Applications*. Prentice Hall of India, New Delhi, fifth printing. 2005.

[16] Raul Rojas. *Neural Network- A Systematic Introduction*. Spriger-Verlag, Berlin. 1996.

[17] Zerikat M and Chekroun S. *Adaptation Learning Speed Control for a High-Performance Induction Motor using Neural Networks*. Proceedings of World Academy of Science, Engineering and Technology. 2008; 35: 294-299.

[18] Kim Seong-Hwan, Park Tae-Sik, Yoo Ji-Yoon and Park Gwi-Tae. Speed-Sensorless Vector Control of an Induction Motor Using Neural Network Speed Estimation. *IEEE Transaction on Industrial Electronics*. 2001; 48(3): 609-614.

[19] Khalid Saifullah, Dwivedi Bharti. *Comparison of Control Strategies for Shunt Active Power Filter under balanced, unbalanced and distorted supply conditions*. Proceedings of IEEE Sponsored

National Conference on Advances in Electrical Power and Energy Systems (AEPES-2013). 2013: 37-41.

[20] Mauricio Aredes, Jurgen Hafner, Klemens Heum GA. "Three-Phase Four-Wire Shunt Active Filter Control Strategies". *IEEE Transactions on Power Electronics.* 1997; 12(2): 311-318.

[21] Khalid Saifullah, Dwivedi Bharti. Power quality improvement of constant frequency aircraft electric power system using Fuzzy Logic, Genetic Algorithm and Neural network control based control scheme. *International Electrical Engineering Journal (IEEJ).* 2013; 4(3): 1098-1104.

[22] Khalid Saifullah, Dwivedi Bharti. Power Quality Issues, Problems, Standards & their Effects in Industry with Corrective Means. *International Journal of Advances in Engineering & Technology (IJAET).* 2011; 1(2): 1-11.

[23] Khalid Saifullah, Dwivedi Bharti. A Review of State of Art Techniques in Active Power Filters and Reactive Power Compensation. *National Journal of Technology.* 2007; 1(3): 10-18.

[24] RC Dugan, MF Mc Granaghan and HW Beaty. *Electrical Power Systems Quality.* New York: McGraw-Hill. 1996.

[25] Khalid Saifullah, Dwivedi Bharti. Power Quality: An Important Aspect. *International Journal of Engineering, Science and Technology.* 2010; 2(11): 6485-6490.

[26] *IEEE Recommended Practices and Requirements for Harmonic Control in Electrical Power Systems*, IEEE Standard 519-1992. 1992.

[27] A Ghosh and G Ledwich. *Power Quality Enhancement Using Custom Power Devices.* Boston, MA: Kluwer. 2002.

[28] S Khalid, N Vyas. Application of Power Electronics to Power System. University Science Press, INDIA. 2009.

[29] Khalid Saifullah, Dwivedi Bharti. *Comparative Critical Analysis of SAF using Soft Computing and Conventional Control Techniques for High Frequency (400 Hz) Aircraft System.* Proceeding ofIEEE-CATCON Conference. 2013: 100-110.

5

Fast Algorithm for Computing the Discrete Hartley Transform of Type-II

Mounir Taha Hamood
Department of Electrical Engineering
Collage of Engineering, Tikrit University, Tikrit P O Box 42, Iraq
e-mail: m.t.hamood@tu.edu.iq

Abstract

The generalized discrete Hartley transforms (GDHTs) have proved to be an efficient alternative to the generalized discrete Fourier transforms (GDFTs) for real-valued data applications. In this paper, the development of direct computation of radix-2 decimation-in-time (DIT) algorithm for the fast calculation of the GDHT of type-II (DHT-II) is presented. The mathematical analysis and the implementation of the developed algorithm are derived, showing that this algorithm possesses a regular structure and can be implemented in-place for efficient memory utilization.Theperformance of the proposed algorithm is analyzed and the computational complexity is calculated for different transform lengths. A comparison between this algorithm and existing DHT-II algorithms shows that it can be considered as a good compromise between the structural and computational complexities.

***Keywords**: Generalized discrete Hartley transforms (GDHTs), discrete Hartley transform of type-II (DHT-II), fast transform algorithms*

1. Introduction

Over last years, the discrete Hartley transform (DHT) has gained popularity in the fields of digital signal and image processing, as it possesses many desirable properties, used in many applications [1-3]. In general, there are four types of this transform known as type-I, -II, -III and -IV DHTs respectively. In the literature, the type-I DHT is just the DHT and other types are called the generalized discrete Hartley transforms (GDHTs), because their definition contains shifts in either time, frequency index or both indices. The GDHTs are also known as the W transforms that were introduced by Z. Wang [4, 5], the only difference between them is that they use different scaling factors in their definitions. If these scaling factors are ignored, the same fast algorithms can be used for both GDHTs and W transforms. The (GDHT/W) transforms have proven to be important tools in the signal processing and related fields, such as for fast computation of different types of convolutions [6-9], filter banks, signal representations [10] and many other applications.

The direct computation of the GDHTs is intensive and requires large arithmetic operations of order N^2, where N is the transform length; therefore fast GDHTs (GFHTs) algorithms have been introduced to reduce the arithmetic complexity and implementation costs. Among them, Hu *et al.* [11] proposed several fast algorithms for computing the GDHTs, Bi and Chen [12] derived a split-radix algorithm for the computation of GDFTs and GDHTs. Chiper [13] introduced an algorithm for decomposing DHT-II of length N using two adjacent $N/2$ sets of coefficients. The aforementioned algorithms are focused on the sequences with length N being power of two. Other fast algorithm for calculating DHT-II with length N being power of three is introduced by Shu*et al.* [14] and for composite transform lengths is developed by Bi *et al.* [15]. Moreover, Shu*et al.* [16] developed a new fast algorithm for the direct computation of the (DHT-II) based on decomposition of DHT-II into two DHTs of type-II of length $N/2$ that can be used to solve such a problem.

While Shu's algorithm is based on decimation-in-frequency (DIF) approach; however, for any transform to stand as a good candidate for real time applications, its complete fast algorithms need to be developed, such as the decimation-in-time (DIT) approach. Therefore, it is the aim of this paper to develop such an algorithm for fast computation of the DHT-II.

The rest of the paper is organized as follows. Section 2 presents the derivation of the new fast DIT algorithm for computing the DHT-II. The analysis of the computational complexity for the developed algorithm is given and comparison with the Hu's algorithm is also provided in Section 3. Finally, Section 4 concludes the paper.

2. Algorithm Derivation

The DHT-II transform for a real-valued sequence $x(n)$ of length N is defined as [1]:

$$X(k) = \sum_{n=0}^{N-1} x(n) cas\left(\theta n\left(\tfrac{2k+1}{2}\right)\right) \qquad k = 0,1,\ldots,N-1 \tag{1}$$

and the corresponding inverse transform (called type-III DHT) is given by:

$$x(n) = \frac{1}{N}\sum_{k=0}^{N-1} X(k) cas\left(\theta k\left(\tfrac{2n+1}{2}\right)\right) \qquad n = 0,1,\ldots,N-1 \tag{2}$$

where $\mathrm{cas}(\theta) = \cos(\theta) + \sin(\theta)$, $\theta = 2\pi/N$ and N is the transform length.

The decimation-in-time algorithm derivation begins by dividing the input sequence $x(n)$into its even $X_e(k)$ and odd $X_o(k)$ parts. Therefore (1)can be decomposed as:

$$X(k) = X_e(k) + X_o(k) \tag{3}$$

Where

$$X_e(k) = \sum_{n=0}^{\frac{N}{2}-1} x(2n) cas\left(2\theta n\left(\tfrac{2k+1}{2}\right)\right) = X_{2n}(k) \tag{4}$$

and

$$\begin{aligned} X_o(k) &= \sum_{n=0}^{\frac{N}{2}-1} x(2n+1) cas\left(\theta(2n+1)\left(\tfrac{2k+1}{2}\right)\right) \\ &= \sum_{n=0}^{\frac{N}{2}-1} x(2n+1) cas\left(2n\theta\left(\tfrac{2k+1}{2}\right) + \theta\left(\tfrac{2k+1}{2}\right)\right) \end{aligned} \tag{5}$$

Using the following *cas*property

$$\mathrm{cas}(\alpha+\beta) = \cos(\alpha)\mathrm{cas}(\beta) + \sin(\alpha)\mathrm{cas}(-\beta) \tag{6}$$

cas(.) term in (5) can be simplified to:

$$cas\left(2\theta n\left(\tfrac{2k+1}{2}\right) + \theta\left(\tfrac{2k+1}{2}\right)\right) = \cos\left(\theta\tfrac{2k+1}{2}\right) cas\left(2\theta n\left(\tfrac{2k+1}{2}\right)\right) + \sin\left(\theta\tfrac{2k+1}{2}\right) cas\left(-2\theta n\left(\tfrac{2k+1}{2}\right)\right) \tag{7}$$

Therefore, $X_o(k)$ can be decomposed further to yield:

$$X_o(k) = \cos\left(\theta\tfrac{2k+1}{2}\right)\sum_{n=0}^{\frac{N}{2}-1} x(2n+1) cas\left(2\theta n\left(\tfrac{2k+1}{2}\right)\right) + \sin\left(\theta\tfrac{2k+1}{2}\right)\sum_{n=0}^{\frac{N}{2}-1} x(2n+1) cas\left(-2\theta n\left(\tfrac{2k+1}{2}\right)\right) \tag{8}$$

The second summation of (8) can be simplified further to

$$\begin{aligned}\sum_{n=0}^{N\text{-}1} x(2n+1)\,cas\left(-2\theta n\frac{(2k+1)}{2}\right) &= \sum_{n=0}^{N\text{-}1} x(2n+1)\,cas\left(2\theta n(-k-\tfrac{1}{2})\right)\\ &= \sum_{n=0}^{N\text{-}1} x(2n+1)\,cas\left(2\theta n(N-k-\tfrac{1}{2})\right)\\ &= \sum_{n=0}^{N\text{-}1} x(2n+1)\,cas\left(2\theta n\frac{2(N-k-1)+1}{2}\right)\end{aligned} \tag{9}$$

Substituting (9) into (8) we get:

$$X_o(k) = \cos\left(\theta\tfrac{2k+1}{2}\right)X_{2n+1}(k) + \sin\left(\theta\tfrac{2k+1}{2}\right)X_{2n+1}(N\text{-}k\text{-}1) \tag{10}$$

where$X_{2n+1}(k)$ and $X_{2n+1}(N-k-1)$ can be identified as two *N*/2 point DHT-II for odd part $x(2n+1)$ of $x(n)$, given by:

$$\begin{aligned}X_{2n+1}(k) &= \sum_{n=0}^{\frac{N}{2}\text{-}1} x(2n+1)cas\left(2\theta n(\tfrac{2k+1}{2})\right)\\ X_{2n+1}(N\text{-}k\text{-}1) &= \sum_{n=0}^{\frac{N}{2}\text{-}1} x(2n+1)cas\left(\text{-}2\theta n(\tfrac{2k+1}{2})\right)\end{aligned} \tag{11}$$

Replacing (4) and (10) into (3), we obtain the following recursive formula:

$$X(k) = X_{2n}(k) + \left[\cos(\theta\tfrac{2k+1}{2})X_{2n+1}(k) + \sin(\theta\tfrac{2k+1}{2})X_{2n+1}(N\text{-}k\text{-}1)\right] \tag{12}$$

For radix-2 algorithm, another point $X\left(k+\frac{N}{2}\right)$needs to be computed. This point can be derived using trigonometric identities and the periodicity property of DHT-II, we get:

$$X(k+\tfrac{N}{2}) = X_{2n}(k) - \left[\cos(\theta\tfrac{2k+1}{2})X_{2n+1}(k+\tfrac{N}{2}) + \sin(\theta\tfrac{2k+1}{2})X_{2n+1}(N\text{-}k\text{-}1)\right] \tag{13}$$

Examine (12) and (13), we realize that the in-place property of these decompositions is not possible, due to the fact that the computation of $X(k)$ and $X\left(k+\frac{N}{2}\right)$ points require $X_{2n+1}(k)$ as well as $X_{2n+1}(N-k-1)$. However this difficulty *can* solve by additionally considering decompositions for $X(N-k-1)$ and $X\left(\frac{N}{2}-k-1\right)$ points, as follows:

$$X(N\text{-}k\text{-}1) = X_{2n}(\tfrac{N}{2}\text{-}k\text{-}1) + \left[\cos(\theta\tfrac{2k+1}{2})X_{2n+1}(N\text{-}k\text{-}1) - \sin(\theta\tfrac{2k+1}{2})X_{2n+1}(k+\tfrac{N}{2})\right] \tag{14}$$

$$X(\tfrac{N}{2}\text{-}k\text{-}1) = X_{2n}(\tfrac{N}{2}\text{-}k\text{-}1) - \left[\cos(\theta\tfrac{2k+1}{2})X_{2n+1}(N\text{-}k\text{-}1) - \sin(\theta\tfrac{2k+1}{2})X_{2n+1}(k+\tfrac{N}{2})\right] \tag{15}$$

Equations (12)-(15) can be implemented using an in-place butterfly structure shown in Figure 1. An example for calculating a 16-point DHT-II using the developed algorithm is shown in Figure 2.

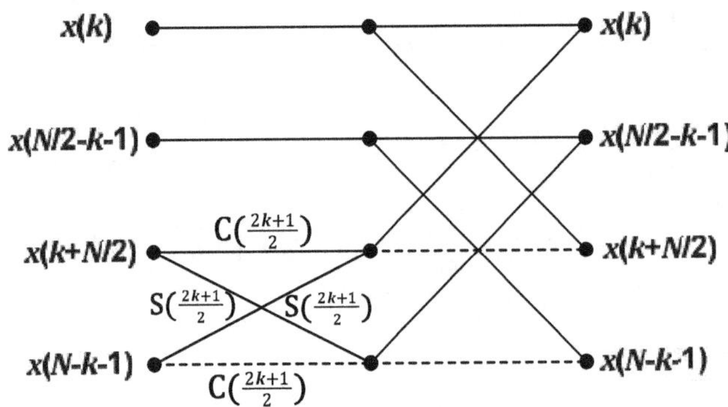

Figure 1. An in-place butterfly of radix-2 DHT-II DIT algorithm; where $C(i) = cos(2\pi i/N)$ and $S(i) = sin(2\pi i/N)$, solid and dotted lines stand for additions and subtraction respectively

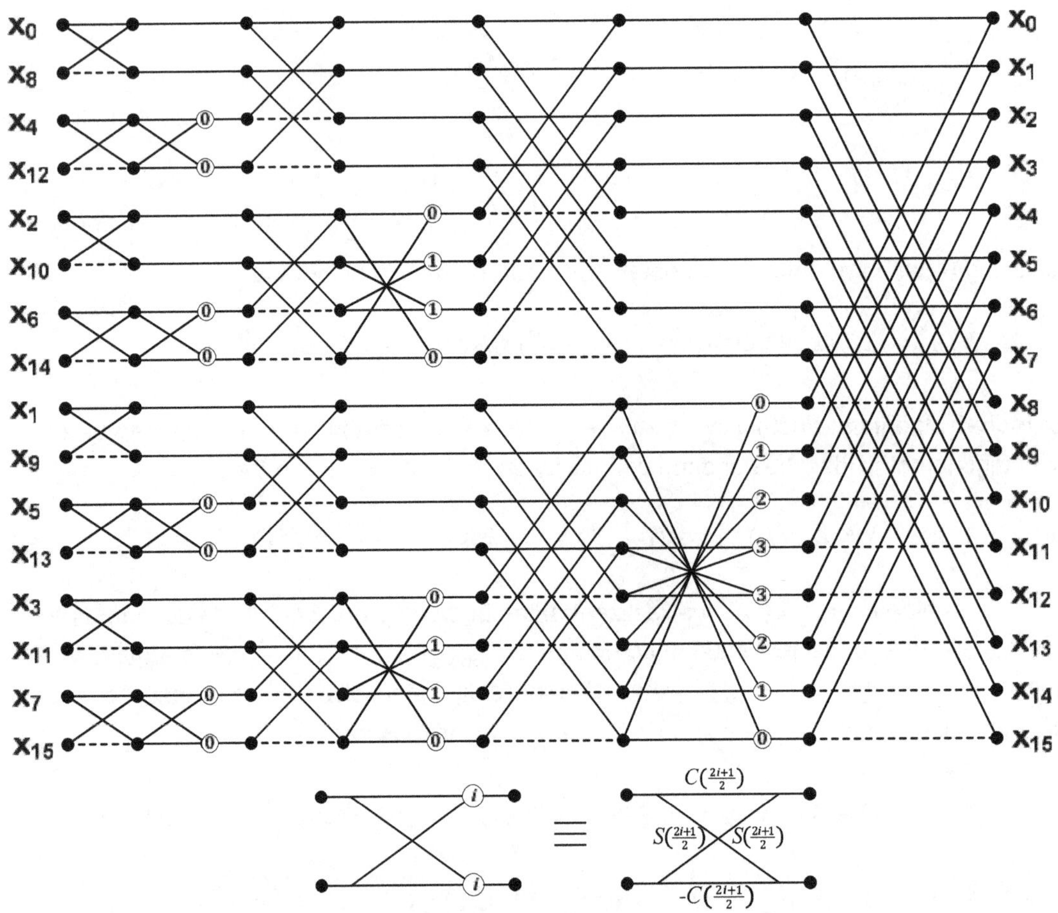

Figure 2. The calculation of 16-point DHT-II using radix-2 DIT approach; where $C(i) = cos(2\pi i/N)$and $S(i) = sin(2\pi i/N)$, solid and dotted lines stand for additions and subtraction respectively

3. Computational Complexity

The radix-2 DHT-II DIT algorithm combines four points together to formulate the in-place butterflyshown in Figure 1. Each butterfly calculates four points and requires four multiplications and six additions. In general, this algorithm requires $\log_2 N$ stages of butterfly computations in which each stage uses N multiplications and $3N/2$ additions. Therefore, the calculation of the whole transform is satisfies:

$$M(N) = 2M(\tfrac{N}{2}) + N$$
$$A(N) = 2A(\tfrac{N}{2}) + \tfrac{3N}{2} \quad (16)$$

where$M(N)$ and $A(N)$ stand for the number of multiplications and additions respectively.

It should be noted that (16) is for general calculation of the arithmetic complexity of radix-2 DIT algorithm using single butterfly; thus ±1 are considered as a twiddle factor that is counted as a multiplication. The total number of multiplications and additions could be reduced further using more than one butterfly.

The arithmetic complexities given by (16) are recursive. To obtain complexity in a closed form, the initial values of these complexities are required. In this case, the initial values can be the number of operations that are needed by length-4 DHTs-II, which are equal to $M(4)=2$ for multiplications and $A(4)=6$ for additions. Solving (16) by repeated substitution of the initial values and using multiple butterflies, we get:

$$M(N) = N\log_2 N - \tfrac{3N}{2}$$
$$A(N) = \tfrac{3N}{2}\log_2 N - \tfrac{3N}{2} \quad (17)$$

A comparison has been made between this algorithm and Hu's radix-2 DIT algorithm [11] in terms of number of multiplications and additions, as shown in Table 1. The result of this comparison reveals that the total number of multiplications and additions of the proposed algorithm is better than Hu's algorithm. Furthermore, comparison in terms of structural complexity for the signal flow graphs of the latter [Figure 1 of 11] with the former shown in Figure 2, we can easily deduce a high regularity and structural simplicity of the developed algorithm in contrast to Hu's algorithm.

Table 1. Comparison of computational complexity for decimation-in-time DHT-II algorithms

Transform length N	Proposed Algorithm			Hu's Algorithm [11]		
	Mults.	Adds.	Total	Mults.	Adds.	Total
8	10	22	32	8	28	36
16	34	66	110	28	84	112
32	98	178	276	84	220	304
64	258	450	708	228	540	768
128	642	1090	1732	580	1276	1856
256	1538	2562	4100	1412	2940	4352
512	3586	5890	9476	3332	6652	9984
1024	8194	13314	21508	7684	14844	22528

4. Conclusion

This paper has been focused on a new fast algorithm for direct computation of the DHT-II transform. The presented radix-2 decimation-in-time GFHT algorithm has a regular signal flow graph that provides flexibility for different transform lengths, substantially reducing the arithmetic complexity as compared with the indirect algorithms. The developed algorithm has been implemented through the DIT approach, and its computational complexity is analyzed and compared with existing algorithms, showing its significantly reduce the structural complexity with a better indexing scheme and ease implementation.

References

[1] G Bi and Y Zeng. *Transforms and Fast Algorithms for Signal Analysis and Representations*. Birkhauser Boston. 2004.

[2] V Britanak and KR Rao. "The fast generalized discrete Fourier transforms: A unified approach to the discrete sinusoidal transforms computation". *Signal Process.* 1999; 79: 135-150.
[3] G Bongiovanni, P Corsini and G Frosini., "One-dimensional and two-dimensional generalized discrete Fourier transforms". *IEEE Trans. Acoust., Speech, Signal Process.* 1976; 24: 97-99.
[4] Z Wang. "Fast algorithms for the discrete W transform and for the discrete Fourier transform". *IEEE Trans. Acoust., Speech, Signal Process.* 1984; 32: 803-816.
[5] Z Wang and BR Hunt. "The discrete W transform". *Appl. Math. Comput.* 1985; 16: 19-48.
[6] OK Ersoy. *Fourier-related transforms, fast algorithms, and applications*. Prentice Hall PTR. 1997.
[7] Z Wang, GA Jullien and WC Miller. "The generalized discrete W transform and its application to interpolation". *Signal Process.* 1994; 36: 99-109.
[8] O Ersoy. "Semisystolic array implementation of circular, skew circular, and linear convolutions". *IEEE Trans. Comput.* 1985; 34: 190-196.
[9] SA Martucci. "Symmetric convolution and the discrete sine and cosine transforms". *IEEE Trans. Signal Process.* 1994; 42: 1038-1051.
[10] AK Jain. "A Sinusoidal Family of Unitary Transforms". *IEEE Trans. Pattern Anal. Mach. Intell.* 1979; 1: 356-365.
[11] H Neng-Chung, IC Hong and OK Ersoy. "Generalized discrete Hartley transforms". *IEEE Trans. Signal Process.* 1992; 40: 2931-2940.
[12] G Bi and Y Chen. "Fast generalized DFT and DHT algorithms". *Signal Process.* 1998; 65: 383-390.
[13] DF Chiper. "Fast radix-2 algorithm for the discrete Hartley transform of type II". *IEEE Signal Process. Lett.* 2011; 18(11): 687-689.
[14] HZ Shu, J Wu, C Yang and L Senhadji. "Fast radix-3 algorithm for the generalized discrete Hartley transform of type II". *IEEE Signal Process. Lett.* 2012; 19(6): 348-351.
[15] G Bi, Y Chen and Y Zeng. "Fast algorithms for generalized discrete Hartley transform of composite sequence lengths". *IEEE Trans. Circuits Syst. II, Exp. Briefs.* 2000; 47(9): 893-901.
[16] HZ Shu, Y Wang, L Senhadji and LM Luo. "Direct computation of type-II discrete Hartley transform". *IEEE Signal Process. Lett.* 2007; 14(5): 329-332.

A Study on IP Network Recovery through Routing Protocols

K Karthik*[1], T Gunasekhar[2], D Meenu[3], M Anusha[4]
[1,2,4]Dept of Computer Science and Engineering, K L University, Vijayawada, Vaddeswaram 522502, India
[3]Dept of Computer Science and Engineering, Ideal College of Engineering, Kolkata, India
*Corresponding author, e-mail: kkarthik46@gmail.com[1], meenudonka@gmail.com[3]

Abstract

Internet has taken major role in our communication infrastructure. Such that requirement of internet availability and reliability has increasing accordingly. The major network failure reasons are failure of node and failure of link among the nodes. This can reduce the performance of major applications in an IP networks. The network recovery should be fast enough so that service interruption of link or node failure. The new path taken by the diverted traffic can be computed either at the time of failures or before failures. These mechanisms are known as Reactive and Proactive protocols respectively. In this paper, we surveyed reactive and proactive protocols mechanisms for IP network recovery.

***Keywords**: Proactive protocol, Reactive protocol, IP network recovery*

1. Introduction

Network which contains many network components of both hardware and software can incur failures due to one (or even multiple) of its contained components incurs a failure, as shown in Figure 1. Ranging from the largest to the smallest and from hardware to software, network failures can be divided into the following categories [1] [2]:

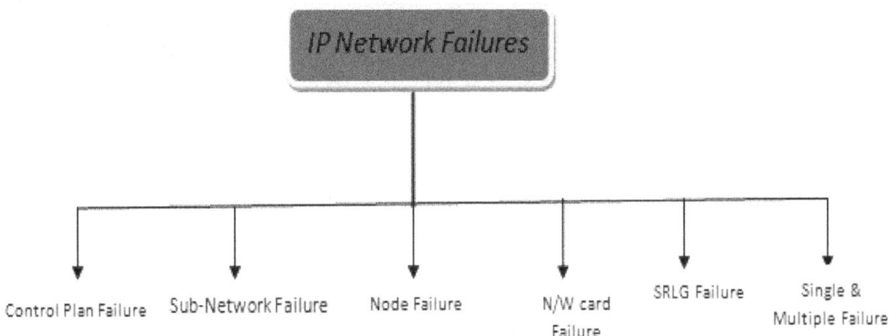

Figure 1. IP Network Failure Classification

Control plane failure

This type of failure is mainly related to software, i.e., network control plane software. For example, in a GMPLS-based network which is made up of a control plane and data plane, the control plane failure would lose the control of the data plane, which means that we cannot establish new service connections, or terminate or modify an existing service connections within the data plane, even though the existing connections can still perform normally to carry user's data [3].

Sub network failure

This is a type failure occurred with a regional sub network that commonly shares a risk, e.g., a region that has high occurring frequency of earthquake. In addition, some large disasters such as flooding, tsunami, etc. can also disable a regional sub network [4].

Node failure

This is a type of failure occurred with a single network node. The reasons for this kind of failure include accidents or disasters at a network operational center, such as power shutdown due to fire, flooding, etc [5].

Network card failure

Network card failure is a type of failure under the umbrella of the node failure type.

Link failure

link failure in general is the most common network failure that occurs due to fiber cut.

SRLG failure

SRLG failure is a generic concept to define all types of network failures whenever a common SRLG incurs a failure. Here a SRLG can be a fiber link, node, sub network, or control plane, etc.

Single failure and multiple failures

In general network failure implies a single network failure because network failure normally seldom occurs. However, under some situations, there can be more than failure occurring with a network. This kind of situation is called multiple failures [6] [7].

2. Proactive Protocols

Number of techniques has been proposed for local and fast protection in IP networks. It doesn't require any notification to neighbor node after failure. A router on detection of failure will redirect traffic to backup paths right away instead of waiting for the completion of network-wide routing convergence [8] [9].

O2 Routing

A network is configured in such a manner that all nodes have two valid next-hops to all destinations. Traffic is split between the next-hops in the normal case, and they function as backup for each other in case of a failure. To avoid loops, some links are excluded from packet forwarding for certain destinations in the normal case, and are only used as backup. O2 requires well connected network topology to give complete protection [10].

Failure Insensitive Routing (FIR)

In FIR mechanism, routers are not explicitly made aware of a failure through notification messages. Instead, they infer that a link failure if a packet for a given destination arrives at an unusual interface.

Loop Free Alternatives

The basic idea behind Loop Free Alternates is to use a precompiled alternate next hop that will not loop the packets back to the detecting node or to the failure in the event of a link failure so that traffic can be routed through this alternate next hop when a failure is detected [20].

NotVia Addresses

To protect against the failure of a component P, a special not via address is created for this component at each of P's neighbors. Forwarding tables are then calculated for these addresses without using the protected component. This way, all nodes get a path to each of P's neighbors, without passing through ("Not-via") P. It is complex because it uses tunneling [19].

Multiple Routing Configurations (MRC)

Multiple Routing Configurations (MRC) is a proactive and local protection mechanism. MRC is based on enervating back up configurations. Back up configurations are generated in such a way that for all links and nodes in the network, there is a configuration where that link or node is not used to forward traffic. Thus, for any single link or node failure, there will exist a configuration that will route the traffic to its destination on a path that avoids the failed element [10] [11].

3. Reactive Protocols

In this type of routing protocol, each node in a network discovers or maintains a route based on-demand. It floods a control message by global broadcast during discovering a route and when route is discovered then bandwidth is used for data transmission [18]. The main advantage is that this protocol needs less touting information but the disadvantages are that it produces huge control packets due to route discovery during topology changes which occurs frequently in MANETs and it incurs higher latency. The examples of this type of protocol are Dynamic Source Routing (DSR) [12] [13].

3.1. Ad-hoc On demand Distance Vector (AODV)

AODV is distance vector type routing where it does not involve nodes to maintain routes to destination that are not on active path. As long as end points are valid AODV does not play its part. Different route messages like Route Request, Route Replies and Route Errors are used to discover and maintain links. UDP/IP is used to receive and get messages.. AODV uses a destination sequence number for each route created by destination node for any request to the nodes. Then Route Reply is sent back to the source node. Finally the animator in any simulation has to be discussed. NAM is used in NS2 [13] [20].

3.2. Dynamic Source Routing (DSR)

This is an On-demand source routing protocol. In DSR the route paths are discovered after source sends a packet to a destination node in the ad-hoc network. The source node initially does not have a path to the destination when the first packet is sent. The DSR has two functions first is route discovery (Figure 2) and the second is route maintenance [14]

Different DSR Algorithms

a) Route discovery.
b) Route maintenance.

Assumptions:

a) X, Y, Z, V and W form ad-hoc network.
b) X is the source node.
c) Z is the destination node.

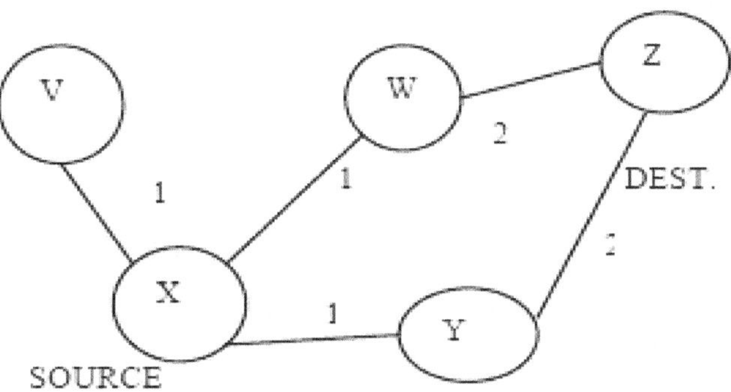

Figure 2. Route discovery

Route discovery algorithm:

a) X broadcasts a Route Request Packet with the address of destination node Z.
b) The intermediate nodes V, W, Y receive the Route Request Packet from X, as shown in Figure 2.

c) The receiving nodes V, W, Y each append their own address to the Route Request Packet and broadcast the packet (as shown in Figure 3).
d) The destination node Z receives the Route Request packet. The Route Request packet now contains information of all the addresses of the nodes from source node X to destination node Y.
e) On receiving the Route Request Packet the destination node Z sends a reply called the Route Reply Packet to the source node X by traversing a path of addresses it has got from the Route Request packet.
f) DSR caches the route information for future use.

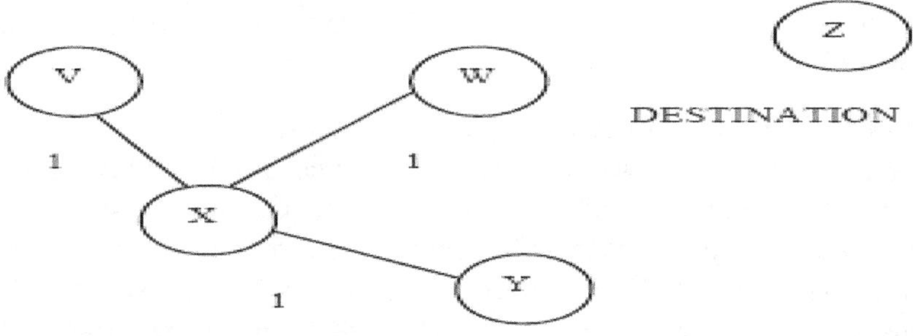

Figure 3. Showing re-broadcasting by nodes V, W, Y

Route Maintenance algorithm [5, 8]

a) In DSR algorithm a link break is detected by a node along the path from node X to node Z, in this case node W.
b) Then node W sends a message to source node X indicating a link break.
c) In this case, node X can use another path like X-Y- Z or it must initiate another route discovery packet to the same destination node, in this case 'Z' [13] [14].

4. Conclusion

The resumption of forwarding after a link failure typically takes seconds. During this period, some destinations could not be reached and the packets to those destinations would be dropped. Proactive mechanism, in which routers compute and store backup paths for potential failures before hand, and once a local link failure is detected, a router will redirect traffic to backup paths right away instead of waiting for the completion of network-wide routing convergence. Proactive routing has short failure recovery time and reduces the overhead of both update propagation and path re-calculation.

References

[1] A Raja, OC Ibe. "A survey of IP and multiprotocol labelswitching fast reroute schemes". *Comput.* 2007; 51(8): 1882-1907.
[2] Amund Kvalbein, Audun Fosselie Hansen, Tarik Ciˇcic, Stein Gjessing, Olav Lysne. "Multiple Routing Configuration for Fast IPNetwork Recovery". *IEEE/ACM Transactions on networking*. 2009; 17(2).
[3] A Markopoulou, G Iannaccone, S Bhattacharyya, C Chuah, C Diot. "Characterization of failures in an IP backbone". in Proc. *IEEE INFOCOM*. 2004: 2307–2317.
[4] C Labovitz, A Ahuja, A Bose, F Jahanian. "Delayed Internetrouting convergence". *IEEE/ACM Trans. Networking*. 2001; 9(3): 293–306.
[5] DD Clark. "The design philosophy of the DARPA internet protocols". *ACM SIGCOMM Comput. Commun. Rev*. 1988; 18(4): 106–114.
[6] A Basu, JG Riecke. "*Stability issues in OSPF routing*". in Proc. ACM SIGCOMM, San Diego, CA. 2001: 225–236.
[7] C Labovitz, A Ahuja, A Bose, F Jahanian. "Delayed internet routing convergence". *IEEE/ACM Trans. Networking*. 2001; 9(3): 293–306.
[8] C Boutremans, G Iannaccone, C Diot. "Impact of link failures on VoIP performance". in Proc. *Int. Workshop on Network and Operating System Support for Digital Audio and Video*. 2002: 63–71.

[9] D Watson, F Jahanian, C Labovitz. "*Experiences with monitoring OSPF on a regional service provider network*". in Proc. 23rd Int. Conf. Distributed Computing Systems (ICDCS'03), Washington, DC, IEEE Computer Society. 2003: 204–213.

[10] P Francois, C Filsfils, J Evans, O Bonaventure. "Achieving sub-second IGP convergence in large IP networks". *ACM SIGCOMM Comput. Commun. Rev.* 2005; 35(2): 35–44.

[11] A Markopoulou, G Iannaccone, S Bhattacharyya, CN Chuah, C Diot. "Characterization of failures in an IP backbone network". in Proc. *IEEE INFOCOM.* 2004; 4: 2307–2317.

[12] S Nelakuditi, S Lee, Y Yu, ZL Zhang, CN Chuah. "Fast local rerouting for handling transient link failures". *IEEE/ACM Trans. Networking.* 2007; 15(2): 359–372.

[13] S Iyer, S Bhattacharyya, N Taft, C Diot. "An approach to alleviate link overload as observed on an IP backbone". in Proc. *IEEE INFOCOM.* 2003: 406–416.

[14] S Rai, B Mukherjee, O Deshpande. "IP resilience within an autonomous system: Current approaches, challenges, and future directions". *IEEE Commun. Mag.* 2005; 43(10): 142–149.

[15] Gunasekhar T et.al. "A Survey on Denial of Service Attacks". *International Journal of Computer Science and Information Technologies.* 2014; 5(2): 2373-2376.

[16] Anusha M, Srikanth Vemuru, T Gunasekhar. "Transmission protocols in Cognitive Radio Mesh Networks". *International Journal of Electrical and Computer Engineering (IJECE).* 2015; 5(4).

[17] Gunasekhar T, Rao KT, Basu MT. "Understanding insider attack problem and scope in cloud". *2015 International Conference on Circuit, Power and Computing Technologies (ICCPCT).* 2015; 1(6).

[18] Anusha M, Vemuru S, Gunasekhar T. "TDMA-based MAC protocols for scheduling channel allocation in multi-channel wireless mesh networks using cognitive radio". *2015 International Conference on Circuit, Power and Computing Technologies (ICCPCT).* 2015: 1, 5.

[19] Gunasekhar T et.al. "Mitigation of Insider Attacks through Multi-Cloud". *International Journal of Electrical and Computer Engineering (IJECE).* 2015; 5.1): 136-141.

[20] M Dileep Kumar, M Trinath Basu, T Gunasekhar. "Meshing VANEMO protocol into VANETs". *International Journal of Applied Engineering Research.* 2015; 10(12): 31951-31958.

[21] R Praveen Kumar, Jagdish Babu, T Gunasekhar, S Bharath Bhushan. "Mitigating Application DDoS Attacks using Random Port Hopping Technique". *International Journal of Emerging Research in Management &Technology.* 2015; 4(1): 1-4.

[22] Gunasekhar T, K Thirupathi Rao. "EBCM: Single Encryption, Multiple Decryptions". *International Journal of Applied Engineering Research.* 2014; 9(19): 5885-5893.

[23] Kalavakolanu Narasimha Sastry, T Gunasekhar. "Novel Approach for Control Data Theft Attack in Cloud Computing". *International Journal of Electrical and Computer Engineering (IJECE).* 2015; 5(6): 2088-8708.

7

Super-Spatial Structure Prediction Compression of Medical Image Sequences

M Ferni Ukrit*[1], GR Suresh[2]
[1]Department of CSE, Sathyabama University, Chennai, Tamilnadu, India
[2]Department of ECE, Easwari Engineering College, Chennai, Tamilnadu, India
*Corresponding author, e-mail: fernijegan@gmail.com

Abstract

The demand to preserve raw image data for further processing has been increased with the hasty growth of digital technology. In medical industry the images are generally in the form of sequences which are much correlated. These images are very important and hence lossless compression Technique is required to reduce the number of bits to store these image sequences and take less time to transmit over the network The proposed compression method combines Super-Spatial Structure Prediction with inter-frame coding that includes Motion Estimation and Motion Compensation to achieve higher compression ratio. Motion Estimation and Motion Compensation is made with the fast block-matching process Inverse Diamond Search method. To enhance the compression ratio we propose a new scheme Bose, Chaudhuri and Hocquenghem (BCH). Results are compared in terms of compression ratio and Bits per pixel to the prior arts. Experimental results of our proposed algorithm for medical image sequences achieve 30% more reduction than the other state-of-the-art lossless image compression methods.

***Keywords**: Lossless Compression, Medical Image Sequences, Super-Spatial Structure Prediction, Interframe Coding, MEMC*

1. Introduction

Medical Science grows very fast and hence each hospital, various medical organizations needs to store high volume of digital me dical image sequences that includes Computed Tomography (CT), Magnetic Resonance Image (MRI), Ultrasound and Capsule Endoscope (CE) images. As a result hospitals and medical organizations have high volume of images with them and require huge disk space and transmission bandwidth to store this image sequences [1].The solution to this problem could be the application of compression. Image compression techniques reduce the number of bits required to represent an image by taking advantage of coding, inter-pixel and psycho visual redundancies. Medical image compression is very important in the present world for efficient archiving and transmission of images [2]. Image compression can be classified as lossy and lossless. Medical imaging does not require lossy compression due to the following reason. The first reason is the incorrect diagnosis due to the loss of useful information. The second reason is the operations like image enhancement may emphasize the degradations caused by lossy compression. Lossy scheme seems to be irreversible. But lossless scheme is reversible and this represents an image signed with the smallest possible number of bits without loss of any information thereby speeding up transmission and minimizing storage requirement. Lossless compression reproduces the exact replica of the original image without any quality loss [3]. Hence efficient lossless compression methods are required for medical images [4]. Lossless compression includes Discrete Cosine Transform, Wavelet Compression [5], Fractal Compression, Vector Quantization and Linear Predictive Coding. Lossless consist of two distinct and independent components called modeling and coding. The modeling generates a statistical model for the input data. The coding maps the input data to bit strings [6].

Several Lossless image compression algorithms were evaluated for compressing medical images. There are several lossless image compression algorithms like Lossless JPEG,JPEG 2000,PNG,CALIC and JPEG-LS.JPEG-LS has excellent coding and best possible compression efficiency [1]. But the Super-Spatial Structure Prediction algorithm proposed in [7] has outperformed the JPEG-LS algorithm. This algorithm divides the image into two regions, structure regions (SRs) and non-structure regions (NSRs).The structure regions are encoded with Super-Spatial Structure Prediction technique and non-structure regions are encoded using

CALIC. The idea of Super-Spatial Structure Prediction is taken from video coding. There are many structures in a single image. These include edges, pattern and textures. This has relatively high computational efficiency. No codebook is required in this compression scheme because the structure components are searched within the encoded image regions [8]. CALIC is a spatial prediction based scheme which uses both context and prediction of the pixel values [9] which accomplishes relatively low time and space complexities. A continuous tone mode of CALIC includes the four components, prediction, context selection and quantization, context modeling of prediction errors and entropy coding of prediction error [10].

Most of the lossless image compression algorithms take only a single image independently without utilizing the correlation among the sequence of frames of MRI or CE images. Since there is too much correlation among the medical image sequences, we can achieve a higher compression ratio using inter-frame coding. The idea of compressing sequence of images was first adopted in [11] for lossless image compression and was used in [12], [13], [14] for lossless video compression. The Compression Ratio (CR) was significantly low (i.e.) 2.5 which was not satisfactory. Hence in [1] they have combined JPEG-LS with inter-frame coding to find the correlation among image sequences and the obtained ratio was 4.8.Super-Spatial Structure Prediction algorithm proposed in [15] has outperformed JPEG-LS. However this ratio can be enhanced using Super-Spatial Structure Prediction technique and Bose, Chaudhuri and Hocquenghem (BCH).Super-Spatial Structure Prediction is applied with a fast block matching algorithm Inverse Diamond Search (IDS) algorithm which include lower number of searches and search points [16]. BCH scheme is used repeatedly to increase the compression ratio [17].

In this paper, we propose a hybrid algorithm for medical image sequences. The proposed algorithm combines Super-Spatial Structure Prediction technique with inter-frame coding and a new innovative scheme BCH to achieve a high mpression ratio. The Compression Ratio (CR) can be calculated by the equation (1) and PSNR by equation (2)

$$\mathrm{CR} = \text{Original Image Size}/\text{Compressed Image Size} \quad (1)$$

$$\mathrm{PSNR} = 20 * \log 10(255/\mathrm{sqrt}(\mathrm{MSE})) \quad (2)$$

This paper is organized as follows:
Section II explains the methodology used which includes Overview, Super-Spatial Structure Prediction, Motion Estimation and Motion Compensation, Motion Vector, Block Matching Algorithm and BCH. Section III discusses the results obtained for the proposed methodology.

2. Research Method

2.1. Overview

The objective of the proposed method is to enhance the compression efficiency using Super-Spatial Structure Prediction (SSP) technique combined with Motion Estimation and Motion Compensation (MEMC). The compression ratio is further enhanced by BCH algorithm an error correcting technique. Figure1 illustrates the complete encoding technique of the proposed method. The steps in the proposed method are discussed.

Step 1: Given an image sequence, input the first image to be compressed

Step 2: The image is classified as Structure Regions (SRs) and Non-Structure Regions (NSRs). SRs are encoded using SSP and NSRs are encoded using CALIC

Step 3: The first image will be compressed by Super-spatial Structure Prediction since there is no reference frame.

Step 4: Now the second frame becomes the current frame and the first frame becomes the reference frame for the second frame.

Step 5: Inter-frame coding includes MEMC process to remove temporal redundancy. Inter-coded frame will be divided to blocks known as macro blocks.

Step 6: The encoder will try to find a similar block as the previously encoded frame. This process is done by a block matching algorithm called Inverse Diamond Search.

Step 7: If the encoder succeeds on its search the block is directly encoded by a vector known as Motion Vector.

Step 8: After MEMC is done the difference of images is processed for compression. The difference is also compressed using SSP compression. MV derived from MEMC is also compressed.
Step 9: BCH converts SSP output code to binary and divide it to 7 bits each
Step 10: Each block is checked if it is valid codeword or not.BCH converts the valid block to 4 bits.
Step 11: This method adds 1 as an indicator for the valid codeword to an extra file called map otherwise if it is not a codeword it remains 7 and adds 0 to the same file.
Step 12: This step is iterated three times to improve CR.
Step 13: Flag bits and the encoded bits are concatenated.
Step 14: Once the compression of the second frame is done it becomes the reference frame for the third frame and this processing will be repeated for the next image until the end of image sequence

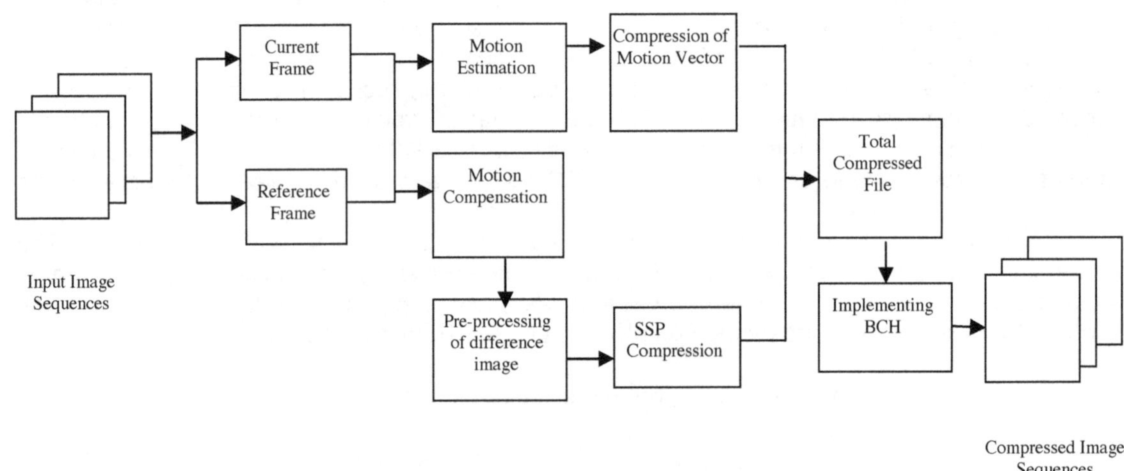

Figure 1. Encoding Technique of the Proposed Method

2.2. Super-Spatial Structure Prediction

Super-Spatial Structure Prediction borrows its idea from motion prediction [18].In SSP an area is searched within the previously encoded image region to find the prediction of an image block. The reference block that results in the minimum block difference is selected as the optimal prediction. Sum of Absolute Difference (SAD) is used to measure the block difference. The size of the prediction unit is an important parameter. When the size is small the amount of prediction and coding overhead will become large. If larger prediction unit is used the overall prediction efficiency will decrease. In this paper, a good substitution between this two is proposed. The image is partitioned into blocks of 4x4 and classifies these blocks to structure and non-structure blocks. Structure blocks are encoded using SSP and non-structure blocks using CALIC.

CALIC is a spatial prediction based scheme in which GAP (Gradient Adjusted Predictor) is used for adaptive image prediction. The image is classified to SRs and NSRs and then SSP is applied to SRs since its prediction gain in the non structure smooth regions will be very limited. This will reduce the overall computational complexity [7].

2.3. Motion Estimation and Motion Compensation

Motion estimation is the estimation of the displacement of image structures from one frame to another in a time sequence of 2D images
The steps in MEMC is stated as

- Find displacement vector of a pixel or a set of pixels between frame
- Via displacement vector, predict counterpart in present frame
- Prediction error, positions, motion vectors are coded & transmitted

Motion estimation can be very computationally intensive and so this compression performance may be at the expense of high computational complexity. The motion estimation creates a model by modifying one or more reference frames to match the current frame as closely as possible. The current frame is motion compensated by subtracting the model from the frame to produce a motion-compensated residual frame. This is coded and transmitted, along with the information required for the decoder to recreate the model (typically a set of motion vectors).At the same time, the encoded residual is decoded and added to the model to reconstruct a decoded copy of the current frame (which may not be identical to the original frame because of coding losses).This reconstructed frame is stored to be used as reference frame for further predictions. The inter-frame coding should include MEMC process to remove temporal redundancy. Difference coding or conditional replenishment is a very simple inter-frame compression process during which each frame of a sequence is compared with its predecessor and only pixels that have changes are updated. Only a fraction of pixel values are transmitted. An inter-coded frame will finitely be divided into blocks known as macro blocks. After that, instead of directly encoding the raw pixel values for each block, as it would be done for an intra-frame, the encoder will try to find a similar block to the one it is encoding on a previously encoded frame, referred to as reference frame. This process is done by a block matching algorithm. [16]. If the encoder succeeds on its search, the block could be directly encoded by a vector known as motion vector, which points to the position of the matching block at the reference frame.

2.4. Motion Vector

Motion estimation is using a reference frame in a video, dividing it in blocks and figuring out where the blocks have moved in the next frame using motion vectors pointing from the initial block location in the reference frame to the final block location in the next frame. For MV calculation we use Block matching algorithm as it is simple and effective. It uses Mean Square Error (MSE) for finding the best possible match for the reference frame block in the target frame. Motion vector is the key element in motion estimation process. It is used to represent a macro block in a picture based on the position of this macro block in another picture called the reference picture. In video editing, motion vectors are used to compress video by storing the changes to an image from one frame to next. When motion vector is applied to an image, we can synthesize the next image called motion compensation [11], [16]. This is used to compress video by storing the changes to an image from one frame to next frame. To improve the quality of the compressed medical image sequence, motion vector sharing is used [14].

2.5. Block Matching

In the block-matching technique, each current frame is divided into equal-size blocks, called source blocks. Each source block is associated with a search region in the reference frame.The objective of block-matching is to find a candidate block in the search region best matched to the source block. The relative distances between a source block and its candidate blocks are called motion vectors. Figure 3 illustrates the Block-Matching technique.

The block-matching process during the function MEMC taken from [1] takes much time hence we need a fast searching method and we have taken Inverse Diamond Search (IDS) method [16] which is the best among methods both in accuracy and speed. In the matching process, it is assumed that pixels belonging to the block are displaced with the same amount. Matching is performed by either maximizing the cross correlation function or minimizing an error criterion.

In the matching process, it is assumed that pixels belonging to the block are displaced with the same amount. Matching is performed by either maximizing the cross correlation function or minimizing an error criterion. The most commonly used error criteria are the Mean Square Error (MSE) as stated in equation (3) and the Minimum Absolute Difference (MAD) as stated in equation (4)

$$MSE = \frac{1}{M * N} \sum_{i=0}^{M} \sum_{j=0}^{N} [M_1(m,n) - M_2(m,n)]^2 \qquad (3)$$

$$MAD = \frac{1}{M*N}\sum_{i=0}^{m}\sum_{j=0}^{n}|M_1(m,n) - M_2(m,n)| \quad (4)$$

The IDS algorithm is based on MV distribution of real world video sequences. It employs two search patterns, Small Diamond Shape Pattern (SDSP) and Large Diamond Shape Pattern (LDSP) In order to reduce the number of search points, use Small Diamond Search Pattern (SDSP) as the primary shape. The entire process is discussed here.

Step 1: It first uses small diamond search pattern (SDSP) and checks five checking points to form a diamond shape.

Step 2: The second pattern consists of nine checking points and forms a large diamond shape pattern (LDSP).

Step 3: The search starts with the SDSP and is used repeatedly until the Minimum Block Distortion Measure (MBD) point lies on the search centre.

Step 4: The search pattern is then switched to LDSP.

Step 5: The position yielding minimum error point is taken as the final MV.

IDS are an outstanding algorithm adopted by MPEG-4 verification model (VM) due to its superiority to other methods in the class of fixed search pattern algorithms.

2.6. Bose, Chaudhuri and Hocquenghem

The binary input image is firstly divided into blocks of size 7 bits each, 7 bits are used to represent each byte and the eighth bits represent sign of the number(most significant bit).BCH checks each block if it is a valid codeword and converts the valid code word to 4 bits. It adds 1 as an indicator for the valid code word to an extra file called map. If it is not a valid code word it remains 7 and adds 0 to the same file. The map is a key for decompression to distinguish between the compressed blocks and non compressed blocks. BCH is repeated for three times to improve the compression ratio. If repeated for more than three times then there will be increase in time and it may affect other performance factor. Hence it is essential for this algorithm not to be repeated for more than three times.

The first frame is decompressed using BCH followed by Super-Spatial Structure Prediction decoder. After the reproduction of the first frame the difference of the rest of the frames are decompressed. The first frame becomes the reference frame for the next frame. After the reproduction of the second frame it becomes the reference frame for the next frame and the process continues until all the frames are decompressed.

3. Results and Discussion

The proposed methodology has been simulated in Microsoft Visual Studio .Net 2005. To evaluate the performance of the proposed methodology we have tested it on a sequence of MRI and CE images. Medical video is taken from Sundaram Medical Foundation (SMF) and MR-TIP database. Input image sequences are taken from these videos. More than 100 image sequences of MRI and CE are taken and tested for compression. The results are evaluated based on Compression Ratio and Bits per pixel. Figure 2 shows five CE Image sequences and Figure 3 shows five MRI Image Sequences. The images in these CE sequences are of dimension 1024×768 and the images in these MRI sequences are of dimension 514x514. Motion Estimation and Motion Compensation is applied to these image sequences using Inverse Diamond Search algorithm and Super-Spatial Structure Compression is applied. The output code of SSP is then taken and BCH is applied to enhance the efficiency of compression ratio. Super-Spatial Structure Prediction significantly reduces the prediction error. SSP outperforms CALIC and saves the bit rate for high frequency image components. With Inverse Diamond Search algorithm the accuracy is 92% and the time saved is 95% on the average. Bose,Chaudhuri and Hocquenghem gives a good compression ratio and keeps the time and complexity minimum.

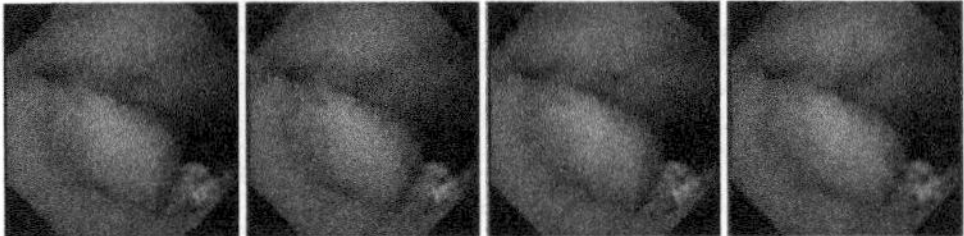

Figure 2. CE Image Sequences

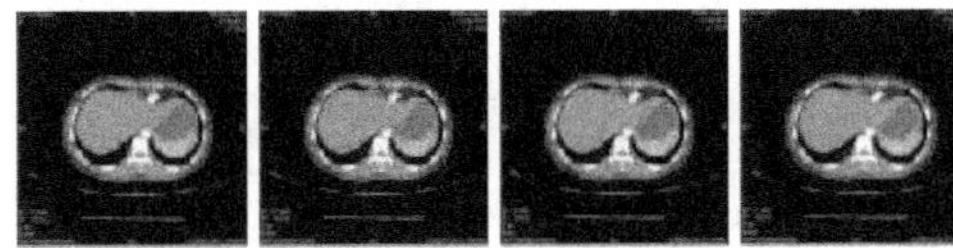

Figure 3. MRI Image Sequences

Table 1 shows the compression ratio of five CE image sequences among the tested image sequences. The results of this proposed methodology are compared with the existing methodology. The result of this algorithm is graphically shown in Figure 4.The results of the proposed algorithm were better.

Table 1. CR of CE Image Sequence

Image Sequences	Compression Ratio(CR)	
	SSP (Existing)	**SSP+IDS+BCH (Proposed)**
FI	5.26	6.81
F2	5.37	6.92
F3	5.62	7.12
F4	5.14	6.75
F5	5.46	7.04
AVG	**5.37**	**6.93**

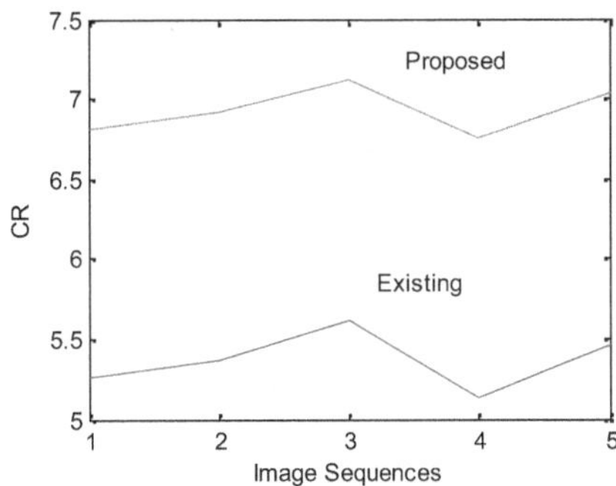

Figure 4. Compression ratio of CE image sequences

From Table 1 and Figure 4 it is easily identified that the proposed methodology has a better compression ratio than the existing one. On the average the proposed has the compression ratio of 6.93 and existing has 5.67.

Table 2 shows average compression ratio of MRI image sequences among the tested image sequences. The results of this proposed methodology are compared with the existing methodology. The proposed has an average compression ratio of 7.52 which outperforms the other state-of-art algorithm and this is illustrated in Figure 5. Experimental results of the proposed methodology gives 30% more reduction than the other state-of-the-art algorithms.

Table 2. Average CR of MRI Image Sequence

Method	CR
JPEG 2000	2.596
JPEG-LS	2.727
JPEG-LS+MV+VAR	4.841
SSP	6.25
Proposed	**7.72**

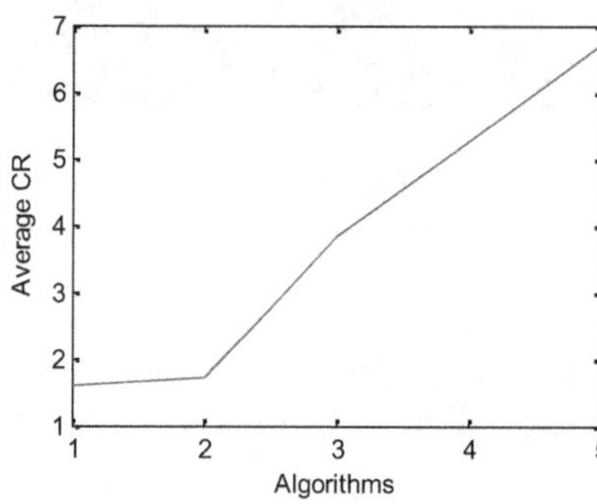

1. JPEG-2000
2. JPEG-LS
3. JPEG-LS+MV+VAR
4. SSP
5. PROPOSED (SSP+IDS+BCH)

Figure 5. Average CR of MRI Image sequences with Existing algorithms

Table 3 shows the Bits per Pixel (BPP) of CE sequences and MRI sequences.

From Table 3 the proposed methodology has 1.503 bpp for CE sequences and 0.82 bpp for MRI sequences. The results show that the proposed methodology produced improved results.

Table 3. Bits per pixel of CE and MRI Image Sequences

Method	Bits Per Pixel (BPP)	
	CE Sequence	MRI Sequence
JPEG 2000	2.757	3.08
JPEG-LS	2.394	2.93
JPEG-LS+MV+VAR	2.115	1.65
SSP	1.827	1.28
Proposed	**1.503**	**0.82**

4. Conclusion

The algorithm given in this paper makes use of the lossless image compression technique and video compression to achieve higher CR. To achieve high CR the proposed method combines Super-Spatial Structure Prediction (SSP) with inter-frame coding along with Bose, Chaudhuri and Hocquenghem (BCH). The technique used in proposed algorithm gives better result than JPEG-LS and SSP. Fast block-matching algorithm is used here. Since the full search block matching was time consuming as proposed in [1] we have taken Inverse Diamond Search (IDS) algorithm for block-matching process. Inverse Diamond Search (IDS) is faster than Diamond Search (DS) as the number of searches and search points are low. Since this paper exploits inter-frame correlation in the form of MEMC the proposed is compared with [1] [15]. To enhance the compression ratio, SSP is combined with BCH. From Table 1, 2 and 3 it is analyzed that proposed is much better than other state-of-the art lossless compression methods.

References

[1] Shaou-Gang Miaou, Fu-Sheng Ke, and Shu-Ching Chen. A Lossless Compression Method for Medical Image Sequences Using JPEG-LS and Interframe Coding. *IEEE Transaction on Information Technology in Biomedicine*. 2009; 13(5).

[2] GM Padmaja, P Nirupama. Analysis of Various Image Compression Techniques. *ARPN Journal of Science and Technology*. 2012; 2(4).

[3] Ansam Ennaciri, Mohammed Erritali, Mustapha Mabrouki, Jamaa Bengourram. Comparative Study of Wavelet Image Compression: JPEG2000 Standart. *TELKOMNIKA Indonesian Journal of Electrical Engineering*. 2015; 16(1).

[4] SE Ghare, MA Mohd Ali, K Jumari and M Ismail. An Efficient Low Complexity Lossless Coding Algorithm for Medical Images. *American Journal of Applied Sciences*. 2009; 6 (8): 1502-1508.

[5] Arikatla Hazarathaiah, B Prabhakara Rao. Medical Image Compression using Lifting based New Wavelet Transforms. *International Journal of Electrical and Computer Engineering (IJECE)*. 2014; 4(5): 741-750.

[6] S Bhavani, Dr K Thanushkodi. A Survey in Coding Algorithms in Medical Image Compression. *International Journal on Computer Science and Engineering*. 2010; 02(5): 1429-1434.

[7] Xiwen Owen Zhao, Zhi hai Henry He. Lossless Image Compression Using Super-Spatial Structure Prediction. *IEEE Signal Processing*. 2010; 17(4).

[8] CS Rawat, Seema G Bhatea, Dr Sukadev Meher. A Novel Algoritm of Super-Spatial Structure Prediction for RGB Colourspace. *International Journal of Scientific & Engineering Research*. 2012; 3(2).

[9] X Wu and N Memon. Context-based, adaptive, lossless image coding. *IEEE Trans.Commun*. 1997; 45(4): 437-444.

[10] X Wu. Lossless Compression of Continuous-tone Images via Context Selection, Quantization, and Modeling. *IEEE Trans. Image Processing*. 1997; 6(5): 656-664.

[11] YD Wang. The Implementation of Undistorted Dynamic Compression Technique for Biomedical Image. *Master's thesis*. Dept. Electr. Eng; Nat.Cheng Kung Univ; Taiwan. 2005.

[12] D Brunello, G Calvagno, GA Mian and R Rinaldo. Lossless Compression of Video using Temporal Information. *IEEE Trans. Image Process*. 2003; 12(2): 132–139.

[13] ND Memon and Khalid Sayood. Lossless Compression of Video Sequences. *IEEE Trans. Commun.* 44(10): 1340-1345.

[14] MF Zhang, J Hu, and LM Zhang. Lossless Video Compression using Combination of Temporal and Spatial Prediction. *In Proc. IEEE Int. Conf. Neural Newt. Signal Process.* 2003; 2: 1193–1196.

[15] Mudassar Raza, Ahmed Adnan, Muhammad Sharif and Syed Waqas Haider. Lossless Compression Method for Medical Image Sequences Using Super-Spatial Structure Prediction and Inter-frame Coding. *International Journal of Advanced TResearch and Technology*. 2012; 10(4).

[16] Wen-Jan Chen and Hui-Min Chen. Inverse Diamond Search Algorithm for 3D Medical Image Set Compression. *Journal of Medical and Biological Engineering*. 2009; 29(5): 266-270.

[17] A Alarabeyyat, S Al-Hashemi, T Khdour, M Hjouj Btoush, S Bani-Ahmad and R Al-Hashemi. Lossless Image Compression Technique Using Combination Methods. *Journal of Software Engineering and Applications*. 2012; 5: 752-763.

[18] T Wiegand, GJ Sullivan, G Bjntegaard and A Luthra. Overview of the H.264/AVC video coding standard. *IEEE Trans.Circuits Systems Video Technology*. 2003; 13(7).

8

Review on Opinion Mining for Fully Fledged System

Asmita Dhokrat*, Sunil Khillare, C Namrata Mahender
Dept. of Computer Science and IT, Dr. Babasaheb Ambedkar Marathwada University
Aurangabad, Maharashtra-431001, India
*Corresponding author, e-mail: asmita.dhokrat@gmail.com

Abstract

Humans communication is generally under the control of emotions and full of opinions. Emotions and their opinions plays an important role in thinking process of mind, influences the human actions too. Sentiment analysis is one of the ways to explore user's opinion made on any social media and networking site for various commercial applications in number of fields. This paper takes into account the basis requirements of opinion mining to explore the present techniques used to develop a fully fledged system. Is highlights the opportunities or deployment and research of such systems. The available tools used for building such applications have even presented with their merits and limitations.

***Keywords**: Opinion Mining, Emotion, Sentiment Analysis*

1. Introduction

Emotions are the complex state of feelings that results in physical and emotional changes that influences our behavior. Emotion is a subjective, conscious experience characterized mainly by psycho-physiological expressions, biological reactions, and mental states. Emotion is often associated and considered commonly significant with mood, nature, personality, disposition, and motivation. It is also influenced by hormones and neurotransmitters such as dopamine, noradrenaline, serotonin, oxytocin, cortisol and GABA [1]. Emotion is a positive or negative experience that is associated with a particular pattern of physiological activity. Humans carry lot of emotions like happiness, sadness, angry, disgust, surprise, fear, panic, scared etc. identifying these emotions are very easy in face to face communication compare to written communication. But now a day's use of social media has increased rapidly and the huge amount of textual data became available on web, mining and managing this vast data has become a crucial task. As the growth of E-facilities have increased lots of people got encouraged to write their emotions, views, opinions about a person, product, place or anything they want.

Opinion Mining or Sentiment analysis involves building a system to explore user's opinions made in blog posts, comments, reviews or tweets, about the product, policy or a topic [2]. Opinion mining is nothing but finding the opinion of person from sentences and classify them on the basis of polarity. As the world changed into E-World the way of expression has dramatically changed for example wide use of smiley's and symbols can be seen as expression while texting. Social communication can be observed on internet and new term has been coined for various ways of communication like texting, twitting, posting etc. people like to communicate with others through internet, they want to share their feelings, likes, dislikes, opinions, views, reviews, emotions etc. people are happy to share their personal life via social media, the use of social media has increased so much and so rapidly that even no body worries about what they are sharing and is this good to share our personal life with unknown persons? Is there any need to share our photos, videos or our daily activities on internet? So finding the sentiment, emotion behind this activity is also an important task for understanding the psycho-socio status. So from that text, mining the opinions of people and finding their views, reaction, sentiments and emotions have become challenging task.

Opinion Mining is the field of study that analyzes people's opinion, sentiments, evaluations, attitudes and emotions from written text. Opinion Mining is one of the most active research areas in Natural Language Processing and is also widely studied in data mining, web mining and text mining this research has spread outside of computer science to the management science and social science due to its importance to business and society. The growing importance of sentiment analysis coincides with the growth of social media such as Reviews, Forums, discussion groups, chatting, blogs, micro-blogs, twitter and social networks.

1.1. Categorization of Text

Sentiment analysis is also called as opinion mining; as it mines the information from various text forms such as reviews, news & blogs and classifies them on the basis of their polarity as positive, negative or neutral [3]. It focuses on categorizing the text at the level of subjective and objective nature. Subjectivity indicates that the text contains/bears opinion content for e.g. Battery life of Samsung mobiles are good. (This sentence has an opinion, it talks about the Samsung mobile phones and showing positive (good) opinion hence it is Subjective). Samsung mobiles are having long battery life. (This sentence is a fact, general information rather than an opinion or a view of some individual and hence its objective) [4].

1.2. Components of Opinion Mining

There are mainly three components of Opinion Mining [3]:

- Opinion Holder: Opinion holder is the holder of a particular opinion; it may be a person or an organization that holds the opinion. In the case of blogs and reviews, opinion holders are those persons who write these reviews or blogs.
- Opinion Object: Opinion object is an object on which the opinion holder is expressing the opinion.
- Opinion Orientation: Opinion orientation of an opinion on an object determines whether the opinion of an opinion holder about an object is positive, negative or neutral.

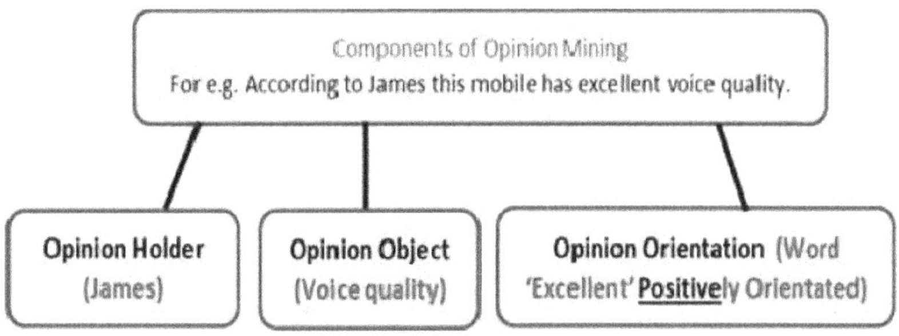

Figure 1. Components of opinion Mining

2. Different Levels of Sentiment Analysis

In general, sentiment analysis has been investigated mainly at three levels [4].

- Document level: The task at this level is to classify whether a whole opinion document expresses a positive or negative sentiment. For example, given a product review, the system determines whether the review expresses an overall positive or negative opinion about the product. This task is commonly known as document level sentiment classification.
- Sentence level: The task at this level goes to the sentences and determines whether each sentence expressed a positive, negative, or neutral opinion. Neutral usually means no opinion. This level of analysis is closely related to subjectivity classification which distinguishes sentences (called objective sentences) that express factual information from sentences (called subjective sentences) that express subjective views and opinions.
- Entity and Aspect level: Both the document-level and sentence-level analyses do not discover what exactly people liked and did not like. Aspect level performs fine-grained analysis. Aspect level was earlier called feature level (feature-based opinion mining and summarization).

3. Challenges in Opinion Mining

There are several challeges in Opinion Mining as follows,

- **Domain-independence:** The biggest challenge faced by opinion mining and sentiment analysis is the domain dependent nature of sentiment words. One features set may give very good performance in one domain, at the same time it perform very poor in some other domain [5].
- **Asymmetry in availability of opinion mining software**: The opinion mining software is very expensive and currently affordable only to big organizations and government. It is beyond the common citizen's expectation. This should be available to all people, so that everyone gets benefit from it [6].
- **Detection of spam and fake reviews:** The web contains both authentic and spam contents. For effective Sentiment classification, this spam content should be eliminated before processing. This can be done by identifying duplicates, by detecting outliers and by considering reputation of reviewer [5].
- **Incorporation of opinion with implicit and behavior data:** For successful analysis of sentiment, the opinion words should integrate with implicit data. The implicit data determine the actual behavior of sentiment words [6].
- **Mixed Sentences:** Suppose the word is positive in one situation may be negative in another situation. For e.g. Word LONG, suppose if customer says the battery life of Samsung mobile is too long so that would be a positive opinion. But suppose if customer says that Samsung mobile take too long time to start or to charge so it would be a negative opinion.
- **Way of Expressing the Opinion:** The people don't always express opinions in the same way. The opinion of every individual is different because the way of thinking, the way of expressing is vary from person to person.
- **Use of Abbreviations and shortforms:** People using social media more and that to for chatting, expressing their views using shortcuts or abbreviations so the use of colloquial words is increased. Uses of abbreviation, synonyms, special symbols is also increase day by day so finding opinion from that is too difficult. For e.g. F9 for fine, thnx for thanks, u for you, b4 for before, b'coz for because, h r u for how are you etc.
- **Typographical Errors:** Sometimes typographical errors cause problems while extracting opinions.
- **Orthographics Words:** People use orthographic words for expressing their excitement, happiness for e.g. Word Sooo..... Sweeetttt....., I am toooo Haappy or if they in hurry they stress the words for e.g. comeeeee fasssssssst I am waittttnggg.
- **Natural language processing overheads:** The natural language overhead like ambiguity, co-reference, Implicitness, inference etc. created hindrance in sentiment analysis too [6].

4. Data Sources and Tools of Opinion Mining

While doing research the collection of data is the biggest issue and for the task like opinion mining, sentiment analysis its too difficult because lots of information is available on internet and collection of that data and extraction of opinion from huge amount of data is too hard. So here we discussed about some available data sources and the tools which is used for extraction the sentiments and opinion of the given text.

4.1. Data Sources Available for Opinion Mining

There are various data sources available on web , i.e. Blogs, Microblogs, online posts, News feeds, Forums, review sites etc.

- **Blogs:** Blogs are nothing but the user own space or diary on internet where they can share their views, opinions about topics they want.
- **Online Reviews:** on Internet various review sites are available through that you can check online reviews of any product before purchasing that.
- **Micro blogging:** Microblogs allow users to exchange small elements of content such as short sentences, individual images, or video links", which may be the major reason for their popularity.
- **Online Posts:** people share their own ideas, opinions, photos, videos, views, likes, dislikes, comments on specific topics etc.

- **Forums:** An Internet forum, or message board, is an online discussion site where people can hold conversations in the form of posted messages.

This table gives you an idea about the available data sources along with the address of sites from which you can download the posts, tweets, reviews for products etc.

Table 1. Available Data sources with web address

Data Sources	Respective Sites/ Source
Blogs	http://indianbloggers.org/,http://www.bloggersideas.com/, http://www.digitaltrends.com/,http://thoughts.com/free-blog,http://blog.com/,https://wordpress.com/, http://blog.hubspot.com/
Review Sites	http://www.sitejabber.com/,http://www.toptenreviews.com/, http://www.trustedreviews.com/,https://in.pinterest.com, http://www.business-edge.com/,http://www.websitemagazine.com/, http://www.yellowpages.com
Micro-Blogging	https://tumblr.com/(Tumblr),http://friendfeed.com/(Frendfeed), http://www.plurk.com/top/(Plurk),https://twitter.com/(Twitter), http://www.jaiku.com/(Jaiku),http://www.qaiku.com/(Quiku), https://www.identi.ca/(Identica),http://www.spotjots.com/(Spotjots), http://www.meetme.com/ (Meet me)
Online Posts	https://www.facebook.com/(Facebook),https://myspace.com/(MySpace), http://www.skype.com/en/(Skype),https://www.linkedin.com/(Linkedin), https://diasporafoundation.org/(Diaspora),https://plus.google.com/(Google Plus),https://www.whatsapp.com/(Whatsapp),https://www.snapchat.com/ (Snapchat),https://telegram.org/(Telegram),https://www.flickr.com/(Flickr)
Forums	http://www.forums.mysql.com,http://www.forums.cnet.com, http://www.forum.joomla.org,https://forums.digitalpoint.com, http://www.bookforum.com,http://www.myspace.com/forums, http://tsrmatters.com/(The Student Room), http://ubuntuforums.org/, https://stackoverflow.com/,

4.2. Tools Available for Opinion Mining

As we discussed in 4.1 there are various data sources are available on web and mining those data is difficult task. Main difficulty is extraction of emotions, structure of text, form of data i.e. image or text, the language used on internet for communication is vary from person to person or state to state. So here are some ready to use tools for opinion mining for various purposes like data preprocessing, classification of text, clustering, opinion mining, sentiment analysis etc.

The table no. 2 shows the name of particular tool as well as uses of these tools.

Table 2. List of available tools

Name of Tools	Uses
STANFORD CORENLP [7]	POS tagging, Named entity recognizer, Parsing, Coreference resolution system, Sentiment analysis, Bootstrapped pattern learning
WEKA [8]	Machine learning algorithm for Data Mining, Data pre-processing, Classification, Regression, Clustering, Association rules, Visualization.
NLTK [9]	Classification, Tokenization, Stemming, Tagging, Parsing, Semantic reasoning, Provides lexical resources such as WordNet
Apache Open NLP [10]	Tokenization, Sentence segmentation, Part-of-speech tagging, Named entity extraction, Chunking, Parsing, Coreference resolution
LingPipe [11]	Entity extraction, POS tagging, Clustering, Classification.
GATE [12]	Tokenizer, Gazetteer, Sentence splitter, POS tagging, Named entities transducer, Coreference tagger
Pattern [13]	Data mining, POS tagging, N-gram search, Sentiment analysis, WordNet, Machine learning, Network analysis, Visualization
Robust Accurate Statistical Parsing [14]	Statistical Parser, Tokenization, Tagging, Lemmatization and Parsing

5. Existing Work in Opinion Mining

As we know the beginning of opinion mining has marked in late 90's but this paper discusses the advances carried out from the year 2002 to 2014. In this section brief tabulated information about the major contribution in the field of opinion mining is shown. The table no. 3

shows details about the author, their work, different techniques used while working on Opinion Mining and brief introduction of that paper as conclusion of that paper.

Table 3. List of available tools

Ref. No.	Author's	Title of the Paper	Techniques used	Conclusion	Year
[15]	B. Pang, L. Lee, S. Vaithyanathan	Thumbs up? Sentiment Classification using Machine Learning Techniques	Naïve Bayes, Maximum entropy classification and SVM	In this paper, discussion on sentiment classification of movie reviews on the basis of positive and negative is given in length. They used three different machine learning algorithms for text classification for increasing accuracy in classification.	2002
[16]	Turney P.D	Thumbs Up or Thumbs Down? Semantic Orientation Applied to Unsupervised Classification of Reviews	Pointwise mutual information (PMI) and Information Retrieval (IR)	In this paper author classified reviews on the basis of thumbs up(recommended) and thumbs down (Not recommended) and classification is predicted by semantic orientation, for this purpose they used unsupervised learning algorithm and PMI-IR uses to measure the similarity of pairs of words or phrases.	2002
[17]	Michael G. Madden	A New Bayesian Network Structure for Classification Tasks	Partial Bayesian Network (PBN) K2algorithm	Author proposed a methodology for induction of a Bayesian network structure for classification and this structure is called Partial Bayesian Network (PBN). It is implemented using the K2 framework. Learning the Partial Bayesian Network essentially reduces to a Bayesian Network learning problem using the K2 algorithm. The complexity of K2 algorithm is exponential to the number of variables; hence, PBN is also feasible only for small data sets.	2002
[18]	Hai Leong Chieu, Hwee Tou Ng	Named Entity Recognition: A Maximum Entropy Approach Using Global Information	Maximum Entropy	In this paper authors presented maximum entropy based named entity recognizer for global information. It uses information from the whole document to classify each word with just one classifiers.	2003
[19]	Turney P.D., Littman M.L	Measuring praise and criticism: inference of semantic orientation from association.	Pointwise mutual information (PMI) and Latent semantic analysis (LSA)	This paper introduced a method for inferring the semantic orientation of a word from its statistical association with a set of positive and negative paradigm words.. They use pointwise mutual information (PMI) and latent semantic analysis (LSA) to measure the relation between a word and a set of positive or negative words and according to this paper LSA gives better results than PMI.	2003
[20]	S.-M. Kim, E. Hovy	Determining the sentiment of opinions	SVM (Support Vector Machine)	Here author discusses about identifying sentiments. Here classification and combination of sentiment at word and sentence levels for identifying opinion holder they used learning techniques like SVM.	2004
[21]	Soo-Min Kim, Eduard Hovy	Identifying Opinion Holders for Question Answering in Opinion Texts	Maximum Entropy	In this paper they used Maximum Entropy for learning the opinion holders automatically with the help of two ways i.e. Classification and Ranking.	2005
[22]	Jack G. Conrad, Frank Schilder	Opinion Mining in Legal Blogs	Language Model and Naive Bayes classifiers.	Here in this paper authors discuss about the scope and opinion mining of blogs which is increased in legal domain. Here they first construct a Weblog test collection containing blog entries that discuss legal search tools. Then they subsequently examine the performance of a language modeling approach deployed for both subjectivity analysis and polarity analysis.	2007

Ref. No.	Author's	Title of the Paper	Techniques used	Conclusion	Year
[23]	Xiaowen Ding, Bing Liu	A Holistic Lexicon-Based Approach to Opinion Mining	Holistic lexicon-based approach	In this paper author discussed about customer reviews of products, they also discussed about the opinion words that show desirable and undesirable states. They used a holistic lexicon-based approach to solving the problem by exploiting external evidences and linguistic conventions of natural language expressions.	2008
[24]	Alec Go, Richa Bhayani, Lei Huang	Twitter Sentiment Classification using Distant Supervision	Naive Bayes, Maximum Entropy, and SVM	Author introduced a novel approach for automatically classifying the sentiment of Twitter messages as either positive or negative with respect to a query term. Author shows that machine learning algorithms like Naive Bayes, Maximum Entropy, and SVM for classifying sentiment, have accuracy above 80% when trained with emoticon data. This paper also describes the preprocessing steps needed in order to achieve high accuracy	2009
[25]	Zhao Yan-Yan Qin Bing, Liu Ting	Integrating Intra- and Inter-document Evidences for Improving Sentence Sentiment Classification	Graph-Based approach, NB (Naive Bayes), SVM(Support Vector Machine)	Here author classify the sentence into positive, negative and objective. Here author propose two such outside sentence features: intra-document evidence and inter-document evidence. Then in order to improve the sentence sentiment classification performance, a graph-based propagation approach is presented to incorporate these inside and outside sentence features.	2010
[26]	Zhongwu Zhai, Bing Liu, Hua Xu, Peifa Jias	Clustering Product Features for Opinion Mining	Expectation–Maximization (EM) algorithm	In this paper author discussed about problem in clustering of product reviews, for the same features people can express their views in different words that are domain synonyms these words need to be grouped under same feature group. For solving this problem author used EM algorithm.	2011
[27]	S. Chandra Kala,C. Sindhu	Opinion Mining and Sentiment Classification: A Survey	Naïve Bayes, Maximum Entropy, Support Vector Machine	In this paper author discussed about opinion Mining, Sentiment Analysis its approaches and various Machine Learning tools like Naïve Bayes, Maximum Entropy, Support Vector Machine has been discussed.	2012
[28]	Anand Mahendran, et. al.	Opinion Mining For Text Classification	Naïve Bayes, Frequency distribution.	Micro blogging services for posting views, images, videos, audios, links etc. author collected this data as raw data. They used Naïve bayes classifier and frequency distribution for classifying the raw data.	2013
[29]	Maqbool Al-Maimani	Semantic and Fuzzy Aspects of Opinion Mining	Fuzzy based logic techniques SVM BN classifiers PMI-IR classifier Rule-based techniques	According to Author Opinions are fuzzy in nature and dealing with the semantic part of the expressed sentiments possesses many challenges and require effective techniques to properly extract and summarize people's views. This paper presents a review covering the semantic and Fuzzy based logic techniques and methods in sentiment analysis.	2014
[30]	Nidhi R. Sharma, Prof. Vidya D. Chitre	Opinion Mining, Analysis and its Challenges	Naïve Bayesian, Expectation Maximization.	In this paper author discussed about reviews of products and services, opinion mining and also its challenges. They discussed about Expectation Maximization and Naïve Bayesian algorithm.	2014

The above table information suggest that statistical techniques have been used mostly used by the researchers for extracting or mining the opinions.

6. Conclusion

Emotions are often associated and considered commonly significant with mood, nature, personality, disposition, and motivation. Opinion Mining or Sentiment analysis refers to extraction of opinion from given text and classify them on the basis of polarity i.e. positive, negative and neutral. In this paper, we discussed about various levels of sentiment analysis and technique used to identify and extract opinions. Here we gave some challenges used while working on opinion mining like some orthographic errors, typographical mistakes, abbreviations, colloquial words etc. are the major challenges. This paper provides a brief review to cover the major challenges, stages, application and advantages of opinion mining. In our study, we find some techniques like Naive Bayes, Maximum Entropy, SVM etc. are very oftenly used in opinion mining and sentiment analysis.

Acknowledgements

We are thankful to the Computational and Psycho-linguistic Research Lab, Dept. of Computer Science & Information Technology, Dr Babasaheb Ambedkar Marathwada University, Aurangabad (MS) for providing the facility for carrying out the research.

References

[1] Cacioppo. "Studyguide for Discovering Psychology: The Science of Mind". *John Publication*. 2014.
[2] S Chandra Kala and C Sindhu. "Opinion Mining and Sentiment Classification: A Survey". *ICTACT Journal On Soft Computing*. 2012; 03(01).
[3] Sharma R et al. "Opinion Mining in Hindi Language: A Survey". *International Journal in Computer Science & Technology(IJFCST)*. 2014 ; 4(2).
[4] Bing L. "Sentiment Analysis and Opinion Mining". *Morgan & Claypool Publishers*. 2012.
[5] Liu B. "Sentiment Analysis and Subjectivity". *Appeared in Handbook of Natural Language Processing, Indurkhya, N. & Damerau, F.J. [Eds.]*. 2010.
[6] Haseena Rahmath. "Opinion Mining and Sentiment Analysis - Challenges and Applications". *International Journal of Application or Innovation in Engineering & Management (IJAIEM)*. 2014; 3(5).
[7] http://nlp.stanford.edu/software/corenlp.html
[8] http://www.cs.waikato.ac.nz/ml/weka/
[9] http://www.nltk.org/
[10] https://opennlp.apache.org/
[11] http://alias-i.com/lingpipe/
[12] https://gate.ac.uk/
[13] http://www.clips.ua.ac.be/pattern
[14] http://www.sussex.ac.uk/Users/johnca/rasp/offline-demo.html
[15] B Pang, L Lee and S Vaithyanathan. "*Thumbs up? Sentiment classification using machine learning techniques*". Proceedings of the Conference on Empirical Methods in Natural Language Processing. 2002: 79-86.
[16] Turney PD. "Thumbs up or down? Semantic orientation applied to unsupervised classification of reviews". *2002 ACL*. 2002: 417-424.
[17] MG Madden. "*A new Bayesian network structure for classification tasks*". Proceedings of 13th Irish Conference on Arti ficial Intelligence & Cognitive Science. 2002;. 2464.
[18] Hai Leong Chieu et al. "Named Entity Recognition: A Maximum Entropy Approach Using Global Information". *http://www.cnts.ua.ac.be/conll2003/pdf/16063chi.pdf*
[19] Turney PD, Littman ML. "Measuring praise and criticism: Inference of semantic orientation from association". *ACM TOIS*. 2003; 21(4): 315-346.
[20] SM Kim and E Hovy. "*Determining the sentiment of opinions*". Proceedings COLING-04, the Conference on Computational Linguistics, Geneva, Switzerland. 2004.
[21] SM Kim and E Hovy. "*Identifying opinion holders for question answering in opinion texts*". Proceedings of AAAI Workshop on Question Answering in Restricted Domains. 2005: 20-26.
[22] Jack G Conrad and Frank Schilder. "Opinion Mining in Legal Blogs". *ACM,ICAIL '07*. 2007.
[23] Xiaowen Ding, Bing Liu. "A Holistic Lexicon-Based Approach to Opinion Mining". *ACM WSDM'08*. 2008.
[24] Alec Go, Richa Bhayani, Lei Huang. "Twitter sentiment classification using distant supervision". *CS224N Project Report, Stanford*. 2009.
[25] Zhao Yan-Yan, Qin Bing, Liu Ting. "Integrating Intra- and Inter-document Evidences for Improving Sentence Sentiment Classification". *ACTA AUTOMATICA SINICA– Elsevier*. 2010.
[26] Zhongwu Zhai et al. "Clustering Product Features for Opinion Mining", *WSDM'11 Hong Kong, China*. 2011.

[27] S Chandra Kala, C Sindhu. "Opinion Mining and Sentiment Classification: A Survey". *ICTACT Journal on Soft Computing*. 2012; 03(01).

[28] Anand Mahendran. "Opinion Mining For Text Classification". *International Journal of Scientific Engineering and Technology (ISSN : 2277-1581)*. 2013; 2(6): 589-594.

[29] Maqbool Al-Maimani et al. "Semantic and Fuzzy Aspects of Opinion Mining". *Journal of Theoretical and Applied Information Technology*. 2014; 63(2).

[30] Nidhi R Sharma et al. "Opinion Mining, Analysis and its Challenges". *International Journal of Innovations & Advancement in Computer Science IJIACS ISSN 2347 – 8616*. 2014; 3(1).

9

Measurement and Analysis of Power in Hybrid System

Vartika Keshri*, Prity Gupta
Department of Electrical & Electronics Engineering, Oriental college of Technology, Bhopal
e-mail: indiadgreat554@gmail.com

Abstract

Application with renewable energy sources such as solar cell array, wind turbines, or fuel cells have increased significantly during the past decade. To obtain the clean energy, we are using the hybrid solar-wind power generation. Consumers prefer quality power from suppliers. The quality of power can be measured by using parameters such as voltage sag, harmonic and power factor. To obtain quality power we have different topologies. In our paper we present a new possible topology which improves power quality. This paper presents modeling analysis and design of a pulse width modulation voltage source inverter (PWM-VSI) to be connected between sources, which supplies energy from a hybrid solar wind energy system to the ac grid. The objective of this paper is to show that, with an adequate control, the converter not only can transfer the dc from hybrid solar wind energy system, but also can improve the power factor and quality power of electrical system. Whenever a disturbance occurs on load side, this disturbance can be minimized using open loop and closed loop control systems.

***Keywords**: power quality, hybrid, harmonic, PWM-VSI, push- pull inverter, solar, wind, open loop, closed loop control.*

1. Introduction

Stand-alone power generation systems are utilized by many communities and remote area around the world that have no access to grid electricity Renewable energy sources are predicted to become competitive with conventional power generation systems in the near future. Unfortunately, they are not very reliable. For example, the PV source is not available during the night or during cloudy conditions. Other sources such as fuel cells may be more reliable but have economic issues associated with them.

Because of this, two or more renewable energy sources are required to ensure a reliable and cost- effective power solution. Such a combination of different types of energy sources into a system is called a hybrid power system. With increasing concern of global warming and the depletion of fossil fuel reserves, many are looking at sustainable energy solutions to preserve the earth for the future generations. Other than hydro power, wind and photovoltaic energy holds the most potential to meet our energy demands. Alone, wind energy is capable of supplying large amounts of power but its presence is highly unpredictable.

Hybrid means utilization of two or more sources for the single load. So many advantages may be derived from Hybrid wind-solar generation. System is using these are:

a. Continuous power can be supplied to the consumers.
b. Environmental pollution can be reduced by using Hybrid solar- wind generation system.
c. Hybrid solar-wind can be made available to the far away consumers at economical rate from the utility saving hydraulic energy, which can be Kept in the dams during the dynamic, to be used at height. That is the dams ma y operate as an energy storage system.
d. The maintenance cost of hybrid solar-wind generation system is less when compared to conventional generation system. We can supply the power with low cost to the consumers.
e. In spite of a very attractive alternative with a reasonable price, the feeding of energy generated by a hybrid solar-wind generation into an existing AC grid poses some problems to the control of the converters that connect the two systems. This is especially true if the AC system cannot supply reactive power and absorb harmonic current.

2. Solar Photo Voltaic System

The physical of PV cell is very similar to that of the classical diode with a p-n junction formed by semiconductor material. When the junction absorbs light, the energy of absorbed photon is transferred to the electron-proton system of the material, creating charge carriers that are separated at the junction. The charge carriers in the junction region create a potential gradient, get accelerated under the electric field, and circulate as current through an external circuit. The solar cell is the basic building of the PV power system it produces about 1 W of power. To obtain high power, numerous such cell are connected in series and parallel circuits on a panel (module), The solar array or panel is a group of a several modules electrically connected in series-parallel combination to generate the required current and voltage. The equivalent circuit of solar cells shown in Figure.1

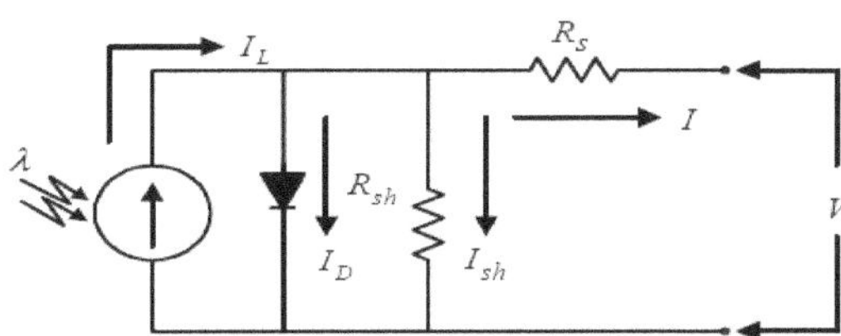

Figure. 1. Equivalent Circuit of Solar Cells

Figure 2 shows the I-V characteristics of the photovoltaic module at different solar illumination intensities. The I-V characteristic of the solar PV decreases gradually as the voltage goes up and when the voltage is low the current is almost constant. The power output of the panel is the product of the voltage and current outputs. The PV module must operate electrically at a certain voltage that corresponds to the peak power point under a given operation conditions.[1]-[5]

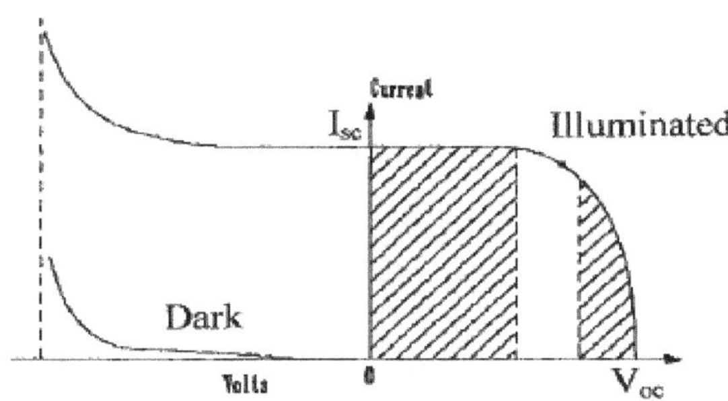

Figure. 2. I-V Characteristics of PV Module

The PV array must operate electrically at a certain voltage which corresponds to the maximum power point under the given operating conditions, i.e. temperature and irradiance. To do this, a maximum power point tracking (MPPT) technique should be applied. Various MPPT techniques like look-up table methods, perturbation and observation (P & O) methods and computational methods have been proposed in the literature. The perturbation and observation (P&O) method has been used in this work. If the array is operating at voltage V and current I, the operation point toward the maximum power point by periodically increasing or decreasing the array voltage, is often used in many PV systems. The advantage of this method is that it works well when the irradiation does not vary quickly with time, however, the P&O method fails to quickly track the maximum power points. In incremental conductance method the maximum power points are tracked by comparing the incremental and instantaneous conductance values of the PV array. Figure. 3 presents the flow of the perturbation and observation technique implemented. [4][5]

For most PV modules, the ratio of the voltage at the maximum power point for different insulation levels to the open circuit voltage is approximately constant. Also, the ratio of the current at the maximum power point for different insulation levels to the short circuit current is constant. If the direction of the perturbation i.e an increase or decrease in the output voltage of a PV array results in a positive change in the output power, then the control algorithm will continue in the direction of the previous perturbation. Conversely, if a negative change in the output power is observed, then the control algorithm will reverse the direction of the pervious perturbation step. In the case that the change in power is close to zero (within a specified range) then the algorithm will invoke no changes to the system operating point since it corresponds to the maximum power point (the peak of the power curves). The MPPT technique proposed in this work makes use of a predetermined relationship between the operating voltage or current and the open circuit voltage/short circuit current to obtain MPPT at any operating conditions.

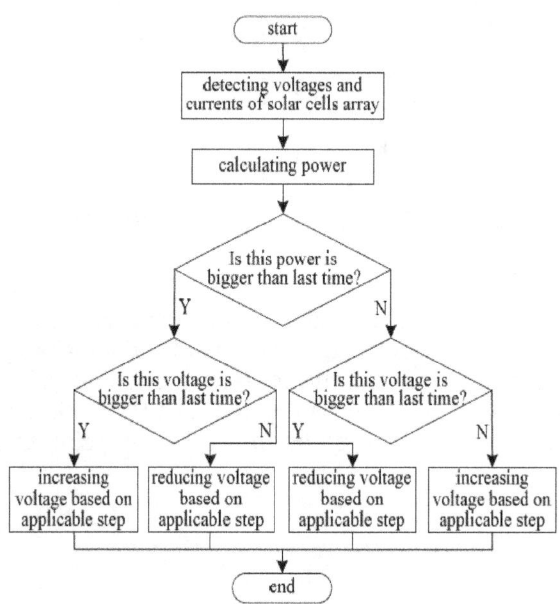

Figure 3. Flow Chart of the MPPT Technique Implemented

3. Wind Energy Conversion System

Power from the wind turbine, real and reactive power, is basically controlled by the wind-side converter and stalled by the wind blade. Below rated wind speeds, the real power from the wind generator is regulated to capture the maximum wind energy from varying wind speed. Reactive power generation is maintained at zero to minimize the thermal rating of the generator and the converter. Above rated wind speeds the maximum power control is overridden by stall regulation for constant power. In this study, the wind blade is assumed to be ideally stall regulated at rated power so that rotor speed can keep constant at rated speed under high wind speeds.

The typical turbine torque vs. rotor speed and power vs. rotor speed characteristics are shown in Figure.5 and Figure.6 respectively. The maximum power for different wind speeds is generated at a different rotor speeds. Therefore, the turbine speed should be controlled to follow the ideal TSR, with an optimal operating point which is different for every wind speed. This is achieved by incorporating a speed control in the system design to run the rotor at high speed in high wind and at low speed in low wind.

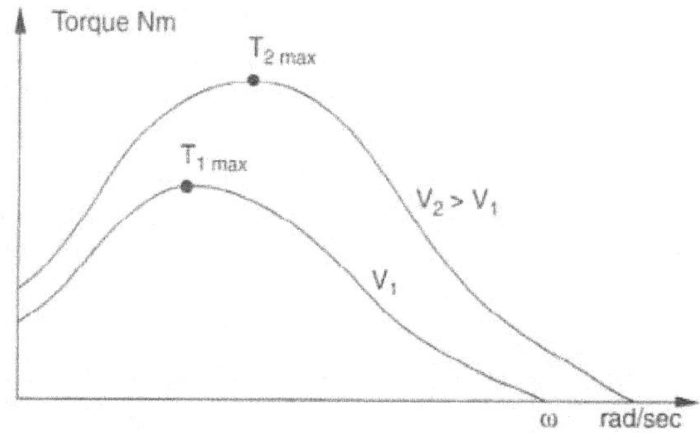

Figure 4. Wind Turbine Torque vs. Rotor Speed

4. Proposed System Configuration

The proposed standalone hybrid wind-solar power generation system is, as shown in Figure,in outline composed off our power sources: a wind power generation system (with a WT converter and a WT inverter), solar power generation system (with a PV inverter), storage battery (with a bidirectional inverter), and engine generator (EG); and a control unit .

The control unit acts to send ON/OFF operation commands to individual power sources and monitor power status via a simple communication line, which is all that is needed since the data traffic volume is small. Once an ON command is sent, each power source is autonomously operated via individual inverters; however, manual setting of inverter operating conditions is also possible if required. The inverters enable redundant parallel operation, making a reliable, stable power supply possible.

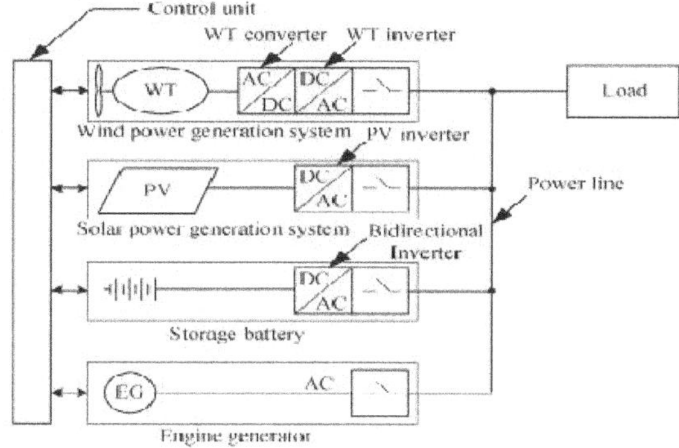

Figure 5. Proposed Standalone Hybrid Wind Solar Power Generation

5. Hardware

Data sheets for all hardware components give operational data as well as limitations of the products and best conditions at which the product operates. The data sheets were very important reference to ensure circuits were being setup correctly.

5.1. Microcontroller

The AT89S52 is a low-power, high-performance CMOS 8-bit microcontroller with 8K bytes of in-system programmable Flash memory. The device is manufactured using Atmel's high-density non volatile memory technology and is compatible with the industry standard 80C51 instruction set and pin out. The on-chip Flash allows the program memory to be reprogrammed in-system or by a conventional non volatile memory programmer. By combining a versatile 8-bit CPU with in-system programmable Flash on a monolithic chip, the Atmel AT89S52 is a powerful microcontroller which provides a highly-flexible and cost-effective solution to many embedded control applications. The AT89S52 provides the following standard features: 8K bytes of Flash, 256 bytes of RAM, 32 I/O lines, Watchdog timer, two data pointers, three 16-bit timer/counters, a six-vector two-level interrupt architecture, a full duplex serial port, on-chip oscillator, and clock circuitry. In addition, the AT89S52 is designed with static logic for operation down to zero frequency and supports two software selectable power saving modes. The Idle Mode stops the CPU while allowing the RAM, timer/counters, serial port, and interrupt system to continue functioning. The Power-down mode saves the RAM contents but freezes the oscillator, disabling all other chip functions until the next interrupt or hardware reset.

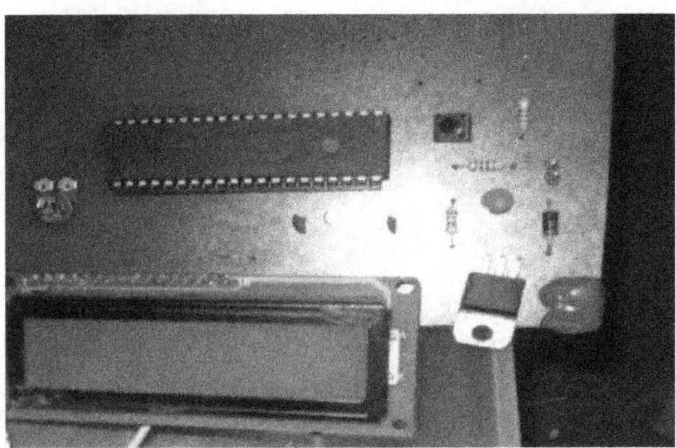

Figure 6. Microcontrollers

5.2. Oscillator Characteristics:

XTAL1 and XTAL2 are the input and output, respectively, of an inverting amplifier which can be configured for use as an on-chip oscillator, as shown in Figure 1. Either a quartz crystal or ceramic resonator may be used. To drive the device from an external clock source, XTAL2 should be left unconnected while XTAL1 is driven as shown in Figure 7. There are no requirements on the duty cycle of the external clock signal, since the input to the internal clocking circuitry is through a divide-by-two flip-flop, but minimum and maximum voltage high and low time specifications must be observed.

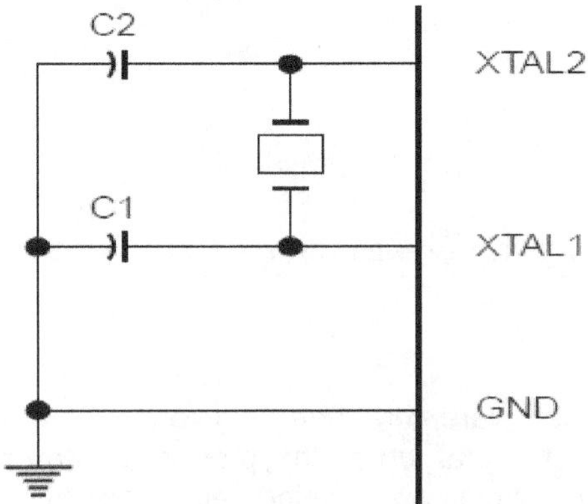

Figure 7. Oscillator Connections

5.3. Relay :

A relay is an electrically operated switch. Many relays use an electromagnet to operate a switching mechanism mechanically, but other operating principles are also used. Relays are used where it is necessary to control a circuit by a low-power signal (with complete electrical isolation between control and controlled circuits), or where several circuits must be controlled by one signal.

Figure 8. Hardware of Hybrid System

6. Results

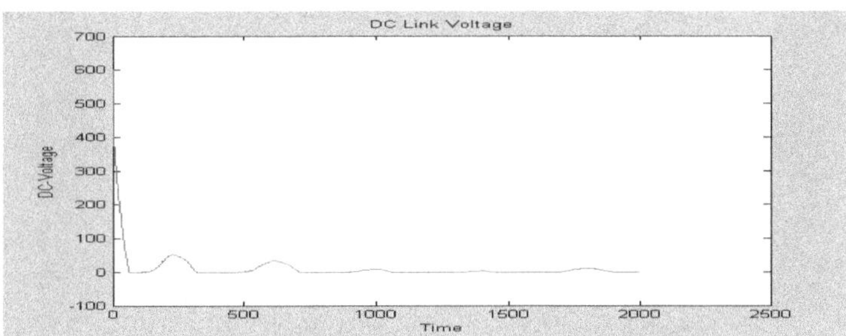

Figure 9. DC Link Voltage

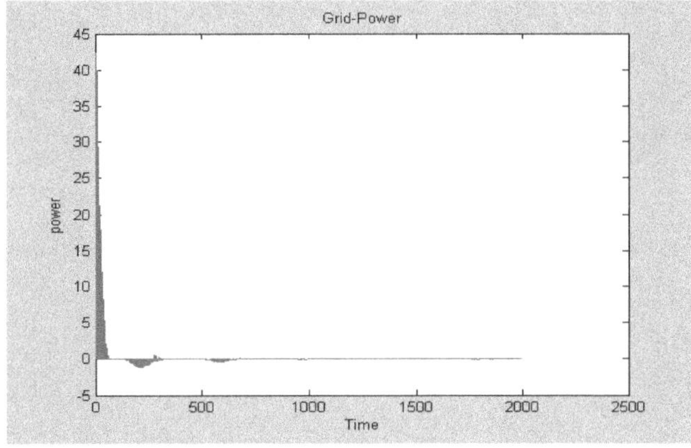

Figure 10. Power Delivered to the Grid

The power stored in battery energy system is as shown in Figure 11.

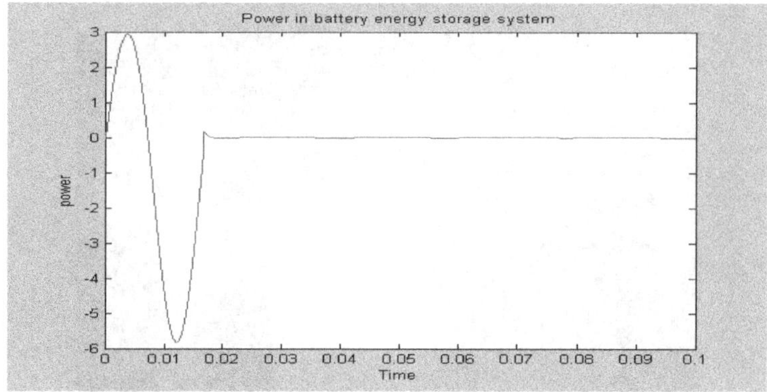

Figure 11. Power in Battery Energy Storage System

The power developed by photo voltaic array is as shown in Figure 12.

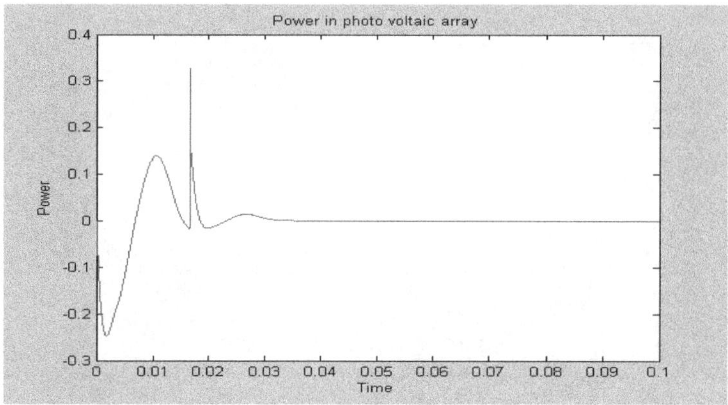

Figure 12. Power In Photo Voltaic Array

The power developed by wind turbine is shown in Figure 13.

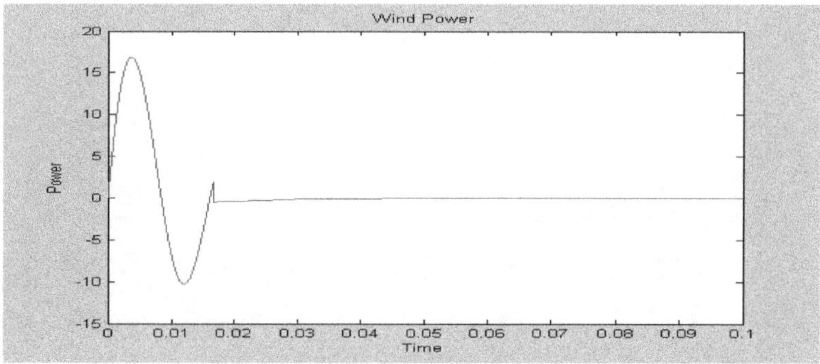

Figure 13. Power Developed by Wind Turbine

From the results it is observed that the variations in output power are more in PV system because the voltage across PV module is changing rapidly whereas the variations in output power in wind energy conversion system is less because the voltage across wind system is almost constant and the battery voltage decreases exponentially and the battery current increases exponentially.

6. Conclusion

Wind and solar power are safe, and do not send emissions or residues to the environment. The production of clean energy, which is harmless and does not aggravate the greenhouse effect, must be promoted. The use of electricity generated from renewable non-pollutant energy sources (green electricity), and all technologies involved must increase the scientific community is also contributing with technological innovations. Nowadays, the development of Power Electronics enables economical solutions for the production of renewable energy based on small power plants. Portugal presents good conditions for the implementation of a large number of these systems, based on wind power and photovoltaic energy. This paper proposes the development of a low-cost high efficiency hybrid system (wind and solar) with an interface to the load that ensures the power quality of the produced energy. The proposed solution may be a contribution to a better, cleaner and safer environment and to a decrease in energy dependence.

References

[1] M. Nayeripour, M. Hoseintabar. Control of Hybrid Renewable Energy System by Frequency Deviation Control. *Intrenational Journal of Engineering and Technology*. 2012; 4(3).

[2] Interfaces for Renewable Energy Sources with Electric Power Systems Environment. 2010. Situation and Perspectives for the European Union 6-10 May 2003. Porto, Portugal.

[3] Barbosa,P.G., Roil, L.G.B, Watanabe E. H. *Control Strategy for Grid-Connected DC-AC Converters with Load Power Factor Correction*. IEEE Proceedings Generation, Transmission and Distribution. 1998; 487-91.

[4] Chowdhury, B. H. *Designing an Innovative Laboratory to Teach Concepts in Grid-Tied Renewable and Other Dispersed Resources*. ASEE Annual Conference and Exhibition. 1999.

[5] S.K. Kim, J.H Jeon, C.H. Cho, J.B. Ahn, And S.H. Kwon. Dynamic Modeling and Control of a Grid - Connected Hybrid Generation System with Versatile Power Transfer. *IEEE Transactions On Industrial Electronics*. 2008; 55: 1677 -1688.

[6] D. Das, R. Esmaili, L. Xu, D. Nichols. *An Optimal Design of a Grid Connected Hybrid Wind/Photovoltaic/Fuel Cell System for Distributed Energy Production*. In Proc IEEE Industrial Electronics Conference2005: 2499-2504.

[7] Wang, Modeling And Control Of Hybrid Wind/Photovoltaic/Fuel Cell Distributed Generation System, A Dissertation Submitted I N Partial Fulfillment of the Requirement for the Degree of Doctor of Philosophy in Engineering. Montreal University 2006

[8] M. H. Nehrir, C. Wang, K. Strunz, H. Aki, R. Ramakumar, J. Bing, Z. Miao, And Z. Salami. A Review of Hybrid Renewable/Altern Ative Energy Systems for Electric Power Generation: Configurations, Control, And Applications. IEEE Trans. Sustain. Energy. 2011; 2(4): 392 –403.

[9] Global Wind 2007 Report, Global Wind Energy Council. [Online]. Available: http://www.gwec.net/index.php?id=90

[10] Wind Power Today—Federal Wind Program Highlights. Nrel,Doe/Go- 102005-2115, Apr. 2005.

[11] M. H. Nehrir, C. Wang, K. Strunz, H. Aki, R. Ramakumar, J. Bing,Z. Miao, And Z. Salameh. A Review of Hybrid Renewable/ Alternative Energy Systems for Electric Power Generation: Configurations, Control, and Applications. *IEEE Trans. Sustain. Energy*. 2011; 2(4): 392 –403.

10

A Secured Cloud Data Storage with Access Privilages

Naresh Vurukonda*[1], B Thirumala Rao[2]

Department of CSE, KLUniversity, Vijayawada, A.P, India
*Corresponding author, email: naresh.vurukonda@gamil.com[1], drbtrao@kluniversity.in[2]

Abstract

In proposed framework client source information reinforcements off-site to outsider distributed storage benefits to decrease information administration costs. In any case, client must get protection ensure for the outsourced information, which is currently safeguarded by outsiders. A configuration and instrument FADE, and a safe overlay distributed storage framework that achieve fine-grained, strategy based methodology control and document guaranteed erasure. It partners outsourced records with document association approaches, and without a doubt erases records to make them unrecoverable to endless supply of document access arrangements. To accomplish such security objectives, FADE is based upon an arrangement of cryptographic key operations that are self-kept up by a majority of key supervisors that are free of outsider mists. In unmistakable, FADE goes about as an overlay framework that works flawlessly on today's distributed storage administrations. Actualize a proof-of-idea model of FADE on Amazon S3, one of today's distributed storage administrations. By behavior broad true studies, and confirm that FADE gives security insurance to source information, while presenting just insignificant behavior and financial cost overhead. My work oversee, esteem included security highlights acclimatize were today's distributed storage administration.

***Keywords**: Cloud Storage, fine grained, policy based access control*

1. Introduction

Distributed storage is a show up administration demonstrates that empowers element and ventures to outsource the storehouse of information reinforcements to remote cloud worker requiring little to no effort. Be that as it may, cloud customers must authorize security confirmation of their outsourced information reinforcements the expanding praise of distributed storage is prevailing associations to analyze moving information out of their own server farms and into the Cloud. It is the long-held long for registering as an adequacy [22], can possibly change over an expansive part of the IT company, making programming considerably additionally beguiling as an administration and build, the way IT equipment is planned and get. Distributed computing alludes to both the applications passed on as administrations over the Internet and the equipment and plan programming in the datacenters that organize those administrations.

A methodology framework that addresses the issues of convoluted strategies is characterized and embellished Based on the necessities of those approaches, cryptographic enhancements that immeasurably propel authorization capacity Of Time-based records, when made, are expressed to have an end time [1]. ABE viewpoint based encryption is to build up the capacity to decrease cryptographic expenses. At the point when the cloud is made open in pay as you go angle to the well known open we call it as open cloud. Self-satisfied Mug a photograph dissemination Website facilitated terabytes of photographs on Amazon S3 in 2006 and spared a great many dollars on proceed with capacity gadgets utilizing distributed storage for far off reinforcement could discover in the system [12].

Drop box-like machine to move sound/video records from their advanced mobile phones to the unhappiness, given that PDAs regularly have characterized capacity assets. Aside from organization and Government Company, people, Third gathering worker security to make substance to the alloted by the substance worker and authorization of endorsement approaches and client consents .we started FADE, The first is selective control key utilized by key controller and the second one is information power key utilized by FADE customer [13]. FADE sums up time-based record ensured cancellation into an all the more fine-grained access called strategy based document settled cancelation, in which documents are join with more pliant document access approach (e.g., time termination, read/compose consents of certify clients) and are totally erased when the consolidate record access strategies are annul and get to be out of date.

2. Related Work on Cloud Security and Access Control

Distributed storage is another business answer for removed reinforcement outsourcing, as it offers a reflection of outright storage room for customers to host information reinforcements in a pay-as you-go way [21]. Time based File guaranteed Deletion is the Existing access [2] [3]. Time-based document settled erasure, which is initially transported in, implies that records can be safely erased and persist for all time remote after a pre-characterized degree. The principle thought is that a record is scrambled with an information key by the proprietor of the document, and this information key is more remote encoded with a control key by a segregated key manager [4] [5]. The key controller is a server that is essential for cryptographic key administration. The control key is time-based, content that it will be totally cleared by the key administrator when a discontinuance time is come to, where the suspension time is portrayed when the record is initially insisted. Without the control key, the information key and thus the information record continue scrambled and are hope to be difficult to reach. In this manner, the fundamental security domain of record guaranteed expunction is that regardless of the fact that a cloud worker does not expel finish up document duplicates from its stockpiling, those documents persist encoded and unrecoverable. Later, the thought of time-based document beyond any doubt cancellation is prototyped in Vanish. Vanish cut an information key into different key shares, which are then accumulated in various hubs of an open Peer-to-Peer Distributed Hash Table (P2P DHT) framework [20].

3. Implementation

We name a distributed storage framework brought secure access benefit over cloud information like FADE, which intends to bear the cost of methodology control settled cancellation for record that are available by today's distributed storage administrations. We collect records with document association strategies that control how documents can be gotten to. We then started arrangement based document settled erasure, in which case are without a doubt cancel and made unrecoverable by anyone when their related record approach strategies are abolish [6] [7] [8] .We portrays the essential operations. On cryptographic keys in order to accomplish approach control and settled erasure [17]. FADE likewise influences real cryptographic strategies, numbering property based encryption (ABE) and a majority of key controller in view of edge arranged sharing. We execute a model of FADE to show its get up and go, and systematically concentrate on its execution flying when it works with Amazon S3. Our exploratory results give bits of knowledge into the execution security exchange off when FADE is sent by and by. In this paper, we characterize the metadata of Fade being joined to individual information records [18] [19]. We then portray how we execute the customer and a majority of key directors and how the customer collaborates with the cloud.

1. Key manager
2. Remote user
3. Cloud admin server
4. Policy based access control
5. Policy based assured deletion

3.1. Key Manager

Fade is based on a majority of key administrators, each of which is a stand-alone substance that keeps up strategy based keys for access control and guaranteed cancellation. Sorts of keys: Data key, control key, access key, remote client. Numerous arrangements, approach recharging. Arrangement cancellation will be finished by key director.

3.2. Remote User

The one is getting to the approaches set by the cloud chief. Client is legitimate on the off chance that he get to just the arrangements set by the cloud administrator or else he will be distinguishing as a misrepresentation client in the cloud organizing. In the event that the client's arrangements are substantial which doled out for him, then the client can get to every one of the benefits in the cloud organizing.

3.2.1. Multiple Policies

* Arrangements are only the entrance benefits being set by the cloud director on the proprietor's information put away in the cloud server.

* Active information documents being put away by the proprietor stay on cloud with related arrangement of client characterized record access strategies (e.g., time termination, read/compose authorizations of approved clients), such that information records are available just to clients who fulfill the document access approaches User keeping in mind the end goal to have entry consent's and for erasure need's sure approaches which are being set by the chief.

3.3. Cloud Admin Server

The cloud, kept up by an outsider supplier, gives storage room to facilitating information records for the benefit of various FADE customers in a pay-as-you-go way. Each of the information documents is connected with a mix of record access policies [12]. FADE is based on the flimsy cloud interface, and accept just the essential cloud operations for transferring and downloading information documents.

3.3.1. Cloud Manager

Typically deals with the proprietor's information/documents from the end clients. Part: Manages the entrance consents for an end client who is looking for access to the proprietor documents put away in the cloud server. Cloud chief makes and includes an end client by getting enrolled, wherein he gives the entrance authorizations to access to the proprietor's document put away in the cloud server. Additionally has the ability to close down the clients framework when he/she tries to get to the documents who has no specific access authorization, wherein they will be obstructed as HACKER/FRAUD.

3.3.2. Cloud Server

Cloud Server gives information storage room to the client/information proprietor to store the information that gives the secured and effective method for putting away the proprietor's information. An asset put away in cloud server has set of access authorizations which are being set by the information proprietor while transferring to the server by means of cloud. Proprietor records put away in cloud server are thusly kept up by the TPA (outsider evaluator), as shown in Figure 1.

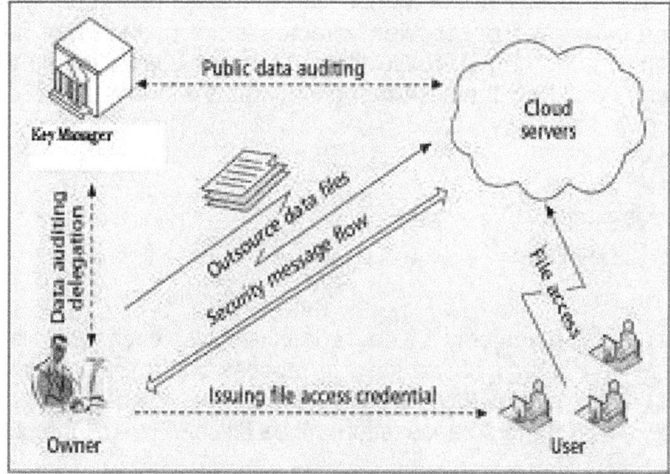

Figure 1. The architecture of cloud data storage service

3.4. Arrangement Based Access Control

A FADE customer is approved to get to just the records whose related strategies are dynamic and are fulfilled by the customer. It gives emit key to the end client for record transferring and downloading [23].

3.4.1. Strategies Renewal

Is the term identified with the entrance authorization's wherein a client solicitations to the cloud supervisor to give the approaches other than which are being distributed to he/her. For the blocked user's(Fraud) keeping in mind the end goal to have admittance to the assets put away in the cloud server need's get to authorization's which are being given by the cloud director when the blocked client goes for asking for the records.

3.5. Arrangement Based Guaranteed Cancellation

A document is erased (or for all time blocked off) if its related approaches are renounced and get to be out of date. That is, regardless of the possibility that a record duplicate that is connected with renounced arrangements, it remains scrambled and we can't recover the relating cryptographic keys to recuperate the document. In this manner, the record duplicate gets to be unrecoverable by anybody (counting the proprietor of the document).

4. Time Performance of FADE

We first measure the time execution of our FADE Prototype. Keeping in mind the end goal to recognize the time overhead of FADE, we isolate the running time of every estimation into three segments:

• File transmission time, the transferring/downloading time for the information record between the customer and the Cloud.

• Metadata transmission time, the ideal opportunity for transferring/Downloading the metadata, which contains the Policy data and the cryptographic keys related. With the record, between the customer and the Cloud.

• Cryptographic operation time, the aggregate time for cryptographic operations, this incorporates the aggregate computational time utilized for performing AES and HMAC on the record, and the ideal opportunity for the customer to organize with the majority of key chiefs on working the cryptographic keys.

5. Results

Figure 2 shows home page. It shows that it is secure overlay cloud storage with file assured deletion. To accomplish security objectives, FADE is based upon an arrangement of cryptographic key operations that are self-kept up by a majority of key supervisors that are free of outsider mists.

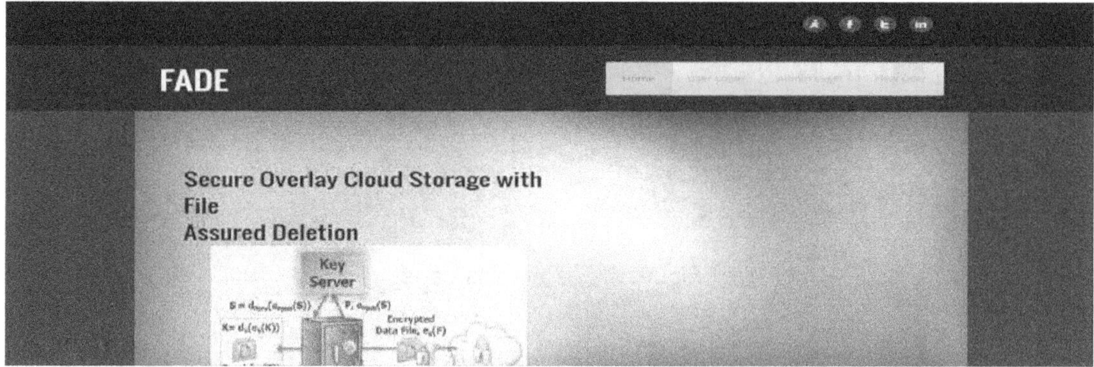

Figure 2. Home page

Figure 3-5 shows admin login, welcome and file access pages, respectively.

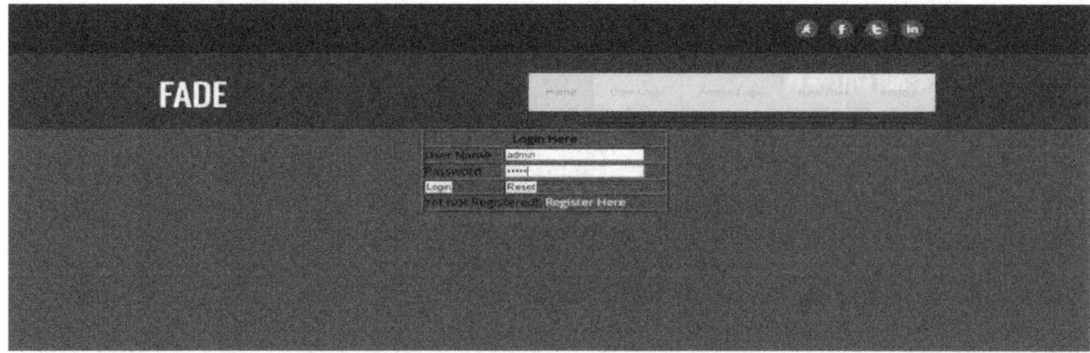

Figure 3. Admin login page

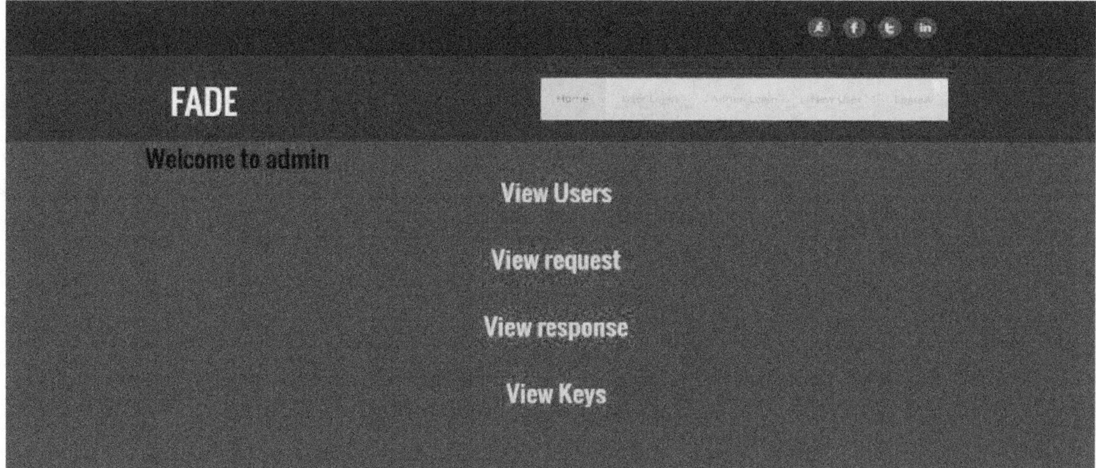

Figure 4. Welcome page

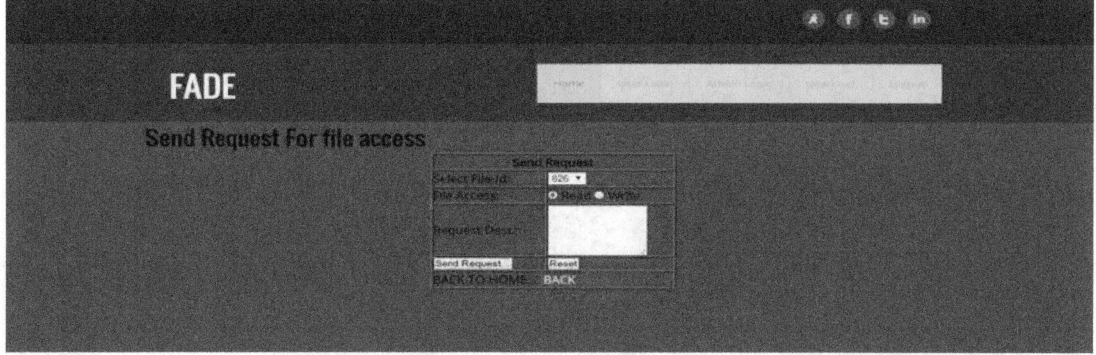

Figure 5. File access page

6. Conclusion

In this System we proposed a commonsense distributed storage framework brought secure access benefit over cloud information like FADE, which means to give access control guaranteed erasure to documents that are facilitated by today's distributed storage administrations. It partner documents with record access arrangements that control how records can be gotten to. And after that present approach based document guaranteed erasure, in which records are definitely erased and made unrecoverable by anybody when their related document access arrangements are denied. Depict the crucial operations on cryptographic keys in order to accomplish access control and guaranteed cancellation. FADE additionally influences existing cryptographic strategies, including property based encryption (ABE) and a majority of key supervisors taking into account edge mystery sharing. Execute a model of FADE to exhibit its common sense, and observationally concentrate on its execution overhead when it works with Amazon S3.Propose exploratory results give bits of knowledge into the execution security exchange off when FADE is conveyed practically speaking.

References

[1] J Bethencourt, A Sahai, B Waters. Cipher text-Policy Attribute-Based Encryption. In Proc. of *IEEE Symp. on Security and Privacy*. 2006.

[2] T Dierks, V Goyal, V Kumar. "Identity based Encryption with Efficient Revocation". In Proc of *ACM CCS*. 2008.

[3] C wang, Q Wang, K Ren, W lou. Privacy-Preserving Public auditing for storage security in cloud computing. In Proc.of *IEEE INFOCOM*. 2010.

[4] W Wang, Z Li, R Owens, B Bhargava. Secure and Efficient Access to Outsourced Data. In *ACM CCSW*. 2009.

[5] S Yu, C Wang, K Ren, W.Lou. Attribute Based Data Sharing with Attribute Revocation. In Proc. of *ACM ASIACCS*. 2010.

[6] A Yun, C Shi, Y Kim. On Protecting Integrity and Confidentiality of Cryptographic File System for Outsourced Storage. In *ACM CCSW*. 2009.

[7] S Ruj, M Stojmenovic, A Nayak. Decentralized access control with anonymous authentication of data stored in clouds. *IEEE Trans. Parallel Distrib. Syst*. 2014; 25(2): 384–394.

[8] Z Wan, J Liu, RH Deng. HASBE: a hierarchical attribute-based solution for flexible and scalable access control in cloud computing. *IEEE Trans. Inform. Forensics Sec*. 2012; 7(2): 743–754.

[9] Y Tang, PP Lee, JCS Lui, R Perlman. Secure overlay cloud storage with access control and assured deletion. *IEEE Trans. Dependable Secure Comput*. 2012; 9(6): 903–916.

[10] R Chandramouli, M Iorga, S Chokhani. Cryptographic key management issues and challenges in cloud services. in: *Secure Cloud Computing*, Springer, New York. 2014: 1–30.

[11] Chaoling, Li, Chen Yue, Zhou Yanzhou. "A data assured deletion scheme in cloud storage". *Communications, China*. 2014; 11(4): 98-110.

[12] Tang, Yang, et al. "FADE: Secure overlay cloud storage with file assured deletion". *Security and Privacy in Communication Networks*. Springer Berlin Heidelberg. 2010: 380-397.

[13] Rahumed, Arthur, et al. "A secure cloud backup system with assured deletion and version control". *2011 40th International Conference on Parallel Processing Workshops (ICPPW)*. IEEE. 2011.

[14] Reardon, Joel, David Basin, and Srdjan Capkun. "Sok: Secure data deletion." *2013 IEEE Symposium on Security and Privacy (SP),* IEEE. 2013.

[15] Xiong, Jinbo, et al. "A secure document self-destruction scheme with identity based encryption". *2013 5th International Conference on Intelligent Networking and Collaborative Systems (INCoS),* IEEE. 2013.

[16] Ali, Mazhar, Samee U Khan, Athanasios V. Vasilakos. "Security in cloud computing: Opportunities and challenges". *Information Sciences*. 2015; 305: 357-383.

[17] Jayalekshmi MB, SH Krishnaveni. "A Study of Data Storage Security Issues in Cloud Computing". *Indian Journal of Science and Technology*. 2015; 8(24).

[18] Rani, NR Anitha, SK Ram Kumar, P Prem Kumar. "A Survey on Data Redundancy Check in a Hybrid Cloud by using Convergent Encryption". *Indian Journal of Science and Technology*. 2016; 9(4).

[19] Saikeerthana R, A Umamakeswari. "Secure Data Storage and Data Retrieval in Cloud Storage using Cipher Policy Attribute based Encryption". *Indian Journal of Science and Technology*. 2015; 8(S9): 318-325.

[20] Shu, Xiao, Xining Li. "A Scalable and Robust DHT Protocol for Structured P2P Network". 2012.

[21] P Mell, T Grance. The NIST definition of cloud computing (draft). *NIST Special Publ*. 2011; 800(145): 7.

[22] Goyal, Vipul, et al. "Attribute-based encryption for fine-grained access control of encrypted data". *Proceedings of the 13th ACM conference on Computer and communications security*. Acm. 2006.

[23] Yang, Tonghao, Junquan Li, Bin Yu. "A Secure Ciphertext Self-Destruction Scheme with Attribute-Based Encryption". *Mathematical Problems in Engineering*. 2015.

11

Electrical Power Generation with Himalayan Mud Soil using Microbial Fuel Cell

Debajyoti Bose[*1], Amarnath Bose[2]
[1,2] Department of Electrical, Power & Energy,
University of Petroleum & Energy Studies, Dehradun 248007, India
e-mail: debajyoti1024@gmail.com*[1], abose@ddn.upes.ac.in[2]

Abstract

Topsoil microbial community primarily consists of bacteria species that can generate electricity if a microbial fuel cell is incorporated with it. Since such electron producing bacteria are abundant in nature, microbial fuel cells can be considered as clean source of electricity generation and a prospect for renewable energy growth. Here, the authors have shown experiments with a real microbial fuel cell, investigating electrical power production from it using the Himalayan top soil of Dehradun in Uttarakhand, India. At the smallest level it can help remote rural areas to power lamps or other less energy intensive devices. Using a setup that includes anode, cathode, and related electrical fittings this work has utilized these bacteria over time and observe the power they produce; also the addition of nutrients to the soil which increases the rate of power production has also been experimented. The setup brings together the concept of energy, electronics and microbiology under one framework and is in line with issues relating to climate change, energy security and sustainability. An attempt has been made to explore the spectrum of scenarios and speculating the possibility of generating renewable power using the Himalayan top soil.

Keywords: *renewable, top soil, microbes, fuel cell, power.*

1. Introduction

During the 1900s the inception of the idea to use microbes to produce electricity started taking shape, which eventually led to the use of microbial fuel cell or MFC technology. This relatively unexplored territory has however gained attention as alternate energy sources are entering the society. Topsoil is filled with bacteria species that decomposes organic matter and release electrons a part of the natural process, and since these soil microbes are abundant in nature, these become an exciting prospect for renewable energy as no toxic chemicals are used and process employing locally sourced material. The MFC setup includes an electrochemical cell with its two electrodes (anode and cathode), and a membrane. For MFCs fuel enters through the anode and leaves from the cathode.

Bacteria that can convert organic matter in the soil to electricity are mainly from the Proteobacteria phylum: Shewanella (Family: Shewanellaceae) and Geobacter (Family: Geobaceraceae). Shewanella appeared in a comic book as "Shewy, the Electric Bacterium" in educational kits that are available for students and hobbyists to understand how microbes contribute to charge density in the soil by giving out electrons after consuming organic matter from it [1]. The fact that these soil microbes live in abundance in all soils makes them an interesting prospect to study their ability to release electrons [2] and see how with varying parameters it is affected. We explore this phenomenon with the Himalayan soil of Dehradun in India and speculate the use of the power it produces.

Present work focuses on the following areas:

Study the sub-Himalayan soil available in Dehradun with a microbial fuel cell and see if it is capable of producing any power.

Observe and record if any power is produced, the peak power and when it starts to drop.

Adding nutrients and secondary chemicals (example: sodium acetate) to soil and check how it affects the microbial growth (positively or negatively)

Speculate the possibility of scaling up the present system in an efficient manner which can then be sold to villagers at an easily affordable price and they can power their own small scale utility devices just by using the local soil.

This work is considered important as it shows that the electro-genic bacteria which can work in a Microbial Fuel Cell exist in Uttarakhand i.e. Indian soil as well. Through our investigation we have opened a plethora of possibilities to experiment and collect more data about the soil in the Himalayan ranges of Uttarakhand. The power of the MFC can be increased by putting the MFC (stacked approach). Renewable and clean forms of energy are one of society's greatest needs. The direct conversion of organic matter to electricity using bacteria is possible in MFC; use of compost is a future prospect. Expensive and toxic chemicals were not needed for mediated electron transfer. Such technology has the possibility to be used even for rural and urban waste management which includes cleaning of river, production of electricity simultaneously.

There is no literature available on this topic for the soil of Uttarakhand, or strictly speaking the soil in this region. There was some interaction with associates from forest research institute (FRI) and Wadia Institute of Himalayan geology, and it is seen that they have data on soil profiles but no studies on microbial profile of the soil. This gives the project an added significance as it would contribute to increase the sphere of knowledge of the soil in terms of power production capacity of microbes in the soil (if any).

2. Background

At Penn State University, Prof. Bruce Logan, one of the most eminent names in MFC research is working on developing MFCs that can generate electricity while accomplishing wastewater treatment. In a project supported by the National Science Foundation (NSF), they are researching methods to increase power generation from MFCs while at the same time recovering more of the energy as electricity [3]. A study conducted by a group at Steven Institute of Technology in New Jersey, USA observed the relationship between organic matter and electrical capacity of MFC fuelled by the sample. High percentage of organic matter in sample resulted in higher electricity production of MFCs powered by that sample [4]. NSN Hishan et al. showed that an MFC has the ability to generate electricity from the wastewater while simultaneously removing carbon and nitrogen [5].

A study by Pranab Barua et al. [6] demonstrated that the first order derivative of voltage generated from an MFC with respect to time is a negative constant. Hence change in voltage with respect to time is independent of time. The key finding of this work has been the mixture of bio-wastes used that resulted in high potential current than any single component. The study used specific quantities of cow dung, sewage water, rice washing water individually and combined with slurry. Cell life decreased after addition of slurry but increased after adding vermicompost, thus confirming to the organic matter consumptions of the microbes.

Fuel cells are considered technologies which are of the need for the 21st century and a potent tool to tackle problems from the combustion technologies. Due to their high efficiency and very little environmental impact, they have emerged as a potent alternative. Space exploration already uses fuel cells and with MFC its use to generate renewable power [7] from soil can be speculated. The work presented in this project highlights an initiative, though small, but if implemented in the right direction can be scaled up to industrial processes working entirely with microbial action. Also it will not be restricted to top soil/mud but waste to energy and other biochemical disciplines as well.

There has been an emphasis on production of fuels through biochemical processes in recent years, to name a significant one would be at IIT-Kharagpur where a group under Prof. Debabrata das is working on production of bio-hydrogen and optimizing the process essentially to bring it from R&D phase to commercialization [11]. The project has the financial support of ONGC and research collaborations from European countries as well. Photo-bioreactors are studied with microbes as part of the carbon sequestration initiative, cultures of chlorella, at UPES [12] at bio-diesel lab and also at Abellon clean energy, a bio-energy company [13] based in Gujarat (to name a few). It must be mentioned that almost all bio-energy projects are in R&D phase nation-wise and are yet to reach commercial viability.

If we look back at the history of soil microbes or simply put bacteria in the soil, they go back all the way to early life (cyanobacteria), these primitive life forms, made our planet green through oxygenic photosynthesis, and set the stage forever more complex life forms to arrive. These microbes millions of years ago figured out how to run those complex cellular processes and they do it even today, and hence are an integral part of nature and life as we know it.

The cyanobacteria made this planet green and today when we suffering from pollution problems from combustion technologies and looking for green alternatives/energy, what better than microbe species to explore and make part of this initiative. The only limitation of such processes is they take time, as almost all natural processes do. The findings of this work also reported the same, for results the time sometimes included a week or sometimes more than a week.

Limitations are coupled with all scientific investigations and it is believed that over time they become a part of the process, given the work develops into a scalable and usable utility. Although it must be mentioned the system that has been used in this work experienced over acidification in the soil between the anode and cathode which is a rare occurrence, yet happened and during that time power output became significantly low, such has been discussed later in this work.

3. Prospect for MFC in Uttarakhand

In the remote locations of the foothills of the Himalayas, there are still many villages with little or no access to electricity. The reason sometimes is extending grid power to remote location comes with a significant price, or otherwise the geo-political scenario. This work reported by the authors ambitiously takes into account the study of Himalayan soil available in a particular region within Uttarakhand and primarily checks the soils capability, i.e. if it can produce power and if so what is the peak power it can produce, which is done by exploiting the microbes in the soil.

The secondary task being to speculate, if enough power is produced, and if so whether it can power small lamps, or less energy intensive utility devices. It is proposed to test MFC on Himalayan soils of Uttarakhand, district Dehradun, situated in NW corner of Uttarakhand state and extends from n latitude 29°58' to 31°02'30" and e longitude 77°34'45" to 78°18'30". In this work we have tested the soil from Bidholi, Dehradun and reported the voltage it generates between the anode and cathode and also the power characteristics.

As for international research on similar topic, Prof. Bruce Logan at the Penn. State University in the US is working on production of hydrogen from microbial fuel cell and also in the same apparatus, purification of water; primarily working on making water infrastructure energy sustainable in the next 20 to 25 years. There are many researches going on with MFCs [9] but specifically understanding the soil microbes and enhancing the process has not gone through significant feats.

4. Methodology

The basic setup used is a microbial fuel cell, the components include, an anode, cathode, an air tight vessel and other secondary materials. As given in Figure 1.

Figure 1. The Anode (with Green Wire), Cathode (Orange Wire), the Vessel and the Electronic Setup.

It must be mentioned that the anode and cathode are made of conductive fibers, which can cause electrical short circuits, to prevent any damage they must be kept away from other electronic devices. The cathode assembly has two parts: cathode felt and the cathode wire which is made of titanium. Similar is for the anode assembly: anode felt and the titanium wire.
The vessel will be filled up with a specific quantity of soil the anode will be placed in the mud where the microbe specie can grow, while the cathode will be on top of the soil layer exposing it to oxygen in the space between the container lid and the felt. Both cathode and anode are made of grapheme connected to titanium wires. This is explained in Figure 2.

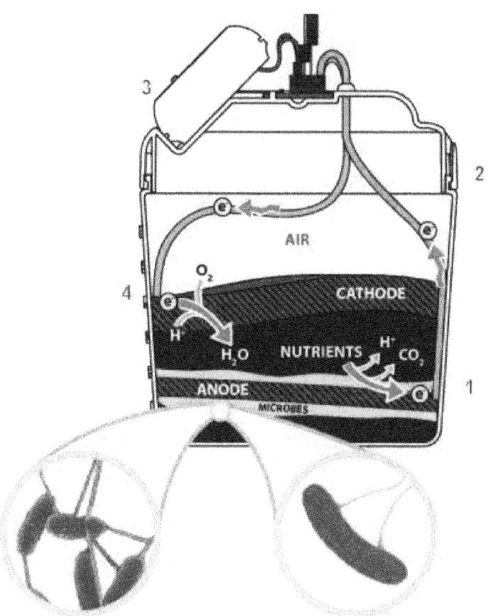

Figure 2. The Flow of Electrons through the System as Adapted From [6]

At Point 1: As the microbes around the anode take up the nutrient in the soil/mud, they deposit electrons onto the anode in one of the three ways:

a. Mediated transfer: using electron-shuttling bio-molecules.
b. Nanowire transfer: using connected appendages developed by the bacteria.
c. Direct transfer: from the cell wall of the bacteria and directly to the anode surface.

At Point 2: Electrons travel through length of the wire to the Board with the power electronics, where they power the small energy intensive devices.

At Point 3: After passing through load, electrons go down through the wire via the cathode.

At Point 4: At the cathode, electrons interact with oxygen inside the closed vessel to form water.

This cycle happens over and over, trillion of times every second. This continuous flow of electrons is the generation of electricity, which power the small electronics. The Blinker Board (part of the electronics, refer to Figure 1) gives a visual indication that the microbial community is producing power, All Blinker Board electronics use lead-free solder. The function of the components:

a. Hacker board: The hacker board takes the low voltage and low current coming from the system and converts into short bursts of higher voltage and higher current. It contains Voltage-boosting microchip and 8-pin hacking socket.
b. Capacitor: Energy storage component, it is able to build up energy as power comes in from the system, and discharge that energy in a quick burst to blink the LED.
c. LED: The Light Emitting Diode takes the electrons being discharged by the capacitor and converts the electron's energy into light energy. The LED will start glowing only when the voltage generated across the terminals is greater than 0.35V.

The Vessel contains a transparent enclosure made of Clear PETE (Polyethylene terephthalate), White ABS (Acrylonitrile butadiene styrene).

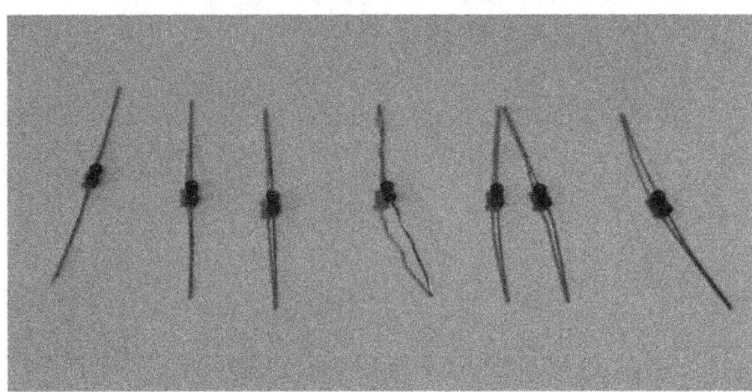

Figure 3. Seven different resistors (47-4700ωs), value can be found by connecting the multimeter

The Power can be calculated directly using the multimeter and for taking power measurements and assessing the health of the microbial community. Seven resistors were used as shown in Figure 3, with values of 47Ω, 100Ω, 220Ω, 470Ω, 1kΩ, 2.2kΩ, and 4.7kΩ respectively.

To find power output (including max, power), power for each resistor is calculated using Ohm's law.

V= I.R also, P = V.I

Therefore,

$$P = V2/R$$

where Power is in Watts, voltage is in Volts, Resistance is in Ohms, and current is in ampere.

As far as analogies are concerned it can be deduced that only when voltage difference exists between terminals, current will be flowing. Similar to pressure drop allowing fluid to flow. Early experiments investigated the possibility of generating power from the local soil available at Bidholi, Dehradun. The geographical details are as follow:

Table 1

Name	UPES, Bidholi
Type	Locality
Latitude	30.3165
Longitude	78.0322
State	Uttarakhand
District	Dehradun

Sample size for all experimental data is:
For sample below anode: 6cm×6cm×2cm (weighing around 84 gm. on average). For sample below cathode and above anode: 6cm×6cm×3cm (weighing around 91 gm. on average).

Detailed graphical analysis of the experiments based on Voltage generated will be discussed now, which demonstrates the fact that soil microbes in this region are capable of producing voltage across terminals using the soil as an electrolyte, now throughout all experiments there were some limitations which will be discussed later.

5. Result and Discussions

For the graphs given in Figures 4(a, b) the voltage developed is studied over time, the soil is prepared, and the whole setup is kept under observation, for the first batch the primary target was not to check power production but to investigate if any voltage is generated across the electrodes, this study was started from 14th Jan, 2016 and was done up to 16th Jan, 2016. A 48 hours period was the total duration of the first experiment.

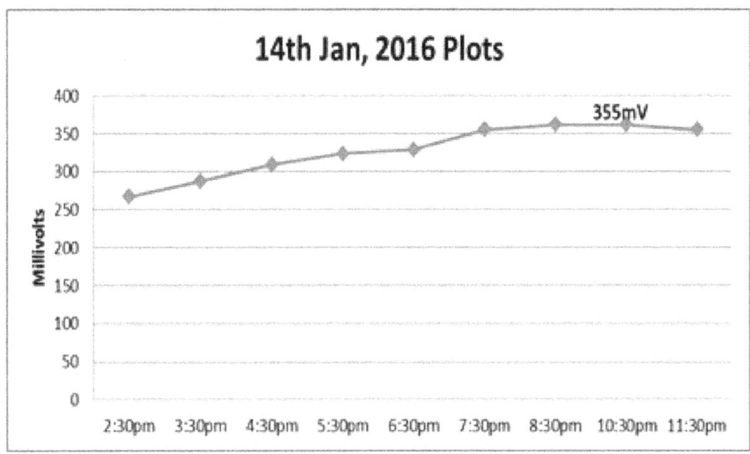

Figure 4a. For the first 12 hour study the peak voltage was 361mV

Now it must be remembered that soil microbes especially bacteria prefer temperatures above 28°C but less than 60°C, at this time of the year in Bidholi, temperature stays around 24°C during the day and drops to around 8°C at night, so this attributes to a limiting factor hence it is assumed that voltage generated (if any) would be significantly low for this study, as temperature is on the colder side and the time given to the microbes to generate electrons is only 48 hours. But however it must be remembered that this experiment is not for investigating peak power but to see if any voltage is generated in this setup by the soil bacteria via the process we have suggested.

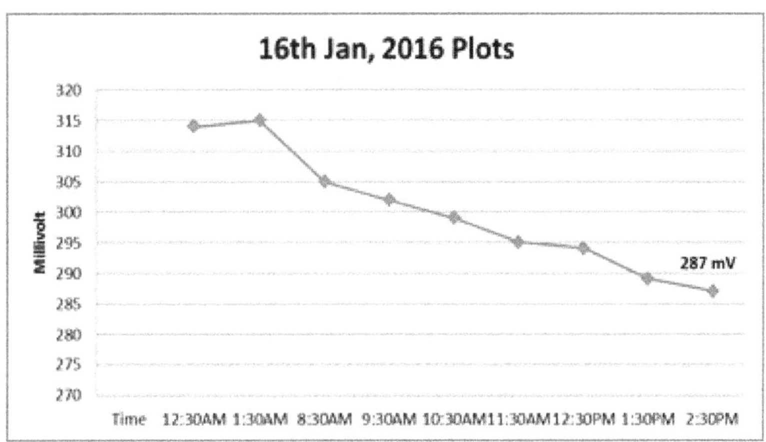

Figure 4b. For the last 12 hours peak value was 299 mV

Figure 4a confirms that potential can exist with the local soil here between the anode and cathode, one hour after charging the system, the value shot up to 267 millivolts (mV). Over time this value increased and for that day the peak voltage value was 361 mV. For figure (b), it is reported the value of voltage starts dropping, now for this there is a very good explanation.

Ideally this system can develop a big community of microbes and can work for a month or even a year, but since in this investigation we only wanted to see the voltage generated across the terminal, we kept it open circuit i.e. the cathode and the anode were not connected to the pins on the Blinker/Hacker board. And the value of voltage was taken by connecting the multimeter with alligator clips and the other end of the clips directly to the titanium wire of the anode and cathode.

Before going on to experiments to see peak power produced and its characteristics it is of interest to see how the soil bacteria reacts to certain materials, for this investigation we have considered two substances, salt and sodium acetate (CH_3COONa).

Sodium chloride or table salt is the most commonly available material in soil and water, we made an investigation where saline water is mixed with the soil and response of the microbes are recorded, in terms of voltage over time, while for sodium acetate some literature from a University in Boston (United States) who made a similar soil based experiment with MFC found that bacteria loves consuming acetate [8], this triggered us to make a study with our local soil and see how the soil microbes react to it.

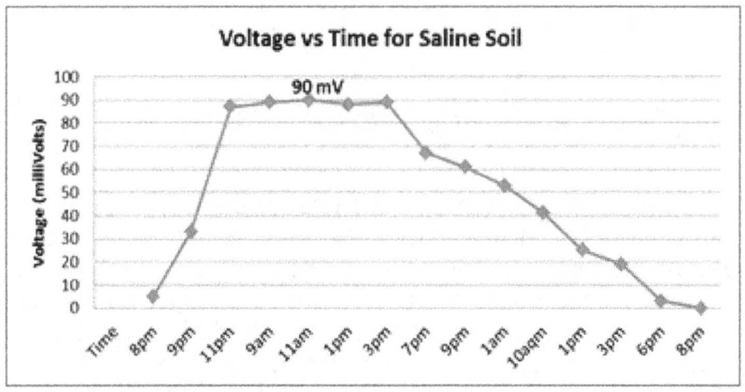

Figure 5a. Voltage variation over time in a saline soil system

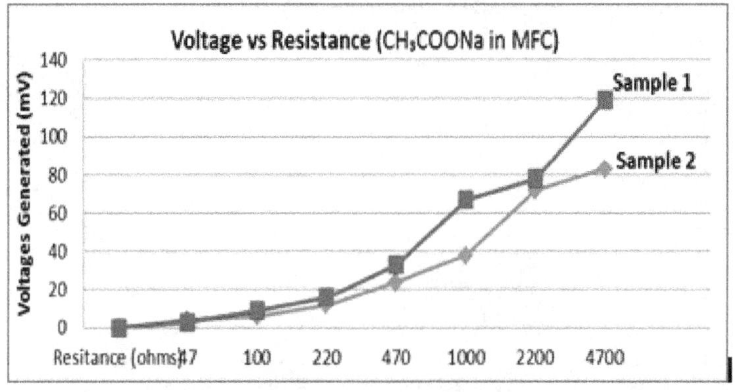

Figure 5b. Voltage variation with resistance over time for two samples of soil with fixed amount of sodium acetate added to it.

For figure 5a, the peak voltage recorded 90mV and then it started dropping within that 24 hours, thus showing that bacteria specie in soil is highly salt intolerant and cannot produce power where the salinity concentration in the soil is high. For this experiment 25gm of salt was added to the soil. This outcome was somewhat expected, as salinization of soil is recognized as one of the most pressing environmental challenges. It is reported that elevated sodium (Na+) decreases plant growth. For figure 5b, some literature has suggested that MFC's running with soil as electrolyte has a remarkable influence on the power produced by soil microbes, but given sample taken for this investigation reports poor microbial growth. Two separate samples were taken and they both peaked up to 83 mV and 119 mV respectively in a 72 hours study.

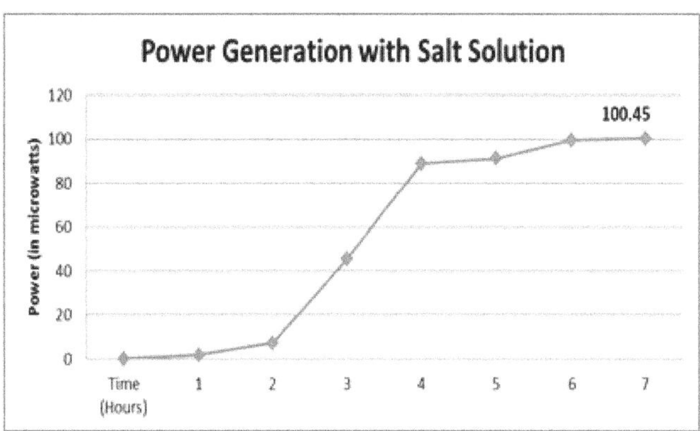

Figure 6a. Power output for saline soil using one resistor (220 ohms)

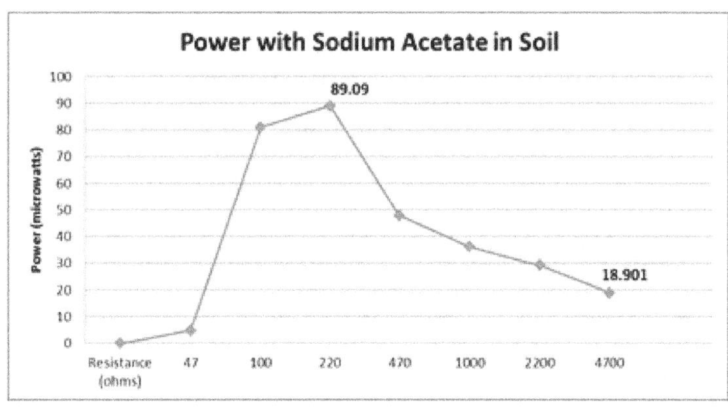

Figure 6b. Power output variations with use of sodium acetate in the soil of MFC.

Figure 6a represents the second system which was setup in the mid of February 2016, when temperature was between 28°C in the morning to 15°C at night, the setup with salt solution was observed for few hours because previous graphical analysis, from Voltage vs. Time plot it was concluded that saline condition do not allow much microbial growth hence this study was completed within 24 hours. This investigation for generating power with salt solution showed that the power stabilized around 100uW. Figure 6b, shows power production with sodium acetate (CH3COONa) solution, now this has been very interesting, in the previous case it is reported that power continued to increase very slowly after 100uW, in this case it is seen after reaching a peak value of 89uW the power production started dropping all the way up to 18uW, and then it was discontinued. This behavior was somewhat unexpected, the authors deduce that Himalayan soil or simply soil microbes in this region do not consume acetate the way some literature suggested.

Now the following discussion will show the maximum power output reported from Himalayan soil using the MFC system, for this system there were significant enrichment of the soil that was done, For the anode, the soil sample was mixed with Nutrient 1 and the sample below cathode was mixed with Nutrient 2. The details of them are as follow:

a. Nutrient 1 (Mixture): Water, Tomato Paste (34.5%), Sugar, Liquid Glucose, Iodized Salt, Thickener (INS415), Onion, Garlic, Spices and Condiments.
b. Nutrient 2 (Mixture): Water, Tomato Paste (34%), Sugar, Edible common salt, Permitted Acid (INS260), Permitted emulsifiers and stabilizer (INS1422, INS415).

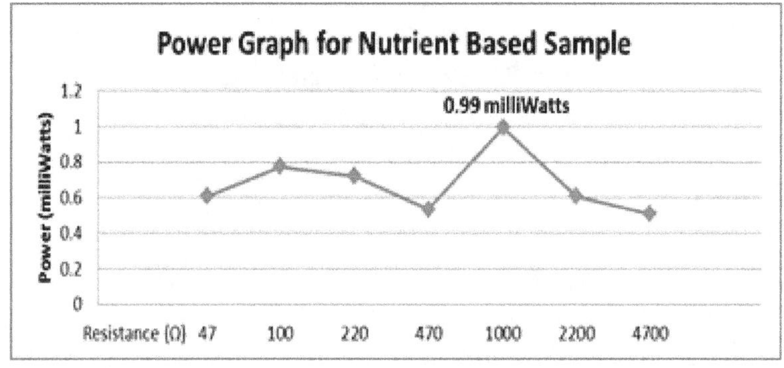

Figure 7a. Power versus resistance

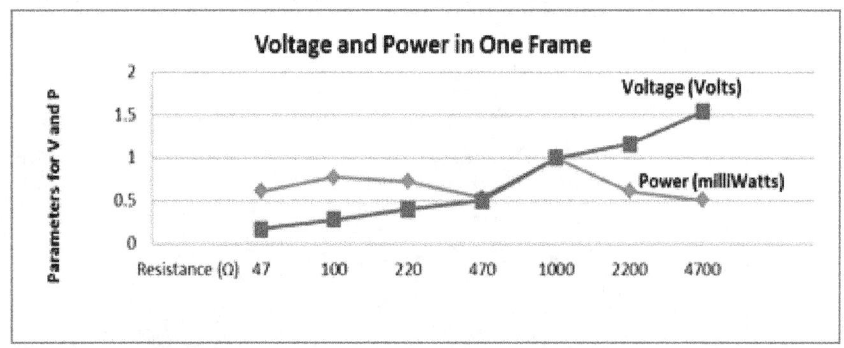

Figure 7b. Bringing voltage and power in one framework to deduce nature of the current produced.

For the graph at Figure 7a, we see the peak power produced by the system in two weeks' time is 0.99 milliwatts, and then it drops, the value fluctuates because different resistors are used to record the voltage which then fed into the equation: P= V2/R gives the power produced. This investigation could have been continued further to see how much power can it produce and given the observation made, it is very clear that power value will again increase.

In the study explained graphically in Figure 7, some very interesting observations are reported. Referring to figure 7b, it is seen that voltage generated across the terminals increases rapidly up to 1.54 Volts, the study was discontinued after that to see the power production, However, the graph given demonstrates the capability of the soil microbes to produce voltage not in the range of milli to micro, but a higher range and if this is allowed to continue (say for a month), there can be some remarkable current produced from the system which we can then speculate to help light less energy intensive devices at rural levels where electricity is still not available. The Microbial fuel cell generates DC or direct current, and with increasing resistance the value of Power produced goes down as both are inversely related to each other.

6. Conclusion

This work investigated the power producing capabilities of soil bacteria, because there was no such literature available on Indian soils for such, this work is considered important in that it has shown that the electrogenic bacteria which can work in a Microbial fuel cell exists here as well, for this work a specific location in India, Uttarakhand, in Dehradun at Bidholi is taken for the soil sample. For sample sizes of 6cm×6cm×3cm this MFC system has produced peak power of 0.99 milliwatts, so it can be speculated with bigger sample size i.e. scaling up this system can produce significant power. Also it must be mentioned because of time limitation most of the investigations in this work were restricted to maximum of two weeks, so the spectrum of possibilities are wide open for this kind of renewable energy technology to grow as this work has shown that the system works. Furthermore, once the power has stabilized for the system we have used, small electronics can be plugged into the Blinker Board and can be operated. The system did show some limitations such as: water build up over the cathode, and the power production is low if temperature is below 24°C but having said that this work has demonstrated the top soil use and also showed the use of salt solution, chemicals like sodium acetate, and tomato pulp. There is literature available, as discussed earlier, which shows use of vermicompost, which makes the soil further enriched with nutrients.

The concept of Microbial Fuel Cells can be used in places lacking proper sanitation and electricity, the process of generating electricity in these can also purify water [10]. Also present work went with direct electron transfer process, hence no cost related to use of chemicals.

7. Acknowledgements

Present work has been funded by the Department of Chemistry at University of Petroleum & Energy Studies, Dehradun, India. We thank them for their inputs and allowing us to carry out this investigation. We would like to acknowledge the continuous support of Dr. Pankaj Kumar, HOD, Chemistry Department and Dr. Kamal Bansal, Dean, CoES, UPES for the RISE initiative which allowed this investigation to meet its goal.

References

[1] Hofstetter, T., K. Pecherl, RP Schwarzenbach. Bioavailability of Clay-Adsorbed U (VI) to Shewanella Putrefaciens. Environ. Sci. Technol. 2002; 36: 1734-1741.

[2] Amann, Rudolf I., Wolfgang Ludwig, Karl-Heinz Schleifer. Phylogenetic Identification and In Situ Detection of Individual Microbial Cells without Cultivation. Microbiological reviews. 1995; 59(1):143-169.

[3] He, W., X. Zhang, J. Liu, X. Zhu, Y. Feng, B.E. Logan. Microbial Fuel Cells With an Integrated Spacer and Separate Anode and Cathode Modules. *Environmental Science: Water Research & Technology*. 2016; 2:186-195.

[4] Jessica, Li. An Experimental Study of Microbial Fuel Cells for Electricity Generating: Performance Characterization and Capacity Improvement. *Journal of Sus. Bioenergy Systems*. 2013; 3: 171-178.

[5] Hisham, Nur Syazana Natasya, et al. Microbial Fuel Cells Using Different Types of Wastewater for Electricity Generation and Simultaneously Removed Pollutant. *Journal of Engineering Science and Technology*; 8(3): 317-326.

[6] Barua, P., Deka, D. *International Journal of Energy, Information and Communications*. 2010; 1(1).

[7] Arulmani, Samuel Raj Babu, Vimalan Jayaraj, Solomon Robinson, David Jebakumar. Long-Term Electricity Production From Soil Electrogenic Bacteria and High-Content Screening Of Biofilm Formation On The Electrodes. *Journal of Soils and Sediments*. 2015: 1-11.

[8] Villarrubia, Narváez, Claudia Wuillma. Integration Of Composite Nanomaterials Into Anode And Cathode Designs. 2011.

[9] Kim, Jung Rae, et al. Power Generation Using Different Cation, Anion, and Ultrafiltration Membranes in Microbial Fuel Cells. *Environmental Science & Technology*. 2007; 41(3): 1004-1009.

[10] Website: http://www.engr.psu.edu/ce/enve/logan/bioenergy/research_mfc.htm (Accessed: 04/22/16)

[11] Das, Debabrata, and T. Nejat Veziroğlu. Hydrogen Production by Biological Processes: A Survey of Literature. *International Journal of Hydrogen Energy*. 2001; 26(1): 13-28.

[12] Sharma, Rohit, et al. Cost Effective and Economic Method for Cultivation of Chlorella Pyrenoidosa for the Simultaneous Treatment of Biogas Digester Waste Water and Biogas Production.

[13] Patel, Beena, Bharat Gami, and Hiral Bhimani. Improved Fuel Characteristics of Cotton Stalk, Prosopis and Sugarcane Bagasse Through Torrefaction Energy for Sustainable Development. 2011; 15(4): 372-375.

12

An Investigative and Comprehensive Study about Deregulation (Restructuring) of Indian Power Market

Ravindrakumar Yadav[1*], **Ashok Jhala**[2]
[1] Babaria Institute of Technology, Varnama, Vadodara, Gujarat, India
[2] RKDF College of Engineering, Bhopal, India
e-mail: rkyadav.bit@gmail.com* , ravindrayadav.ee@bitseducampus.ac.in

Abstract

This paper is a discussion about the introduction of restructuring and deregulation in Indian Power System. In modern era, deregulation has an important impact on power sector. In this paper, recent use of deregulation in Indian Power Sector has been described and measures to be taken in order to improve deregulation are also suggested.

***Keywords**: deregulation, restructuring, Power Sector, CERC, TSO , LDC*

1. Introduction

In current time, due to rise in power demand and supply, it is not an easy task to manage the generation and cost concurrently for one single party. To reduce monopoly of one organization and to provide quality and continue reliable power supply at reasonable cost, it is essential to encourage competition in power market. This can be possible by introducing restructuring and deregulation in electrical power sector.

Deregulation involves unbundling of different components of power system, availability of components for sale and also forming new set of rules for operation and sales of electricity [1].

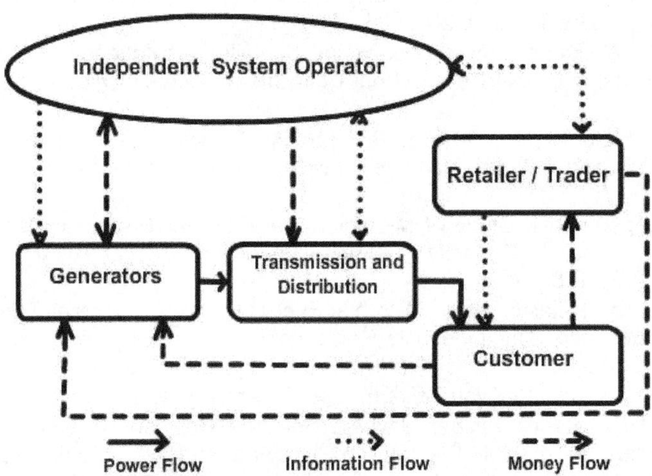

Figure 1. Deregulated power system

An main and important aspect of deregulation is restructuring. Restructuring means unbundling of power system into both horizontal and vertical components. Vertical integrated utilities are mainly broken up into three main components, i.e. Generation, Transmission and Distribution [2]. This introduces competition in generation, transmission open access with retail competition in distribution. Competition in generation reduces cost of power, Transmission open access provides access to transmission grid for the various generations, that enhances reliability of power supply. Retail competition in distribution provides choices for buyers to select between power suppliers, which provides good quality of power. The general structure of deregulated power system is shown in Figure 1.

Many countries like England, United States of America, Canada, Australia, Peru, New Zealand, Chile, Argentina, Colombia and Scandinavian have already adopted deregulated power structure since long[3]. In India, till independence the entire power sector was under the control of private sector. After the enactment of new Electricity Act in 1948, the entire power sector is mostly owned by State Governments and is largely managed by vertically integrated electricity business through State Electricity Boards (SEBs). In 1975, Government of India (GOI) entered in the field of generation and transmission through their Central sector power stations and management. However, distribution sector continued to stay with SEBs as a monopoly business. Over the period of operations, regrettably, the sector developed techno-comm inefficiencies. Restructuring therefore was felt as mandatory option to cure. Accordingly, the power generation was opened up in 1991 followed by transmission in 1998. Electricity Regulatory Commissions Act was enacted in 1998 for establishing regulatory commissions in various States. And then the Electricity Act 2003 has been notified by Government of India in June 2003. The objective of this act is to accelerate the power sector reforms.

2. Power Sector Reforms

The Government of India has been taking several initiatives to invite private sector participation in generation and transmission. Understanding the difficulties faced in the process of reforms, GOI in consultation with the States initiated measures to unblock the difficulties

1. Unbundling of SEB: A number of States have initiated the power sector reform process, starting the unbundling, tariff rationalization and corporatization of generation, transmission and distribution. Practically this brings the accountability at each level of operations in the power business. The results were not much encouraging although not so adverse to consider. Self-sustainability is not achieved through this makeup unless privatization is introduced in stages starting from urban areas.
2. Setting up of Regulatory Mechanism: The Central Electricity Regulatory Commission (CERC) along with State Electricity Regulatory Commissions (SERC) have been established in 19 states under the Electricity Regulatory Commissions Act, 1998, to discharge the following functions:
    a. Regulation of the tariff of generating companies owned or controlled by the Central Government.
    b. Regulation of the tariff for generating companies other than those owned or controlled by the Central Government if such generating companies enter into or otherwise have a composite scheme for generation and sale of electricity in more than one State.
    c. Regulating the Inter-State transmission of energy including tariff of the transmission utilities.
    d. Promoting competition, efficiency and economy in the activities of the electricity industry.
    e. Advising the Central Government in the formulation of tariff policy which shall
        (i) Provide justice to clients
        (ii) Facilitate mobilization of adequate resources for the power sector.
    f. Association with the environmental regulatory agencies to develop appropriate policies and procedures for environmental regulation of the power sector.
    g. Framing of guidelines in matters relating to electricity tariff.
    h. Resolving the disputes involving generating companies or transmission utilities.
    i. Assisting Government of India on any other matter referred to the Central Commission by that Government.
    j. Licensing any person for the construction, maintenance and operation of Inter-State transmission system.

3. Regional Electricity Boards in India

As per the statute, the Central Electricity Authority (CEA) is possible for power planning at the national level. CEA advises the Ministry of Power on the national power policy and planning, whereas the central electricity regulatory commission is looking after the regulatory issues. Day-to-day operation of the regional grid is carried out by Regional Load Dispatch Centers (RLDCs), which are under the operational control of CTU, i.e. PGCIL.

The main function of RLDC is, to carry out the integrated operation of the power system in that region and that of Regional Electricity Board (REB), to facilitate integrated grid operation. Presently five REBs namely, Northern REB, Southern REB, Western REB, Eastern REB and North-Eastern REB exist to promote the integrated operation of the regional power systems. The responsibilities of REBs are to review project progress, to plan integrated operation among the utilities in the region, to co-ordinate the maintenance schedules, to determine the availability of power for inter- state utilities transfer, to prescribe the generation schedule and to determine a suitable tariff for the inter-utility exchange of power. Power sector across the world is undergoing a lot of restructuring; India is no exclusion to this. The need for restructuring the power sector was felt due to the lack of financial resources available with Central and State Governments, and necessity of improving the technical and commercial efficiency.

In some States of India there are multiple private utilities, which are technically and financially in a position to enter the phase of a competitive electricity market. Hence, in 1998 the Regulatory Commissions were formed under the Electricity Regulatory Commissions Act 1998 (Central Law) to promote competition, efficiency and economy in the activities of the electricity industry. Ministry of Power has undertaken Accelerated Power Development and Reform Program (APDRP) from the year 2000-01 with the twin objectives of financial turn-around in the performance of the power sector especially in electric distribution and improvement in quality of supply.

Electricity Act 2003 has come into force from June 2003. As the act allows third party sales, it allows the concept of trading bulk electricity. The act also provides open access to transmission as well as distribution of electricity.

4. Proposed Model for Restructuring in India

In many parts of the world everywhere unbundling taken place, the two models are more established for system operation. The first one is Independent System Operator (ISO) model and the other is Transmission System Operator (TSO) model. In ISO model, transmission companies are also permitted to own, manage and control generation and distribution companies, an independent system operator is created to facilitate open access and competitive markets.

In TSO model, operation of the grid and ownership of the grid are integrated in a single entity, which is responsible for development of transmission system and to provide unbiased open access to all eligible market participants.

Neutrality is an important feature of the TSO to ensure an efficient market. In view of this, TSO model seems to be most suitable for future restructured electricity market in India. This is because the government owned Transmission Company is merely responsible to provide non-discriminatory open access. Some of the developed countries are also moving away from ISO model by formation of Regional Transmission Organizations (RTO), which will finally converge as a TSO model. Even though the conditions in Indian power market are not yet ripe for introducing retail competition, the necessities in a deregulated power market can be summarized below:

a. Non-discriminatory open access to transmission network is a pre-requisite for ensuring competition in wholesale power trading.
b. The system operation functions at the national level can be handled by central transmission utility while state transmission utilities can manage State Load Dispatch Centers (SLDCs) similar to TSO concept.
c. The regional electricity boards will have the responsibility of managing the power exchanges while the Regional Load Dispatch Centers (RLDCs) will manage the overall integrated operation of power system like outage planning, relay coordination, islanding schemes, etc.

5. Competing Models for Restructuring in India

a. Odisha Model:

Orissa was the first state to embark on the reform program after the state Electricity Reform Act became effective in April 1996. Almost immediately, the Orissa State Electricity Board is partially unbundled into three separate entities: Orissa Hydro Power Corporation OHPC, Orissa Power Generation Corporation OPGC and Grid Corporation of Orissa GRIDCO. Generation was first privatized in June 1998, AES purchases 49% stake in OPGC. In the second phase, the distribution assets, properties and personnel of GRIDCO is broken into four distribution companies. BSES purchases three of them (NESCO, WESCO and SOUTHCO) in April 1999 and one (CESCO) is transferred to AES Transpower (joint venture of AES and Jyothi Structures Ltd) in September 1999.

b. Delhi Model:

The Delhi Electricity Reform Act comes into force in March 2001. Two months later, Delhi Vidyut Board DVB, the state's electricity board, establishes six shell companies (holding company, generating company, Transmission Company, three distribution companies) to be operational. Distribution is first privatized. 51% of the equity in three distribution companies is sold to two privately owned Indian power companies, BSES and Tata Power. DVB ceases to exist and is replaced by the holding company, the Generation Company and Transmission Company. Delhi government retains ownership of the generation. Holding company retains all unserviceable liabilities. Existing serviceable DVB liabilities will be paid by successor agency after a four year grace period Introduced concept of aggregate technical and commercial (AT&C) losses, rather than transmission and distribution (T&D) losses. Private investors bid for distribution companies on the basis of a five year AT&C targets, indicative multi-year tariff profile and projected Government assistance, a five year in-between period with some Government support over the period.

c. Andhra Pradesh Model:

State Reforms Act came into force w.e.f. Feb 1999. APSEB unbundled into Andhra Pradesh Generation Company Ltd. (APGENCO) and Andhra Pradesh Transmission Company Ltd. (APTRANSCO for transmission and distribution). Andhra Pradesh Electricity Regulatory Commission has been operational w.e.f. April 1999.

d. Haryana Model:

State Reforms Act came into force w.e.f. 14.8.1998. SERC became operational w.e.f. 17.8.1998. SEB unbundled into Haryana Vidyut Prasaran Nigam Ltd., a Trans Co. (HVPNL) and Haryana Power Corporation Ltd. On 14.8.1998. Two Government owned distribution companies viz. Uttar Haryana Bijli Vitaran Nigam Ltd.(UHBVNL) and Dakshin Haryana Bijli Vitaran Nigam Ltd.(DHBVNL) have been established. Till these two companies become independent licensees, they will operate as subsidiaries of HVPNL.

e. Uttar Pradesh Model:

State Reforms Act has been notified on 15.1.2000.As per the decision of the Government of Uttar Pradesh, the activities of generation, transmission and distribution of erstwhile UPSEB have been transferred to: Uttar Pradesh Rajya Vidyut Utpadan Nigam Ltd. (UPRVUNL) ,Uttar Pradesh Jal Vidyut Nigam Ltd. (UPJVNL) ,Uttar Pradesh Power Corporation Ltd. (UPPCL)- UPPCL took over the transmission and distribution functions of erstwhile UPSEB. Re-organization Committee set up to study the State Power Sector has submitted its recommendations to State Govt. The State Government has set up State Rural Energy Development Corporation as an independent company under the Companies Act to manage distribution for rural and agricultural consumer segments with assistance of Rural Energy Cooperatives. Consultants have submitted the final report of the tariff rationalization study which has been financed by PFC. Four Task Forces have been formed to initiate the implementation of reform program. The areas covered are:

1. HR,
2. Identification, Valuation and Transfer of Assets,
3. Identification and segregation of urban and rural feeders and zones and earmarking operational areas of WBREDC and WBSEB UD System and

4. Metering, Billing, Collection, Electricity Accounting and System Loss Reduction of WBSEB

6. Indian Electrricity Market

India being a very vast country, several independent electricity markets may co-exist having their area of operation clearly demarcated from each other. In India, at the state level, the state power markets and at the regional/ national level, the regional national power markets may emerge. In the immediate future, after the enactment of Electricity Act 2003, it is felt that the prevailing conditions in the country are ripe only for wholesale competition and not for the retail competition.

7. Evolution of Electricity Industry

In the pre-independence time power sector consists of small private players to meet the local needs of the smaller area around them under the provision of Indian Electricity Act 1910. In 1947, the electricity industry in the UK was nationalized. India followed suit in 1948 and except for some pockets such as Mumbai, Kolkata, Ahmedabad and Surat. The entire industry was nationalized by virtue of the aforesaid Act of 1948, which laid down the structure of electricity industry in the independent India. This Act triggered the formation of State Electricity Boards (SEBs) to handle generation, transmission and distribution of electricity within the states. Subsequently central sector steeped in to support the National Load Dispatch Center (NLDC) is also planned to facilitate inter-regional transfer of power and for optimum scheduling and dispatch of electricity among the Regional Load Dispatch Centers (RLDCs). Given these various changes, the industry structure will be transformed from the current 'single-buyer model' to 'multi-buyer model'. In a multi-buyer model, the distribution entities are totally autonomous in procuring and dispatching their supply. This model would lead to better operation and lower cost to the end consumers. On an overall basis, Electricity Act 2003 is comprehensive and provides for progressive development of a market-based regime in Indian power sector through competition. The benefits will, however, start reflecting after a period of 4-5 years. Over the past two decades a number of countries have restructured their electricity industry by significantly reducing the government's role in the ownership and management of domestic electricity industries. It has seen as necessary conditions for increasing the efficiency of electric energy production and distribution, offering a lower price, higher quality and secured supply. The forces behind electric sector deregulation taking place worldwide are different in different countries.

8. Recent Initiatives

The Electricity Act 2003 makes thermal power are-licensed activity, freely permits captive generation and makes the setting up of state regulatory commissions mandatory. It recognizes the trading as a distinct commercial activity and suggests measures like preparation of National Electricity Policy for planned development of the sector. In line with the policy objectives, the act provides a drive to complete rural electrification and provide for management of rural distribution by cooperative societies, non-government organization, franchisees etc. The progressive policies would open new opportunities for setting up merchant generators, utilization of captive generation and electricity market development. A load dispatch center at the national level i.e. layers such as SEBs and utilities. It is a predetermined contracted transaction and there is non-existent of spot market. Currently, State load dispatch Centers (SLDCs) are carrying out the optimum scheduling of the state generating units and the RLDCs are responsible for scheduling of central sector generating units only. SLDCs send the requisition to the RLDCs against their entitlements out of available power from central sector generation (CGS) and the RLDCs allocate total available power to various states in the ratio of their entitlements. Day-to-day operation of the regional grid is carried out by RLDCs, which are under the operational control of Central Transmission Utility (CTU), i.e. Power Grid Corporation of India Ltd. all amount of power (about 2.5% of total generation).

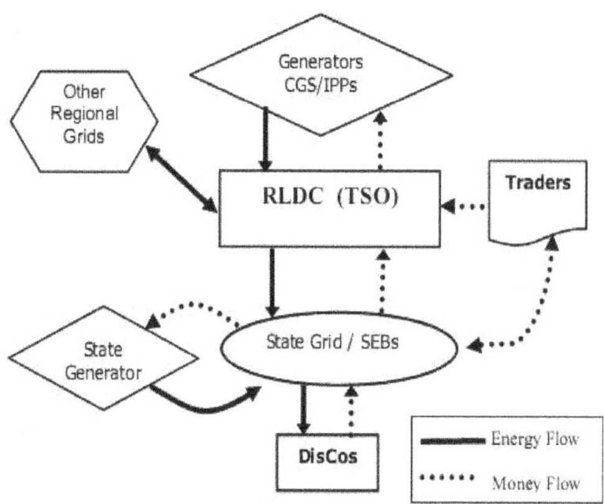

Figure 2. Current electricity market

9. Recent Initiatives

Apart from above, a small amount of generation is being traded at wholesale level through either bilaterally or with the help of power traders. But presently, trading is mostly restricted to players such as SEBs and utilities. Power trading has generated considerable interest among power players, as it is evident from the brisk line-up for licenses at Central Electricity Regulatory Commission (CERC). Presently there are seventeen trading licensees to whom CERC has granted license for inter-state trading in electricity. These traders apply for open access on behalf of suppliers and buyers to the nodal RLDC depending upon transaction requirement. CERC has made the regulations for open access in inter-state transmission and inter-state trading. The market structure of current Indian electricity industry is shown in Figure1, depicting energy flow and money flow separately.

10. Key Issues

The wholesale electric power trading in India, though a scent activity, is poised to develop a market and pave the way for creation of power exchange for economic pricing and optimal utilization of generation capacity. However, power trading has barely got off the ground in India, there are some major issues related to power trading, which need to be addressed for structural development of electricity market in India. These key issues include:

Non-existence of Power traders are just matchmakers. They are lacking formal market and real-time information because at present, no power exchange for power trading in the pattern of stock exchange really exists in the country. Trading is mostly restricted to Exchange: Presently power layers such as SEBs and utilities. It is a predetermined contracted transaction and there is non-existent of spot market.

a. Lack of Pricing Mechanism: CERC has fixed the trading profit margin as 0.02 Rs/kWh but there is no uniform method for price calculation of traded power. Traders simply negotiate the energy price between suppliers and purchasers and then add the trading margin. There is a need of market driven pricing mechanism.
b. Reliability of Supply and Off-take: Contractual obligations are not always honored. There is need of an institution/authority which can guarantee supply side delivery and buy side payment. A clearing system based on pledge accounts can minimize the risk for market participants.
c. Lack of Information: Traders are lacking equal and same time information about the market. Participants need information to evaluate his options and pick the most competitive deal. Setting up an online bulletin board offering real-time information for market participants would help.
d. Risk Management: Electricity market is highly price volatile because of dependence on fuel prices and network constraints during peak hours. Spot price variation in electricity market can be hedged with the help of risk hedging derivative instruments, which can be traded in financial markets for electricity. Competition is necessary for eradication of

inefficiency but it should be introduced gradually in the phased manner. With the enactment of EA 2003, along with other recent initiatives, Government of India has outlined the counters of a suitable enabling framework.

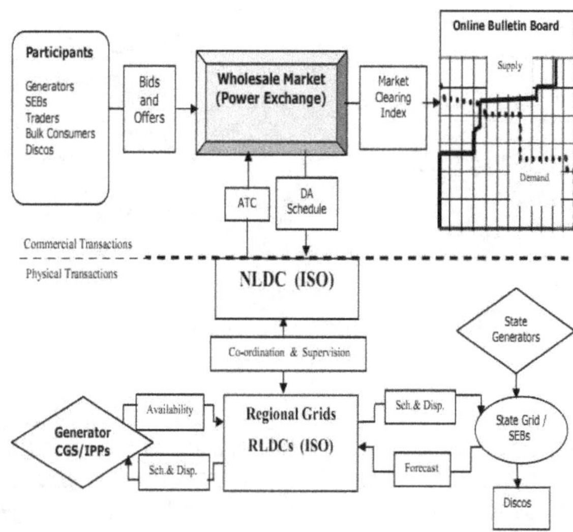

Figure 3. Proposed national electricity market

Figure 3 illustrates a simplified, theoretical model of the proposed national power market having different interactions between the relevant actors that participate in electricity market. The financial flows that result from electricity trade referred as 'commercial transactions' and the physical electricity flows referred as 'physical transactions' are depicted in upper and lower part of Figure 2 respectively.

In decentralized trading arrangement, PX has to be a separate entity from system operator, as an independent market operator. But PX should have strong co-ordination with system operators (NLDC/RLDCs) for ATC information, dispatch of DA schedules and imbalance settlement through UI mechanism.

11. Conclusion

Electricity reform process in India is already in action although at a slow pace. Several state electricity boards are being unbundled into three distinct corporations namely Generation, Transmission and distribution. The distribution system are being horizontally broken down into manageable Discos with separate accountability and privatized for better efficiency in metering, billing and revenue collection. The system operation functions at the regional/national level can be with central transmission utility, while state transmission utilities may manage load dispatch centers inline with TSO.

References

[1] Loi Lei Laai. Power System Restructuring and Deregulation. John Wiley & Sons. England. 2002.

[2] Malik OP. Control Considerations in a Deregulated Electric Utility Environment. *IEEE Canadian Review.* 2000: 9-11.

[3] Yog Raj Sood, Narayana Prasad Padhy, HO Gupta. Wheeling of Power Under Deregulated Environment of Power System – A Bibliographical Survey. *IEEE Transaction on Power Systems.* 2001; 109(1).

[4] Khaparde SA, Kulkarni SV, Karandikar RG, Agalgaonkar AP. *Role of Distributed Generation in Indian Scenario.* Proceedings of South Asia Regional Conference. New Delhi, India. 2003.

[5] Vindal SS, Saxena NS, Srivastava SC. *Industry Structure Under Deregulated Wholesale Power Markets in India.* Proceedings of International Conference on Present and Future Trends in Transmission and Convergence. New Delhi, India. December 2002.

[6] Government of India. The Electricity Act, 2003. New Delhi: The Gazette of India. 2003. Part I1 Section 3 Sub-section (ii).

[7] Kothari DP, Nagrath IJ. *Power System Engineering*. 2nd Edition.

13

Improvement of Quality of Service Parameters in Dynamic and Heterogeneous WBAN

Madhumita Kathuria *, Sapna Gambhir
Department of Computer Engineering, YMCA University of Science and Technology
Faridabad, India
e-mail: *madhumita.fet@mriu.edu.in; sapnagambhir@gmail.com

Abstract

With growth in population and diseases, there is a need for monitoring and curing of patients with low cost for various health issues. Due to life threatening conditions, loss-free and timely sending of data is an essential factor for healthcare WBAN. Health data needs to transmit through reliable connection and with minimum delay, but designing a reliable, and congestion and delay free transport protocol is a challenging area in Wireless Body Area Networks (WBANs). Generally, transport layer is responsible for congestion control and reliable packet delivery. Congestion is a critical issue in the healthcare system. It not only increases loss and delay ratio but also raise a number of retransmissions and packet drop rates, which hampers Quality of Service (QoS). Thus, to meet the QoS requirements of healthcare WBANs, a reliable and fair transport protocol is mandatory. This motivates us to design a new protocol, which provides loss, delay and congestion free transmission of heterogeneous data. In this paper, we present a Dynamic priority based Quality of Service management protocol which not only controls the congestion in the network but also provides a reliable transmission with timely delivery of the packet.

***Keywords**: congestion, delay, dynamic priority, packet drop rate, quality of service, reliability, wireless body area network*

1. Introduction

With the help of Wireless Body Area Network, monitoring of the patient is done remotely anytime from anywhere. WBAN consists of tiny and intelligent sensors. The main job of WBAN is to continuously monitor health data and send it to healthcare server. Dynamic and heterogeneous nature of healthcare WBAN makes Quality of Service (QoS) provisioning very inspiring and essential research facet [1-2]. In healthcare WBAN system reliable data transmission [10-11] with low delay [15] is very important, to improve the quality of life and to reduce treatment cost. The main motive of the proposed protocol is to offer reliable transmission of heterogeneous packets [3-6] within the time bound. It is also deals with duplicate packets along with performing congestion control by adjusting data sending rate dynamically. In addition to the challenges for reliable data transmission, there exist additional challenges due to the unique requirements of the healthcare system such as bounded delay and delay variation [16-17], fair resource allocation, drop rate [7-10] and retransmission control [12-14]. As argued earlier, the traditional transport protocols cannot be directly implemented for WBAN, and the already existing protocols are not considered all issues up to mark. Hence, it motivates us to develop a new transport protocol for WBAN, which ensures QoS requirements of healthcare systems.

This paper has been organized as follows: Section I signifies a brief introduction to the subject matter. Section II illustrates the related work. Section III provides proposed protocol with new techniques. Section IV provides the experimental results. Section V scrutinizes the conclusion part.

2. Related Work

Rate-Controlled Reliable Transport (RCRT) [7] is a new transport protocol for wireless sensor networks. RCRT consists of four major components including: congestion detection, rate adaptation, rate allocation and end-to-end retransmission. RCRT uses the length of retransmission list as the congestion indicator. When there are too many packets in retransmission list, it means that the congestion density is high. In this case, the RCRT tries to adapt the transmission rate of each sensor node, using an AIMD rate control mechanism. RCRT implements a NACK-based end-to-end loss recovery scheme. The sink detects packet losses and repairs them by requesting end-to-end retransmissions from source nodes.

The Congestion Detection and Avoidance (CODA) protocol [8] uses both a hop-by-hop and an end-to-end congestion control scheme. It avoids the congestion by simply dropping packets at the node preceding the congestion area and employing additive increase and multiplicative decrease (AIMD) scheme to control data rate. It uses a static threshold value for detecting the onset of congestion even though it is normally difficult to determine a suitable threshold value that works in dynamic channel environments. In CODA, nodes use a broadcast message to inform their neighboring nodes the onset of congestion though this message is not guaranteed to reach the sources. The CODA only partially minimizes the effects of congestion, and as a result retransmissions and consumption of resources still occur.

Wang et al. [9] proposed "A Prioritization Based Congestion Control Protocol for Healthcare Monitoring Application in Wireless Sensor Networks" for sensing and categorizing physiological signals into different classes. In this architecture, the node with high priority and low congestion get more network bandwidth than the others. When the central node detects any abnormal changes in any signal, it assigns high priority to the correspondence node and sends an alert message to the medical server. It considers a fixed service time for all and adjusts the sending rate according to node's priority without considering the conditions of the system.

The Event to sink reliable transport (ESRT) [10] protocol is transport protocol that achieves reliable event detection with minimum energy consumption. The features provided by ESRT are self-configuration, energy awareness, congestion control, collective identification, biased implementation. In self-configuration, ESRT adjusts the reporting rate according to required condition. If the reliability is higher than required then, sink reports to the sensor to reduce reporting rate which leads to energy awareness. ESRT uses the congestion control mechanism that conserves energy of nodes and simultaneously maintains desirable reliability. The collective identification in ESRT means sink refers only collective information provided by a number of nodes. In the biased implementation, ESRT runs on the sink which is high powered compare to the sensor node. ESRT algorithm runs in different reliability and congestion condition they are as NCHR (No Congestion High Reliability), NCLR (No Congestion Low Reliability), CHR (Congestion High Reliability), CLR (Congestion Low Reliability), and OOR (Optimal Operating Region). In NCHR, sink decreases frequency to achieve required reliability. In NCLR, sink increases frequency rate of sensor nodes. In CHR, sink decreases frequency aggressively which leads to NCHR condition and then it performs action in NCHR to achieve required reliability. In CLR, sink decreases frequency exponentially. In OOR, frequency remains unchanged. Applications of ESRT are bounded to Signal Estimation and Signal Tracking Event Detection only.

In Real Time Reliable Transport Protocol (RT2) [11], Wireless Sensor Actor Network (WSAN) consists of sensors and actors. Actors are resource rich and having better processing capabilities than sensors. RT2 protocol achieves congestion control and also transport event reliably. RT2 protocol mainly works in 2 stages, Sensor-Actor communication, and Actor-Actor communication. Here the sensor nodes sense the information about the environment and deliver this information to the actor nodes in Sensor-Actor communication. RT2 has different reliability and congestion conditions in the Sensor-Actor communication like ERNCC (Early Reliability No Congestion Condition), ERCC (Early Reliability Condition Condition), LRNCC (Low Reliability No Congestion Condition), LRCC (Low Reliability Congestion Condition), Adequate Reliability and No Congestion Condition. In ERNCC, actor node decreases the reporting rate of sensor nodes to conserve unnecessary wastage of energy of the sensor nodes and to maintain reliability. In ERCC, actor node decreases reporting rate of sensors more aggressively to avoid congestion as soon as possible. In LRNCC, actor nodes increase the reporting rate of sensors by using multiplicative strategy to achieve required reliability. In LRCC, actor node decreases reporting rate of sensors until required reliability is achieved. In Adequate Reliability and No Congestion Condition, reporting rate of sensors remains as it is. In the Actor-Actor Communication actor nodes communicate with other actors in the network to take a decision and send this decision to the sink node which acts as the base station.

The Learning Automata-Based Congestion Avoidance Scheme for Healthcare Wireless Sensor Networks (LACAS) [12], tries to make packet arrival rate and packet service rate equal, by avoiding queuing at the nodes for a longer period. Although it focuses to choose better data rates, but these rates are defined randomly and will not change during simulation, hence resulting inefficient channel utilization. Here the source nodes are not provided with feedback by the intermediate nodes to slow down their rates, this leads to increase the drop rate. It does not consider different types of vital signal and treats all nodes as same.

The Optimized Congestion Management Protocol for Healthcare Wireless Sensor Networks (OCMP) [13], designed with serviced prioritization policy. It employs a single physical queue which is divided into several virtual queues and assigns dynamic weights to each child node. If any child node's queue is likely to be full, then it can use the free space of other child node's queue. It minimizes the packet loss rate for high priority traffic classes, reduces starvation for low priority traffic, provides fair scheduling by applying weighted fair scheduling. It does not provide up to mark performance in unusual critical situations and does not focus on heterogeneous traffic flow handling.

3. Proposed Protocol

The proposed DWBAN [18] architecture is consists of three units: i) Wireless Body Area Network Unit (WBANU), ii) Controller Unit (CU), and iii) Medical Server Unit (MSU). The WBANU consists of sensor nodes that sense vital signals and sends them to CU. The CU aggregates and classifies the packets accordingly in packet handling unit using Dynamic Priority based Packet Handling protocol (DPPH). The other job of CU is to improve QoS parameters i.e. Reliability, Delay, and Congestion in its QoS management unit. The MSU receives packets from CU and diagnosed the health condition accordingly.

QoS management Unit:

The purpose of QoS management unit in dynamic and heterogeneous WBAN applications like healthcare system is to achieve high reliability and to reduce delay. The main motive of the proposed QoS management protocol is to reduce loss and drop, avoid unnecessary retransmission, minimize duplicate transmission and make an effort to reduce transmission delay and its variance. To improve the QoS parameters in WBAN, heterogeneous packet delivery, packet loss, packet transmission delay and congestion degree are considered as the key parameters.

3.1. Reliability Unit

Reliability is measured using the packet loss rate (PLR). The main reason for packet loss or drop includes congestion, bad channel conditions, and link breakage. The main functionality of reliability unit is given below:

a. Flow Control: The proposed protocol follows a quick start based flow control policy. According to the sensor priority, the data sending rate is calculated and notifies to the controller unit at the time of three-way handshake connection establishment phase. The initial *Data Sending Rate (DSR)* is calculated exponentially by considering the sensor node priority as given in equation (1) in the beginning phase. The subsequent data sending rate of a sensor node is increased or decreased only by a fractional amount as given in equation (2), depending on the degree of congestion (CD), hence called Fractional Increased and Fractional Decreased (FIFD).

$$DSR_{Sn} = 2^{(N-n)} \quad (1)$$

$$DSR_{Sni} = \begin{cases} DSR_{Sni-}^{1+2floor(k*(N-n))}, \text{ if } CD <= Th_{min} \\ Ceil\ (DSR_{Sn}^{i-1} / 2^{k*n}), \text{ if } Th_{min} < CD < Th_{max} \\ 1, \text{ if } CD >= Th_{max} \end{cases} \quad (2)$$

where N=total number of sensors, n= priority of the sensor.

b. Loss Minimization: Minimization of loss is an essential factor for reduction of unnecessary retransmission and wastage of limited resources.
    1. Loss detection: The gap in sequence number of packets is the indicator of packet loss. In the first two steps of the three-way handshake, both source and sink exchange the initial sequence number along with the maximum sequence number.
    2. Loss ratio calculation: Proposed protocol calculates the packet loss ratio. The packet loss ratio is denoted as the ratio of the total number of lost packets in an interval with respect to the total number of packets transmitted in that interval.
    3. Loss Notification: Unlike ACK/NACK, a novel Duplicate Selective Negative Acknowledgment (DSNACK) based loss notification policy is introduced here. It provides the benefits of multiple loss notifications with the help of two consecutive SNACKs. Here the SNACK packet is having three index fields in its header i.e. Current Sequence Index (CSI), Previous Sequence Index (PSI) and Successive Sequence Index (SSI). The CSI provides the sequence number of very first lost packets in the current SNACK packet, PSI provides the sequence number of very first lost packets in the previous SNACK packet and the SSI provides the total number of consecutive lost packets in current SNACK packet. The CSI of previous SNACK and PSI of current SNACK are used to detect loss of SNACK packets. The SNACK based protocol consumes fewer network resources.
    4. Loss Recovery: The sensor node discovers packet loss after analyzing the DSNACK packet or expiration of the timer. Recovery of loss packets are done in the proposed selective and topical Fast retransmit and Fast recovery phase. The sensor nodes are allowed to select the number of packets it needs to retransmit and are permitted to retransmit latest loss packets according to their dynamic priority. By doing so, it reduces the number of unnecessary retransmission.

c. Reordering: Unlike single ended priority queue, here Double Ended Priority Queue(DEPQ) is used. It is made up of min_max heap-based data structure. A min-max heap is a complete binary tree containing alternating min and max levels to store packets according to their priority and sequence number. This heap structure implicitly reorders the packets.

d. Mitigation of Duplicate transmission: Here the packet sequence number will help to identify and remove duplicate packets. In the proposed protocol, controller unit (CU) maintains a packet loss table (i.e. cuckoo hash table) to store lost packet's sequence numbers. When sink received an out of order packet, it stores the lost packet's sequence number in this loss table. When CU receive a retransmitted lost packet, it first searches the entry of the incoming packet's sequence number in its loss table, if the sequence number is found, then it marks it as a duplicate packet and drops it, otherwise, it inserts this packet into the appropriate queue. Here Cuckoo hashing determines two places for the loss packet by applying hash functions on some header fields of incoming packet P_i^{Sn}. The working principle of this technique given below.
    i. Calculate two places $h1(P_i^{Sn})$ and $h2(P_i^{Sn})$.
    ii. Then it checks both these places for their vacancy.
    iii. If both or any one place is empty, then it inserts packet's sequence number to the empty place.
    Else if neither of these places is empty, then
        a. it selects one of the candidate places,
        b. kicks out the existing sequence number (i.e. re-inserts the victim sequence number to its own alternate place or follows step (ii) until it gets its right place)
        c. insert PiSn's sequence number into this empty place.

Unlike traditional and existing reliability protocols the proposed protocol transmit selected and latest loss packets, which reduce retransmission rate, increase packet delivery ratio, and improves bandwidth utilization. Duplicate packet detection and rejection utilizes buffer or memory in a more efficient way.

3.2. Delay Unit

In a healthcare system, health data recognized after lapsed time, does not reflect the actual condition of the patient and may cause serious problems. Hence, one of the most important metrics for QoS is delay in these kinds of time critical applications. The delay and delay variance is highly dependent on the communication link, topology, and resource used in the network. In this unit, the total transmission delay is evaluated by summing all types of delay, i.e., Total Elapsed Time (T_{EL}), Total delay variance (T_{VA}), and Total Loss Time (T_{LE}). The Following equations are used to calculate total transmission delay from source to destination.

a. Total Elapse time (TEL): It calculates the total elapsed time due to the late arrival of packet Pi at the receiver end.

$$T_{EL}^{Pi} = \begin{cases} T_{AD}^{Pi} - T_{ED}^{Pi}, & \text{if } T_{AD}^{Pi} > T_{ED}^{Pi} \\ 0, & \text{if } T_{AD}^{Pi} <= T_{ED}^{Pi} \end{cases} \quad (3)$$

$$T_{EL}^{Sn} = \sum_{li=1} T_{EL}^{Pi} \quad (4)$$

$$T_{EL}^{Total} = \sum_{n=1}^{N} T_{EL}^{Sn} \quad (5)$$

where TELSn denotes the total elapsed time for a particular sensor node or for one link, TELTotal denotes the total elapsed time for all sensors or all links, TAD denotes the time when packet Pi actually arrived at the receiver, TED denotes the expected deliver time of a packet Pi of sensor node Sn at the receiver end and calculated as:

$$T_{ED}^{Pi} = T_{ED}^{Pi-1} + TTG^{Sn} \quad (6)$$

where TTG^{Sn} denotes the transmission time gap between two consecutive packets.

b. Delay variation time (TVA): Variation in the elapsed time (i.e difference between two consecutive elapsed time).

$$T_{VA}^{Pi} = T_{EL}^{Pi} - T_{EL}^{Pi-1} \quad (7)$$

$$T_{VA}^{Sn} = \sum_{li=1} T_{VA}^{Pi} \quad (8)$$

$$T_{VA}^{Total} = \sum_{n=1}^{N} T_{VA}^{Sn} \quad (9)$$

where T_{VA}^{Sn} denotes the total delay variance time for a particular link, T_{VA}^{Total} denotes the total delay variance time for all links.

c. Loss error time (T_{LE}): Time interval between last delivered in-order packet ($T^{AD}{}_{Pb}$) and current delivered out-of-order packet (i.e. T_{AD}^{Pi}).

$$T_{LE}^{Pi} = T_{AD}^{Pi} - T_{AD}^{Pb} \quad (10)$$

$$T_{LE}^{Sn} = \sum_{li=1} T_{LE}^{Pi} \quad (11)$$

$$T_{LE}^{Total} = \sum_{n=1}^{N} T_{LE}^{Sn} \quad (12)$$

where T_{LE}^{Sn} denotes the total loss error time for a particular link, T_{LE}^{Total} denotes the total loss error time for all links.

d. Total transmission Delay (T_D): Total time is taken for delivery of all generated packets in a given time interval.

$$T_D = (\alpha * T_{EL}) + (\beta * T_{VA}) + (\gamma * T_{LE}) \quad (13)$$

where α, β, γ are small coefficient values with constraints

$0 < \gamma <= \beta <= \alpha < 1$.

This calculated transmission delay value lowers the end-to-end delay tremendously, as it used to calculate the congestion level of the network for next time interval.

3.3. Congestion Unit

Congestion means over-crowding, occurs mainly in many-to-one point topology, burst data rates, and low resources. In the healthcare system, having critical data for transmission, it is essential to avoid congestion as much as possible. Congestion occurs when offered load exceeds available capacity or the link bandwidth is reduced due to fading channels. Network congestion causes channel quality to degrade and loss rates rise. It leads to packets drops at the buffers, increased delays, wasted energy, and requires retransmissions.

So in the proposed protocol, a dynamic congestion mitigation protocol is designed. The proposed congestion unit is consisting of three sub-units: i) Congestion Detection and Notification (CDN) unit, ii) Congestion Control (CC) unit, iii) Congestion Avoidance (CA) unit. In the Congestion detection unit, it calculates congestion degree (CD) from various parameters like packet loss ratio, packet drop rate, transmission delay and current queue length. In the congestion notification unit, it activates CN bit in the header field of the control packet and notifies to all source. The congestion control unit employs the dynamic rate adjustment policy, which computes a new sending rate that is a reflection of the current sending rate and the dynamic priority of the sensor node as already mentioned in equation (2). The congestion avoidance unit is having an Active Queue Managements (AQM) policy. It finds and drops selected amount of low priority packets from the low priority queue.

a. Calculate Queue length (QL)

```
If (QL <=THmin), then
            Set Drop rate =0
   else if (THmin < QL  < THmax), then
            Drop rate: Ceil (QL/ i)
   else if (QL >= 1), then
            Drop rate: Ceil (QL/ 2*i)
```

The congestion unit tries to minimize the packet drop rate and resource consumption, and reduced retransmission rate.

4. Experimental Result

The protocol used for QoS management is the extension of our proposed DPPH protocol so termed as Modified DPPH (MDPPH) protocol. The experimental results show the comparison of MDPPH with DPPH and OCMP (i.e. existing protocol). The performance these protocols are implemented using the NS-2.35 simulator and graphs are generated using Matlab. It considers Packet Delivery Ratio (PDR), End-to-End Delay, and Throughput as the key performance metrics.

4.1 Packet Delivery Ratio (PDR)

The PDR is defined as the total number of packets delivered to the CU. The graph in Figure 1 shows that the dynamic flow control along with data sending rate in MDPPH increases the number of packets delivery ratio.

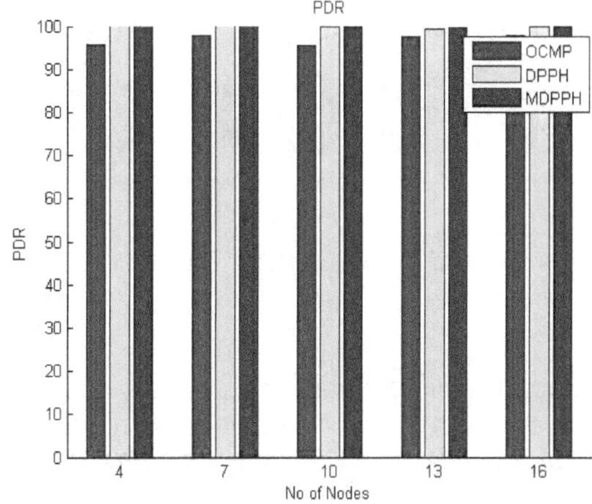

Figure 1. The impact of OCMP, DPPH, MDPPH on PDR.

4.2. Delay

It is defined as the total time required for transmitting a packet from source to destination. The Figure 3 shows that MDPPH protocol tries to minimize delay with an increase in a number of nodes.

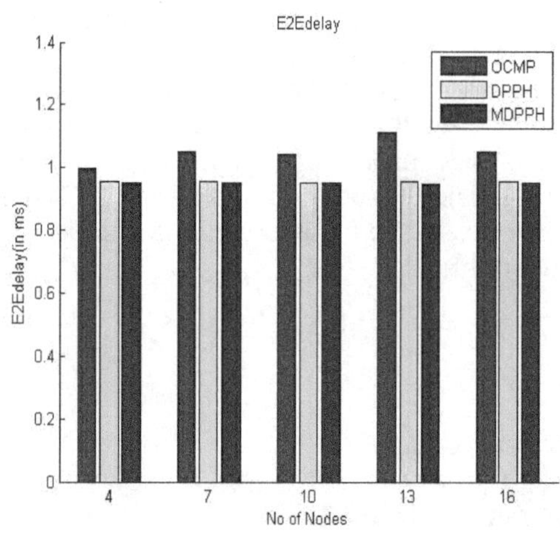

Figure 2. The impact of OCMP, DPPH, MDPPH on E2E Delay.

4.3. Throughput

It indicates the data rate. It denotes the speed of the received data in bits per seconds or data packets per second. The Figure 4 shows that the throughput for proposed MDPPH protocol is raised due to proposed loss recovery and congestion mitigation methods.

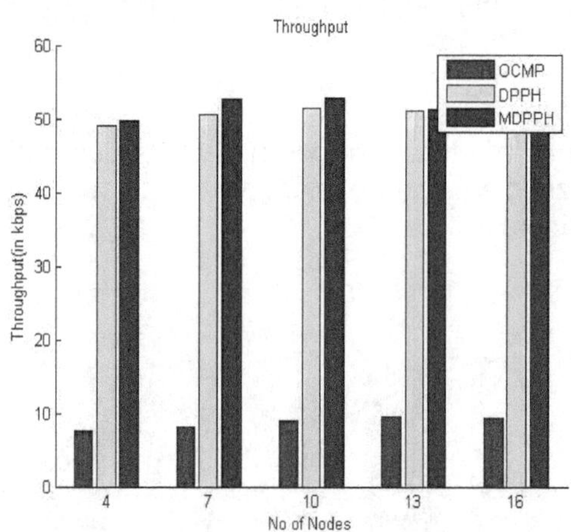

Figure 3. The impact of OCMP, DPPH, MDPPH on Throughput.

5. Conclusion

Congestion may occur due to limited resources in WBANs, which further hampers QoS parameters i.e. increases loss and delay and wastes energy. In this paper, we proposed a QoS provisioning transport protocol for WBANs. The proposed protocol enables QoS guarantees with end-to-end reliable data transmission within the critical time interval. Its main purposes are flow and congestion control, loss recovery, duplicate mitigation, loss, drop, retransmission and delay minimization. The proposed protocol uses the SNACK policy to decrease the number of ACK and NACK packet. It provides a dynamic rate adjustment method. It also overcomes the problem of critical time delivery of heterogeneous data flow. The experimental results validate the performance of the proposed protocol with respect to packet delivery ratio, delay, and throughput.

References

[1] G. Zhou et al. *BodyQoS: Adaptive and Radio-Agnostic QoS for Body Sensor Networks.* In Proceedings of IEEE INFOCOM. 2008.

[2] Madhumita Kathuria, Sapna Gambhir. *Quality of Service Provisioning Transport Layer Protocol for WBAN System.* International Conference on Optimization, Reliability and Information Technology (ICROIT, IEEE). 2014: 222-228.

[3] Madhumita Kathuria, Sapna Gambhir. *Leveraging Machine Learning for Optimize Predictive Classification and Scheduling E-Health Traffic.* IEEE International Conference on Recent Advances and Innovations in Engineering. 2014: 1-7.

[4] Venki Balasubramanian, Andrew Stranieri. *Performance Evaluation of the Dependable Properties of a Body Area Wireless Sensor Network.* International Conference on Optimization. Reliability and Information Technology, IEEE. 2014: 229-234.

[5] Madhumita Kathuria, Sapna Gambhir. Genetic Binary Decision Tree based Packet Handling Schema for WBAN System. *Recent Advances in Engineering and Computational Sciences (RAECS, IEEE).* 2014: 1-6.

[6] Venki Balasubramanian. *Critical Time Parameters for Evaluation of Body Area Wireless Sensor Networks in a Healthcare Monitoring Application.* IEEE Ninth International Conference on Intelligent Sensors, Networks and Information Processin. 2014; 1-7.

[7] Jeongyeup Paek, Ramesh Govindan. RCRT: Rate-Controlled Reliable Transport Protocol for Wireless Sensor Networks. *ACM Transactions on Sensor Networks.* 2010; 7(3): 1-43.

[8] C. Y. Wan, S. Eisenman, A. Campbell. *CODA: Congestion Detection and Avoidance in Sensor Networks.* in Proc ACM SenSys'03. 2003: 266–279.

[9] C. Wang et al. *Priority-based Congestion Control in Wireless Sensor Networks.* Proceedings of the IEEE International Conference on Sensor Networks, Ubiquitous, and Trustworthy Computing (SUTC'06). 2006.

[10] Y. Sankarasubramaniam, O. B. Akan, I. F. Akyildiz. ESRT: Event-To-Sink Reliable Transport in Wireless Sensor Networks. *IEEE/ACM Transactions on Networking.* 2005; 13(05): 1003-1016.

[11] Vehbi Cagri Gungor, Özgür B. Akan, Ian F. Akyildiz. A Real-Time and Reliable Transport (RT2) Protocol for Wireless Sensor and Actor Networks. *IEEE/ACM Transactions On Networking.* 2008; 16(2); 359.

[12] S. Misra et al. LACAS: Learning Automata-Based Congestion Avoidance Scheme for Healthcare Wireless Sensor Networks. *IEEE Journal on Selected Areas in Communications.* 2009: 466-479.

[13] Abbas Ali Rezaee et al. Optimized Congestion Management Protocol for Healthcare Wireless Sensor Networks. *Wireless Pers Commun.* 2014: 11-34.

[14] Sapna Gambhir, Vrisha` Tickoo, Madhumita Kathuria. *Priority Based Congestion Control in WBAN.* Eighth International Conference on Contemporary Computing (DBLP). 2015; 428-433.

[15] N. Javaid, M. Yaqoob, M. Y. Khan, M. A. Khan, A. Javaid, Z. A. Khan. Analyzing Delay in Wireless Multi-hop Heterogeneous Body Area Networks. *Research Journal of Applied Sciences, Engineering and Technology.* 2014; 7(1): 123-136.

[16] J. C. Bolot. *End-to-end Packet Delay and Loss Behavior in The Internet.* ACM SIGCOMM Conference on Communications Architectures, Protocols and Applications. 2005: 289-298.

14

Link Adaptation for Microwave Link using both MATLAB and Path-Loss Tool

Jide Julius Popoola*[1], Damian E. Okhueleigbe[1], Isiaka A. Alimi[1,2]
[1]Department of Electrical and Electronics Engineering, Federal University of Technology, Akure, Nigeria
[2]Instituto de Telecomunicações, DETI, Universidade de Aveiro, Aveiro, Portugal
e-mail: jidejulius2001@gmail.com*; damianeee2008@yahoo.com; compeasywalus2@yahoo.com

Abstract

The inherent multipath transmission on wireless channels usually leads to signal fading which eventually degrades the system performance. In mitigating this problem, link adaptation has been identified as a promising scheme that helps in maximizing the system spectral efficiency (SE) in dispersive wireless channels. In this paper, link adaptation based on adaptive modulation and coding was used to study the performance of M-ary quadrature amplitude modulation radio system subjected to multipath fading. MATLAB® scripts and Simulink model were developed to compare the effect of wireless channel on different constellation sizes. Also, transmission link on Federal University of Technology Akure campus' path terrain was designed with the aid of path-loss® tool software in order to further analyze the effect of using different modulation formats on the system performance. The results show that, employment of link adaptation scheme offers better performance regarding the system availability and SE.

Keywords: *wireless channels, multipath transmission, link adaptation, spectral efficiency, constellation sizes*

1. Introduction

In recent years, radio or wireless communication system has replaced almost all wired communication systems in telecommunications. This is as a result of tremendous advantages that wireless communication systems have over the fixed-wired communication systems. Two of these major advantages are its ability to be rolled out rapidly and its mobility capability. With this shift from wired connection to wireless connection, it is obvious that an acceptable quality of service (QoS) on wireless channels can only be achieved through higher information transfer rates than the currently available rates in today's wireless system. In order to increase the current information or data transfer rates to the next level, a simple approach is to increase the allocated bandwidth. However, this approach is practically infeasible due to the fact that radio spectrum is extremely scarce due to proliferation of wireless services, devices and application. In addition, the growth in information transmission via wireless technologies has led to increase in both the demand for the radio spectrum and consumption of more bandwidth for services such as video-on-demand, high-speed Internet access, video conferences and applications with multimedia contents.

Furthermore, the fact that wireless channels is currently the most commonly used channel for signal transmission, observations show that signals transmitted over wireless channels are often being impaired due to undesirable effects on wireless channels [1]. This makes wireless channels to be characterized by multipath transmission of signal. Thus, results in the variation of the signal strength which leads to signal fading when replicas are combined at the receiver (Rx). With signal fading, there is high degradation in the link carrier-to-noise ratio (CNR) as well as high bit error rate (BER) in the radio link [2].

These detrimental effects of signal fading call for an appropriate measure that can mitigate it. Different measures such as space diversity [3-5], automatic transmit power control (ATPC) [6], adaptive equalization, multilevel coded modulation (MLCM) [7,8], forward error correction (FEC), and cross-polarization interference cancellers (XPICs) [9] have been proposed in the literature to address the challenges of signal fading. However, aiming at the inherent capacity of wireless channel, it is obvious that technology which adapt and adjust transmission parameters in real-time based on the link quality will be the appropriate solution to both the problems of channel impairments and bandwidth in wireless communication. Thus, the primary objective of the study presented in this paper is to investigate the effectiveness of adaptive modulation scheme in enhancing both the transmission rates and SE in wireless communication.

For sequential and logical presentation of the study presented in this paper, the remaining parts of this paper are organized as follows. Section 2 presents a brief overview on link adaptation scheme and SE. Section 3 focuses on methodology involved in carrying out the study with emphasis on different simulation and link design parameters using FUTA as case study. In Section 4, the results of the study are presented and discussed while the conclusion is presented in Section 5.

2. Link Adaptation

Link adaptation, also known as adaptive modulation (AM) or adaptive modulation and coding (AMC), is a technique employed in wireless communication system to denote the matching of the modulation coding as well as the signal and protocol parameters to the conditions on the radio link. Thus, the process of link adaptation is a dynamic technique that brings about changes in both the signal and protocol parameters as the link conditions change. According to [10], link adaptation is a powerful technique for improving the SE in wireless transmission over fading channels. It works by exploiting the channel state information (CSI) at the transmitter. This capability enhances its performance when compared with systems that do not exploit channel knowledge at the transmitting end.

Link adaptation based on AMC as reported in [11] is one of the promising adaptive schemes to counteract fading and enhance the performance of wireless system. It is an efficient link adaptation technique in which transmission parameters such as modulation format, code-rate, and power are regulated in accordance with the CSI. The concept of the scheme is based on monitoring the channel variations in order to determine the error rate at the receiver (Rx). When the error rate increases, the Rx sends feedback information about the nature of the CSI to the transmitter (Tx). Consequently, the Tx dynamically changes the modulation and coding formats in order to achieve better throughput by transmitting higher information rates in good CSI conditions. However, in poor CSI conditions, the Tx automatically shifts its transmission to a more robust but less efficient modulation and coding formats in response to the channel degradation [12]. In essence, in a good CSI condition, highest sustainable data rate is transmitted; whereas, when the CSI is poor, lower data rate is transmitted [13]. The activities involve in both the Tx and Rx in an AMC scheme using an additive white Gaussian noise (AWGN) channel is shown in Fig. 1. Fig. 1 shows the interaction involves in both the Tx and Rx in order to achieve the desired SE and high transmission that AMC scheme offers in wireless communication system [14].

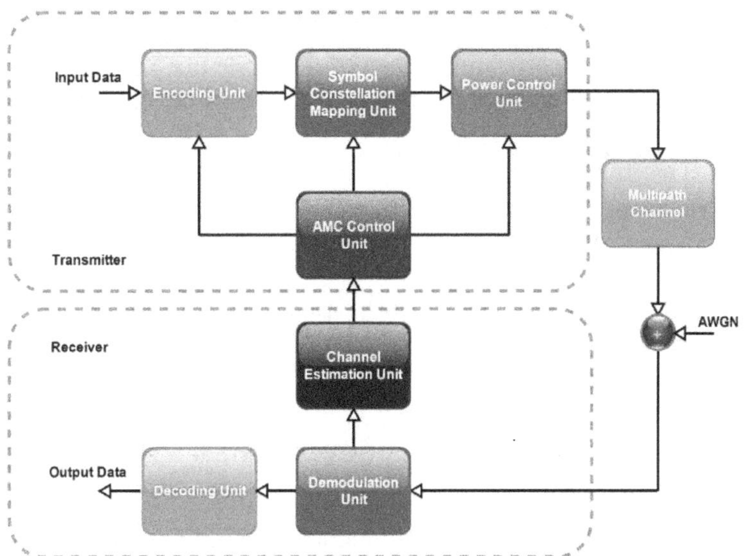

Figure 1. Block Diagram of Adaptive Modulation and Coding Scheme (Adapted from [13])

Spectral efficiency, SE, by simple definition is the amount of information or data that can be transmitted over a given bandwidth in a specific communication system. It is also defined as the ratio of the data rate to the bandwidth of the modulation signal. Thus, SE means getting more bits per hertz and being expressed in bps/Hz. It is the optimized use of the spectrum such that the maximum amount of data can be transmitted with minima transmission errors. It is one of the major factors in the design of wireless communication system [15]. As reported in [15], frequency re-use is one of the ways of achieving SE. However, according to [16-18], application of frequency re-use usually introduces unavoidable co-channel interference in wireless communication system. However, as reported by [15], another technique of increasing spectral efficiency in wireless communication systems is the application of multilevel or M-ary quadrature amplitude modulation (M-QAM) schemes. This is because M-QAM increases the link SE by sending multiple bits per symbol as reported in [15]. QAM is able to achieve this because the two carrier waves in it, $(\cos 2\pi f_c t)$ and $(\sin 2\pi f_c t)$, which are out of phase with each other by 90^0 called quadrature carriers or quadrature components are algebraically summed [19]. The algebraic sum of these modulated waves gives a single signal to be transmitted, containing the in-phase (I) and quadrature (Q) information [20]. Thus, the resulting M-QAM signal is defined mathematically in [19] as;

$$\begin{aligned} s(t) &= I(t).\cos(2\pi f_c t) + Q(t).\sin(2\pi f_c t) \\ &= A_m^I.g(t).\cos(2\pi f_c t) + A_m^Q.g(t).\sin(2\pi f_c t) \qquad m = 1,2,3,\cdots,M \end{aligned} \tag{1}$$

where $I(t)$ and $Q(t)$ are the modulating signals, A_m^I and A_m^Q are the sets of the amplitude levels for the in-phase and quadrature phase respectively, and $g(t)$ is the real valued signal pulse, whose shape influences the spectrum of the transmitted signal .

In digital formats of M-QAM, two or more bits are usually grouped together to form symbols and one of M possible signals is transmitted during each symbol period. Normally, the number of possible signals is expressed mathematically in [19] as;

$$M = 2^n \tag{2}$$

where n is an integer. Therefore, possible M-QAMs are: 4-QAM, 8-QAM, 16-QAM, 32-QAM, 64-QAM, and so on. The number of 4, 8, 16, 32 and 64 is corresponding to 2^2, 2^3, 2^4, 2^5 and 2^6 in which the superscript number 2, 3, 4, 5 or 6 is the bits per symbol respectively. Similarly, survey literature reviews that SE achievable is considerably high when a low order constellation such as quaternary phase shift keying (QPSK) is employed in wireless communication system. Mathematically, QPSK signal can be represented as:

$$s(t) = A_c \cos\left(2\pi f_c t + (i-1)\pi/2\right) \quad i = 1,2,3,4 \tag{3}$$

Using trigonometric identity, $\cos(x+y) = \cos x \cos y - \sin x \sin y$, (3) can be re-written as:

$$s(t) = A_c \cos(2\pi f_c t)\cos\{(i-1)\pi/2\} - A_c \sin(2\pi f_c t)\sin\{(i-1)\pi/2\} \quad i = 1,2,3,4 \tag{4}$$

In microwave radio systems, adaptive modulation is employed for point to point digital communication in order to offer better capacity to the user. This is based on the CSI feedback that enables dynamic radio link adaption. An example of adaptive technique being implemented in the microwave radio links to combat fading is ATPC which is also known as power diversity

[6]. Conceptually, a threshold power is set to achieve a given BER under good channel conditions. The Tx power can be adjusted (increased or decreased) dynamically according to the link condition. However, the increase in power when the CSI is poor should not exceed the set threshold value so as to prevent co-channel interference (CCI) [6,22]. Moreover, an alternative approach of achieving a desired QoS is by the use of adaptive modulation scheme in which the power and modulation level can be dynamically altered in order to regulate the link ability to conform to the set transmission conditions [22]. This simulation and link design carried out in achieving the objectives of the study presented in this paper is presented in the next section.

3. Methodology

In this section, the methodology involved in carrying out this study is presented. This section presents the simulation, design, and analysis of the adaptive modulation system for the microwave transmission link. The simulation and modeling are implemented in MATLAB®/Simulink®. The Simulink for the microwave transmission link is shown in Fig. 2. Furthermore, the design of transmission link on FUTA path terrain is achieved with the aid of PATHLOSS® 4.0 simulation tool. This is done using factors and parameters such as the geographical coordinates of the terrain, path profile from Shuttle Radar Topography Mission database, the International Telecommunication Union (ITU) recognized transmit and receive frequency, a standard Microwave Networks radio model, AMT/07/16E1/14M, an antenna code model, a traffic code of 16E1-16QAM and 16E1-QPSK with horizontal polarization for worst condition in the path design modules of the program.

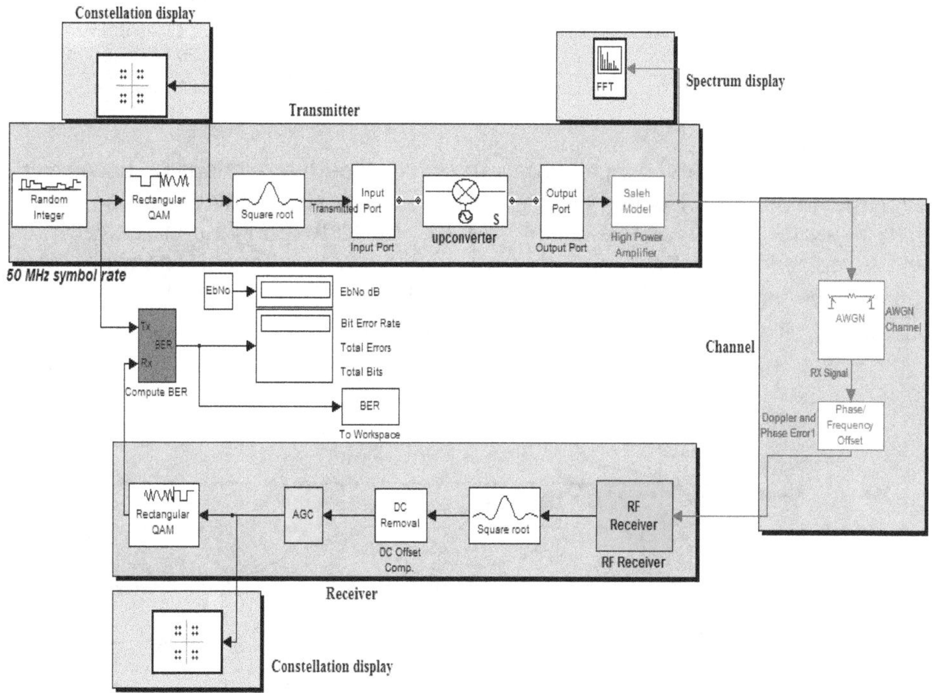

Figure 2. Simulink for Microwave Transmission Link Model

4. Results and Discussion

This section is divided into two subsections. The first subsection presents the effect of link adaptation or adaptive modulation on FUTA wireless link. This was done by presenting the results of MATLAB® scripts and Simulink model that were developed to compare the effect of wireless channel on different constellation sizes. On the other hand, the second subsection presents results of transmission link on FUTA path terrain that was designed with the aid of

Pathloss® tool application software. The results obtained and the discussions are presented as follows in the following sub-sections.

4.1. Effect of Link Adaptation on different Constellation Sizes

This subsection presents results of the link adaptation on both the data or information transmission rate and the SE. Simulink model was employed in varying the constellation sizes and SNR in order to study the effects of adaptive modulation by varying the constellation parameters on the microwave transmission system. In the analysis, the effect of additive white Gaussian noise (AWGN) is observed at a symbol rate of 50 MHz and transmits frequency of 5.29 GHz. The BER generated and number of bits received for different M-ary quadrature amplitude modulation (M-QAM) and SNR are illustrated in Table 1. The results of the study show that out of all constellation sizes considered, 16-QAM has the least BER while 256-QAM has the highest BER. Moreover, for all constellation sizes considered, the number of bits received increase with increase in SNR on the AWGN channel employed. This shows that with the adaptive modulation and coding more bits are transmitted with increase in SNR. The result of the study thus shows that with adaption of link adaptation on wireless link results in more data or information transmission and high SE.

Table 1. BER Values and Number of Bits at Different SNR and M-QAM

SNR	MODULATTION SCHEMES									
	16-QAM		32-QAM		64-QAM		128-QAM		256-QAM	
	BER	No of Bit	BER	No of Bit	BER	No of Bit	BER	No of Bit	BER	No of Bit
0	0.0208	1504	0.125	920	0.1495	1104	0.1929	840	0.2411	448
3	0.0208	5088	0.0673	1560	0.0847	1488	0.0286	840	0.1563	960
6	0.0034	29664	0.0212	4760	0.051	2256	0.0947	1288	0.0505	960
9	0.0001	883936	0.0036	28120	0.0162	6480	0.0495	3184	0.0734	1472
12	1.9E-07	10000020	0.0001	818200	0.0032	31440	0.0146	7112	0.0382	3008

Furthermore, MATLAB® scripts were used to analyse the relationship between SNR and the BER for the five M-QAM schemes considered. The result obtained, as shown in Fig. 3, shows that increase in the constellation size led to corresponding increase in the BER. Therefore, to maintain high transmission quality in a given channel condition, an effective modulation format was employed. For instance, to achieve a BER of 10^{-4}, with M=16, 32, and 128, about 13 dB, 15 dB, and 20 dB are required, respectively. This illustrates that, reducing the constellation size is a good option for reducing the effect of impairments caused by both fading and noise.

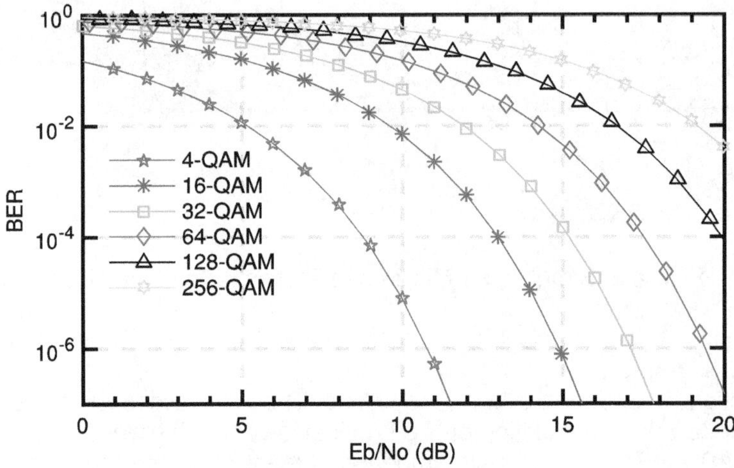

Figure 3. Plot of BER against SNR for Different Constellation Sizes

In an attempt to study the implementation of adaptive modulation in microwave radio system, threshold levels were set in the scripts to achieve a system BER that is less than 10^{-4} for each modulation formats. This allows an automatic switching between modulation formats based on the SNR. In general, for fixed radio-link, modulation formats such as 64, 128, 256, and 512-QAM are normally employed whereas low-order constellations such as 4, 16, and 32-QAM, are implemented in adaptive links [23].

Furthermore, in evaluating the performance of the adaptive modulation, the switching results for 16-QAM and 32-QAM were compared. The compared results include the scatter plots, power spectral density, and in-phase components of the eye diagrams of the analysed signal. Also, the results of the channel impairment effects on the received signal that lead to the switching for the two modulation formats were presented for comparison. The scatter plot and the eye diagrams for the transmitted and received signal for 16-QAM and 32-QAM are shown in Fig. 4 and Fig. 5 respectively for the switching. The eye diagrams confirm that there is significant difference between the received signal and the transmitted signal. The difference is due to the channel impairments on the signal which make the eye to be constrained. Also, the power spectral density of transmit and received signal for 16-QAM and 32-QAM are shown in Fig. 6 and Fig. 7, respectively. The effect of channel impairments is clearly shown on each received signal with 16-QAM having high power efficiency compared with the 32-QAM.

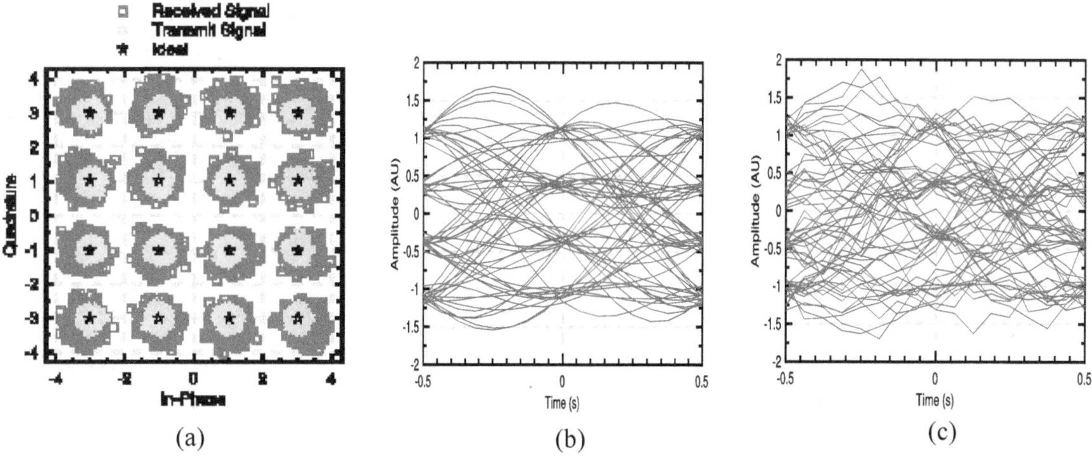

Figure 4. Plot for 16-QAM (a) scatter plot (b) eye diagram for the transmit signal (c) eye diagram for the received signal

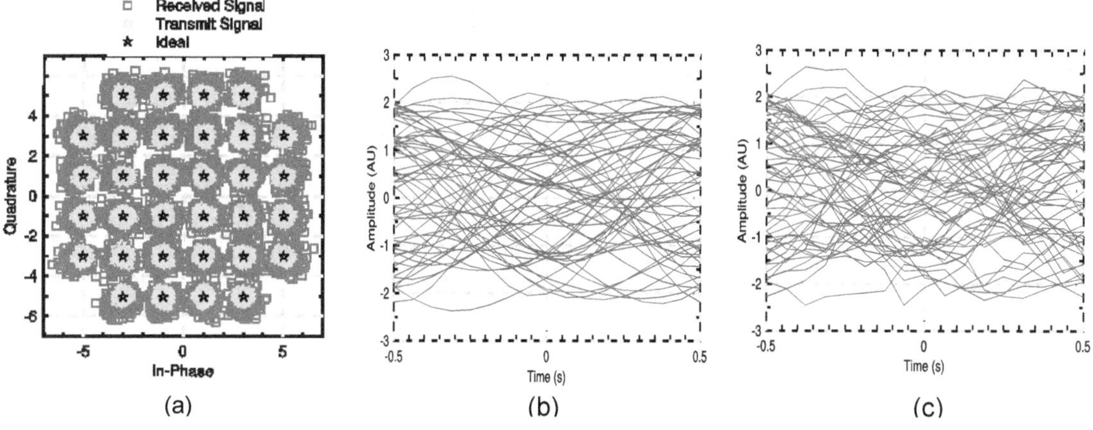

Figure 5. Plot for 32-QAM (a) scatter plot (b) eye diagram for the transmit signal (c) eye diagram for the received signal

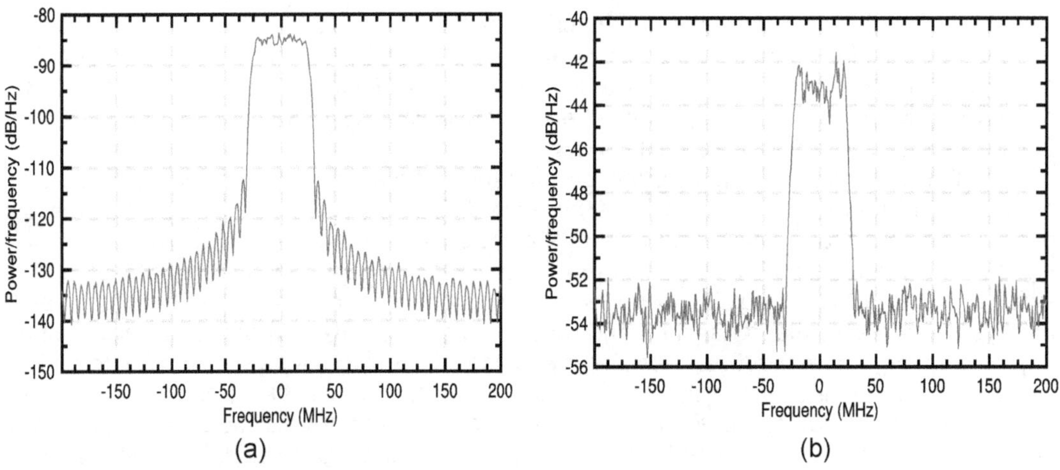

Figure 6. Power Spectral Density of (a) Transmit and (b) received signal for 16-QAM.

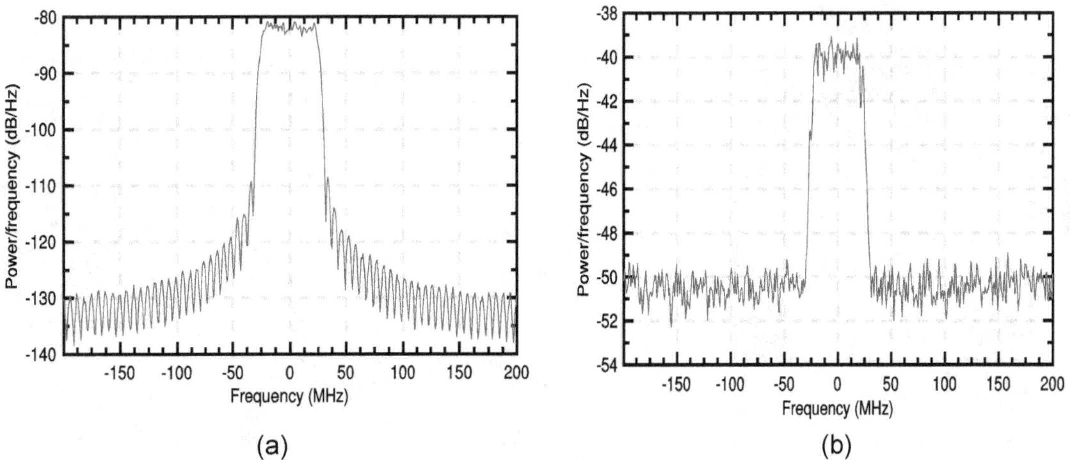

Figure 7. Power Spectral Density of (a) Transmit and (b) received signal for 32-QAM.

4.2. Performance of the Transmission Link Budget

This subsection presents results of design of link budget with respect to the microwave transmission link being designed for Site 1 and Site 2. The effects of using modulation schemes such as 16E1- QPSK and 16E1-16QAM on the link budget are analysed. Table 2 and Table 3 show the link budget designed for 16E1-16-QAM and 16E1-QPSK modulation schemes respectively. The effective isotropic radiated powers (EIRPs) of 51.27dBm and 55.27dBm are obtained for 16-QAM and QPSK modulation schemes respectively. This shows that, lesser power is transmitted with16-QAM so as to meet the transmit mask criterion. Moreover, the fade margins of 44.14dBm and 52.14dBm are achieved for 16-QAM and QPSK modulation schemes, respectively. This result indicates that, the link availability is higher when QPSK is employed. This is due to the fact that the percentage of link availability increases with the fade margin value.

Table 2. Designed Link Budget of 16E1-16QAM Modulation Scheme

Parameter	Site 1	Site 2
Elevation (m)	371.00	373.00
Latitude	07 18 00.00 N	07 18 00.00 N
Longitude	005 08 00.00 E	005 09 00.00 E
True azimuth (^{0})	63.40	243.40
Vertical angle (^{0})	-0.04	0.02
Antenna model	VHP4-71	VHP4-71
Antenna height (m)	22.32	19.21
Antenna gain (dBi)	36.40	36.40
TX line type	EWP77	EWP77
TX line length (m)	100.00	100.00
TX line unit loss (dB/100 m)	6.13	6.13
TX line loss (dB)	6.13	6.13
Frequency (MHz)	7200.00	7200.00
Polarization	Horizontal	Horizontal
Path length (km)	2.06	2.06
Free space loss (dB)	115.88	115.88
Atmospheric Absorption loss (dB)	0.02	0.02
Net path loss (dB)	55.36	55.36
Radio model	AMT/07/16E1/14M	AMT/07/16E1/14M
TX power (watts)	0.13	0.13
TX power (dBm)	21.00	21.00
EIRP (dBm)	51.27	51.27
Emission designator	14MOD7W	14MOD7W
TX Channels	4h 7310.0000H	41 7149.0000H
	41 7149.0000H	4h 7310.0000H
RX threshold criteria	BER 10-6	BER 10-6
RX threshold level (dBm)	-78.50	-78.50
RX signal (dBm)	-34.36	-34.36
Thermal fade margin (dB)	44.14	44.14
Dispersive fade margin (dB)	52.00	52.00
Dispersive fade occurrence factor	1.00	1.00
Effective fade margin (dB)	43.48	43.48
Geoclimatic factor	1.98E-04	1.98E-04
Path inclination (mr)	0.54	0.54
Fade occurrence factor (Po)	8.46E-05	8.46E-05
Average annual temperature (^{0}C)	24.00	24.00
Worst month - multipath (%)	100.00000	100.00000
(sec)	9.98E-03	9.98E-03
Annual - multipath (%)	100.00000	100.00000
(sec)	0.04	0.04
(% - sec)	100.00000 – 0.09	100.00000 – 0.09
Rain region	ITU Region F	ITU Region F
0.01% rain rate (min/hr)	28.00	28.00
Flat fade margin – rain (dB)	44.14	44.14
Rain rate (mm/hr)	8268.82	8268.82
Rain attenuation (dB)	44.14	44.14
Annual rain (%-sec)	100.00000 – 0.00	100.00000 – 0.00
Annual multipath + rain (%-sec)	100.00000 – 0.09	100.00000 – 0.09

futa sites_16QAM.p14
Reliability Method – ITU-R P.530-7/8
Rain – ITU-R P530-7

Table 3. Designed Link Budget of 16E1-QPSK Modulation Scheme

Parameter	Site 1	Site 2
Elevation (m)	371.00	373.00
Latitude	07 18 00.00 N	07 18 00.00 N
Longitude	005 08 00.00 E	005 09 00.00 E
True azimuth (^{0})	63.40	243.40
Vertical angle (^{0})	-0.04	0.02
Antenna model	VHP4-71	VHP4-71
Antenna height (m)	22.32	19.21
Antenna gain (dBi)	36.40	36.40
TX line type	EWP77	EWP77
TX line length (m)	100.00	100.00
TX line unit loss (dB/100 m)	6.13	6.13
TX line loss (dB)	6.13	6.13
Frequency (MHz)	7200.00	7200.00
Polarization	Horizontal	Horizontal
Path length (km)	2.06	2.06
Free space loss (dB)	115.88	115.88
Atmospheric Absorption loss (dB)	0.02	0.02
Net path loss (dB)	55.36	55.36
Radio model	AMT/07/16E1/28M	AMT/07/16E1/28M
TX power (watts)	0.32	0.32
TX power (dBm)	25.00	25.00
EIRP (dBm)	55.27	55.27
Emission designator	28MOG7W	28MOG7W
TX Channels	4h 7310.0000H	41 7149.0000H
	41 7149.0000H	4h 7310.0000H
RX threshold criteria	BER 10-6	BER 10-6
RX threshold level (dBm)	-82.50	-82.50
RX signal (dBm)	-30.36	-30.36
Thermal fade margin (dB)	52.14	52.14
Dispersive fade margin (dB)	49.00	49.00
Dispersive fade occurrence factor	1.00	1.00
Effective fade margin (dB)	47.28	47.28
Geoclimatic factor	1.98E-04	1.98E-04
Path inclination (mr)	0.54	0.54
Fade occurrence factor (Po)	8.46E-05	8.46E-05
Average annual temperature (^{0}C)	24.00	24.00
Worst month - multipath (%)	100.00000	100.00000
(sec)	4.16E-03	4.16E-03
Annual - multipath (%)	100.00000	100.00000
(sec)	0.02	0.02
(% - sec)	100.00000 – 0.04	100.00000 – 0.04
Rain region	ITU Region F	ITU Region F
0.01% rain rate (min/hr)	28.00	28.00
Flat fade margin – rain (dB)	52.14	52.14
Rain rate (mm/hr)	9844.15	9844.15
Rain attenuation (dB)	52.14	52.14
Annual rain (%-sec)	100.00000 – 0.00	100.00000 – 0.00
Annual multipath + rain (%-sec)	100.00000 – 0.04	100.00000 – 0.04

futa sites_QPSK.p14
Reliability Method – ITU-R P.530-7/8
Rain – ITU-R P530-7

5. Conclusion

This paper presents microwave transmission model employed in simulating the effects of different constellation sizes and the SNR on the implementation of adaptive modulation scheme. The symbol rate of 50 MHz and transmit frequency of 5.29 GHz are employed in the analysis. In addition, the paper presents designs for transmission link on FUTA path terrain in which the effect of using different modulation schemes are analyzed. The results obtained show that implementation of adaptive modulation scheme offers better performance with regard to system availability as well as spectral efficiency. Furthermore, the overall result of this study has shown clearly that adaptive modulation scheme as a technology that adapts and adjusts transmission parameters in real-time based on the link quality is the appropriate solution to both the problems of channel impairments and bandwidth in wireless communication.

References

[1] Popoola, J.J. Computer Simulation of Hata's Equation for Signal Fading Mitigation. *Pacific Journal of Science and Technology*. 2009; 10(2): 462-470.

[2] Hole, K.J., Holm, H., Oien, G.E. Adaptive Multidimensional Coded Modulation over Flat Fading Channels. *IEEE Journal on Selected Areas in Communications*. 2000; 18(7): 1153-1158.

[3] Dinc, E., Akan, O.B. Fading Correlation Analysis in MIMO-OFDM Troposcatter Communications: Space, Frequency, Angle and Space-Frequency Diversity. *IEEE Transactions on Communications*. 2015; 63(2): 476-486.

[4] Renzo, M.D., Haas, H. Space Shift Keying (SSK) MIMO over Correlated Rician Fading Channels: Performance Analysis and a New Method for Transmit-Diversity. *IEEE Transactions on Communications*. 2011; 59(1): 116-129.

[5] Zhang, Y.Y., Yu, H.Y., Zhang, J.K., Zhu, Y.J., Wang, J.L., Wang, T. *Full Large-Scale Diversity Space Codes for MIMO Optical Wireless Communications*. IEEE International Symposium on Information Theory (ISIT), Hong Kong, 2015; 1671-1675.

[6] Morelos-Zaragoza, R.H., Suh, K.W., Lee, J.H. *Automatic Transmit Power Control of a Digital Fixed Wireless Link with Co-Channel Interference*, 2nd International Conference on Communications and Networking. Shanghai, 2007; 1178-1184.

[7] Farhoudi, R., Rusch, L.A. Multi-Level Coded Modulation for 16-ary Constellations in Presence of Phase Noise. *Journal of Lightwave Technology*, 2014; 32(6): 1159-1167.

[8] Beygi, L., Agrell, E., Karlsson, M. On the Dimensionality of Multilevel Coded Modulation in the High SNR Regime. *IEEE Communications Letters*. 2010; 14(11): 1056-1058.

[9] Proença, H., Carvalho, *N.B. Cross-Polarization Interference Cancelation (XPIC) Performance in Presence of Non-Linear Effects*. Workshop on Integrated Nonlinear Microwave and Millimeter-Wave Circuits (INMMIC), Goteborg, 2010; 93-96

[10] Falahati, S., Svensson, A., Ekman, T., Sternad, M. Adaptive Modulation Systems For Predicted Wireless Channels, *IEEE Transactions on communication*, 2004; 52(2): 307-316.

[11] Tan, P.H., Wu, Y., Sun, S. Link adaptation based on adaptive modulation and coding for multiple-antenna OFDM. *IEEE Journal on Selected Areas in Communications*. 2008. 26(8): 1599-1606.

[12] Salih, S.H.O., Saliman, M.M.A. *Implementation of Adaptive Modulation and Coding Techniques using Matlab*. Proceedings ELMAR, Zadar, 2011; 137-139.

[13] Shami, A., Maier, M., Assi, C. Broadband Access Networks: Technologies and Deployments, Springer US, 2010, 379 pages

[14] Reddy, B.S.K., Lakshmi, B. *Adaptive Modulation and Coding for Mobile-WiMAX using SDR in GNU Radio.* International Conference on Circuits, Systems, Communication and Information Technology Applications (CSCITA), Mumbai, 2014; 173-178.

[15] Bao, L., Hansryd, J., Danielson, T., Sandin, G., Noser, U. *Field Trial On Adaptive Modulation Of Microwave Communication Link At 6.8 GHz*. 9th European Conference on Antennas and Propagation (EuCAP), Lisbon, 2015; 1-5.

[16] Chinnici, S., Decanis, C. *Channel Coding And Carrier Recovery For Adaptive Modulation Microwave Radio Links.* 5th IEEE Mediterranean Electrotechnical Conference, Valletta, 2010; 6-11.

[17] Ekpenyong, M.E., Isabona, J. *Improving Spectral Efficiency of Spread Spectrum Systems under Peak Load Network Conditions*. 6th International Conference on Systems and Networks Communications (ICSNC), Barcelonia, 2011; 1-8.

[18] Nagata, Y., Akaiwa, Y. Analysis for Spectrum Efficiency in Single Cell Trunked and Cellular Mobile Radio. *IEEE Transactions on Vehicular Technology*, 2006, 36(3): 100-113.

[19] Yao, Y., Sheikh, A. Investigations in Co-Channel Interference in Microcellular Mobile Radio Systems. *IEEE Transactions on Vehicular Technology*. 2002; 41(2): 14-123.

[20] Prasad, R. Kegel, A. Improved Assessment of Interference Limits in Cellular Radio Performance. *IEEE Transactions on Vehicular Technology*, 2002; 40(2): 412-419.

[21] Popoola, J.J. Sensing and Detection of a Primary Radio Signal in a Cognitive Radio Environment Using Modulation Identification Technique. PhD Thesis. Johannesburg, University of the Witwatersrand.
Online[Available]: http://mobile.wiredspace.wits.ac.za/handle/10539/11600. Accessed on 23 August 2016.

[22] Hannan, M.A., Islam M., Samad, S.A., Hussain A. QAM in Software Defined Radio for Vehicle Safety Application. *Australian Journal of Basic and Applied Sciences*, 2010; 4(10): 4904-4909.

[23] Barbieri, A., Fertonani, D., Colavolpe, G. Time-Frequency Packing for Linear Modulations: Spectral Efficiency and Practical Detection Schemes. *IEEE Transactions on Communications*, 2009; 57(10): 2951-2959.

15

Voltage Stabilization of a Wind Turbine with STATCOM using Intelligent Control Techniques

SA Gawish, SM Sharaf, MS El-Harony
Department of Electrical Power and Machines Engineering, Faculty of Engineering of Helwan, University of Helwan

Abstract

Application of FACTS controller called Static Synchronous Compensator STATCOM to improve the performance of power grid with Wind System is investigated. The essential feature of the STATCOM is that it has the ability to absorb or inject fastly reactive power with power grid. Therefore the voltage regulation of the power grid with STATCOM FACTS device is achieved. Moreover restoring the stability of the wind system at suddenly step up or down in wind speed is obtained with STATCOM. This paper describes a complete simulation of voltage regulation of a wind system using STATCOM. Conventional control technique as proportional plus integral controller and intelligent techniques as FLC and ANFIS are used in this work. The control technique is performed using MATLAB package software. The dynamic response of uncontrolled system is also investigated under wide range of disturbances. The voltage regulation by using STATCOM whose output is varied so as to maintain or control output voltage in the system. The dynamic response of controlled system is shown and comparison between the uncontrolled system and the controlled system is described to assure the validity of the proposed controller. Also comparison between the proposed control methods scheme is presented. To validate the powerful of the STATCOM FACTS controllers, the studied power system is simulated and subjected to different severe disturbances. The results prove the effectiveness of the proposed STATCOM controller in terms of fast damping the power system oscillations and restoring the power system stability and voltage.

***Keywords**: wind system; STATCOM; Voltage stabilization; PI controller; fuzzy logic controller, ANFIS controller*

1. Introduction

Global warming is one of the most serious environmental problems facing the world community today. It is typified by increasing the average temperature of Earth's surface and extremes of weather both hot and cold. Therefore, implementing smart and renewable energies such as wind power, photo voltaic etc, are expected to deeply reduce heat-trapping emissions. Moreover, wind power is expected to be economically attractive when the wind speed of the proposed site is considerable for electrical generation and electric energy is not easily available from the grid [1]. This situation is usually found on islands and/or in remote localities. However, wind power is intermittent due to worst case weather conditions such as an extended period of overcast skies or when there is no wind for several weeks. As a result, wind power generation is variable and unpredictable. The wind power with rectifier and inverter generation has been suggested [2, 3] to handle the problem above. Wind system with STATCOM is very reliable because the STATCOM acts as a cushion to take care of variation in wind speed and would always maintain an average voltage equal to the set point. However, in addition to the unsteady nature of wind, another serious problem faced by the isolated power generation is the frequent change in load demands. This may cause large and severe oscillation of power. The fluctuation of output power of such renewable sources may cause a serious problem of frequency and voltage fluctuation of the grid, especially, in the case of isolated microgrid, which is the a small power supply network consisting of some renewable sources and loads. In the worst case, the system may lose stability if the system frequency can not be maintained in the acceptable range.

Control schemes to enhance stability in a wind –system have been proposed by much researchers in the previous work. The programmed pitch controller (PPC) in the wind side can be expected to be a cost-effective device for reducing frequency deviation [4, 5]. Nevertheless, under the sudden change of load demands and random wind power input, the pitch controller of the wind side able to effectively control the system frequency due to theirs slow response.

FACTS devices can be a solution to these problems [6]. They are able to provide rapid active and reactive power compensations to power systems, and therefore can be used to provide voltage support and power flow control, increase transient stability and improve power oscillation damping. Suitably located FACTS devices allow more efficient utilization of existing transmission networks. Among the FACTS family, the shunt FACTS devices such as the STATCOM has been widely used to provide smooth and rapid steady state and transient voltage control at points in the network. In this paper, a STATCOM is added to the power network to provide dynamic voltage control for the wind system, dynamic power flow control for the transmission lines, relieve transmission congestion and improve power oscillation damping.

Traditional optimization methods such as mixed integer linear and non linear programming have been investigated to address this issue; however difficulties arise due to multiple local minima and overwhelming computational effort. In order to overcome these problems, Evolutionary Computation Techniques have been employed to solve the optimal parameters of FACTS devices. Fuzzy logic controller (FLC) [11, 12], is an evolutionary computation technique that has been applied to other power engineering problems, giving better results than classical techniques and with less computational effort. In this paper the gains of the controllers with STATCOM have been optimized and optimum transient by trial and error it using to determine the optimal parameters of the PI controller in STATCOM such as PI controller in AC voltage regulator, DC voltage regulator. Simulation results show that the STATCOM devices significantly improve the performance of the wind with rectifier and inverter system, and the grid during transient disturbances.

2. Wind Energy System with STATCOM

Figure 1 presents a schematic diagram of a generalized wind system with STATCOM. The model represents the wind turbine, and the generator (three phase permanent magnet synchronous generator). The PMSG generator is connected in parallel rectifier, inverter, and connected to STATCOM and the utility grid through transmission line.

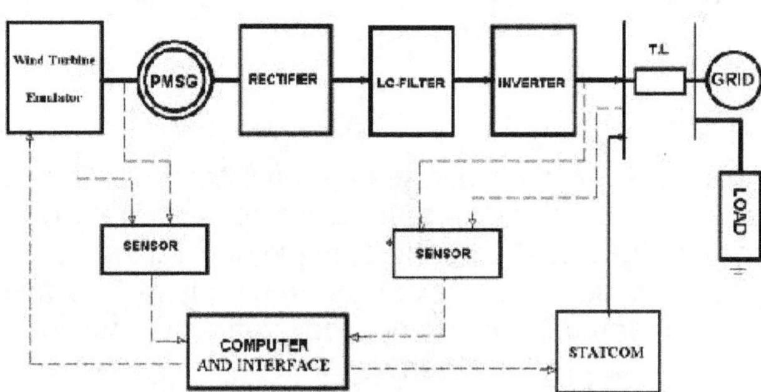

Figure 1. Wind system with STATCOM

3. Mathematical Model of the System

The active power feed to the grid is fulfilled by the permanent magnet synchronous generator. The reactive power required for the operation of permanent magnet synchronous generator and grid is provided by STATCOM and equations for the system shown in Figure 2.1 is given by,

$$P_s = V_i^2 g_s - V_i V_s (g_s \cos(\theta_i - \theta_s) + b_s \sin(\theta_i - \theta_s)) \tag{1}$$

$$Q_s = -V_i^2 b_s - V_i V_s (g_s \sin(\theta_i - \theta_s) - b_s \cos(\theta_i - \theta_s)) \tag{2}$$

Shunt compensators are primarily used for bus voltage regulation by means of providing or absorbing reactive power; they are effective for damping electromechanical oscillations [14, 15]. Different kinds of shunt compensators are currently being used in power systems, of which the most popular ones are Static Var Compensator SVC and STATCOM [16].

In this work, only the STATCOM, which has a more complicated topology than a SVC, is studied. The resulting STATCOM can inject or absorb reactive power to or from the bus to which it is connected and thus regulate bus voltage magnitudes [16]. The main advantage of a STATCOM over a SVC is its reduced size, which results from the elimination of ac capacitor banks and reactors; moreover, a STATCOM response is about 10 times faster than that of a SVC due to its turn-on and turn-off capabilities. Figure 2 illustrate a single-line diagram of the STATCOM and a simplified block diagram of its control system.

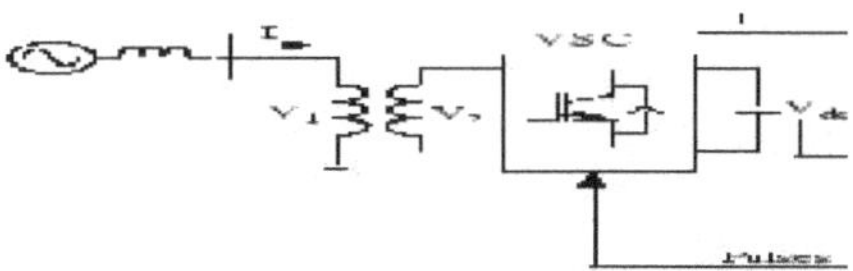

Figure 2. STATCOM Connection

The control system of STATCOM consists of:

1. A phase-locked loop (PLL) which synchronizes on the positive-sequence component of the three-phase primary voltage V1. The output of the PLL (angle θ=wt) is used to compute the direct-axis and quadrature-axis components of the AC three-phase voltage and currents (labelled as Vd, Vq or Id, Iq on the diagram).
2. Measurement systems measuring the d and q components of AC positive-sequence voltage and currents to be controlled as well as the DC voltage Vdc.
3. An outer regulation loop consisting of an AC voltage regulator and a DC voltage regulator. The output of the AC voltage regulator is the reference current Iqref for the current regulator (Iq = current in quadrature with voltage which controls reactive power flow). The output of the DC voltage regulator is the reference current Idref for the current regulator (Id = current in phase with voltage which controls active power flow). Figure 3, and Figure 4 have shown the AC voltage regulator and the DC voltage regulator respectively.
4. An inner current regulation loop consisting of a current regulator. The current regulator controls the magnitude and phase of the voltage generated by the converter (Vd Vq) from the Idref and Iqref reference currents produced respectively by the DC voltage regulator and the AC voltage regulator (in voltage control mode). The current regulator is assisted by a feed forward type regulator which predicts the V2 voltage output (V2d V2q) from the V1 measurement (V1d V1q) and the transformer leakage reactance.

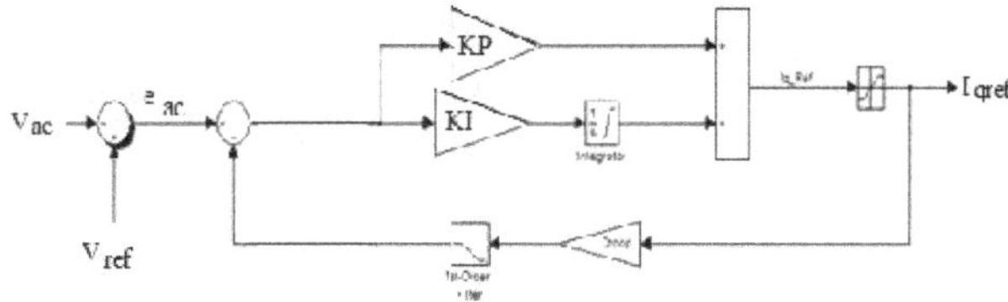

Figure 3. AC voltage regulator

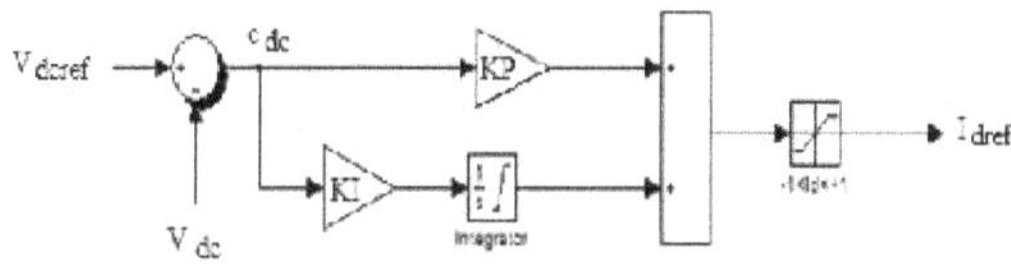

Figure 4. DC voltage regulator

The STATCOM controller in on wind system consists of two controllers, firstly: in AC voltage regulator, secondly: in DC voltage regulator. Errors in system are:

$$e = e_{ac} + e_{dc} \tag{3}$$

$$e_{ac} = V_{ref} - V_{ac} \tag{4}$$

$$e_{dc} = V_{ref} - V_{dv} \tag{5}$$

Figure 5 shows the schematic PI Controller in STATCOM with wind system.

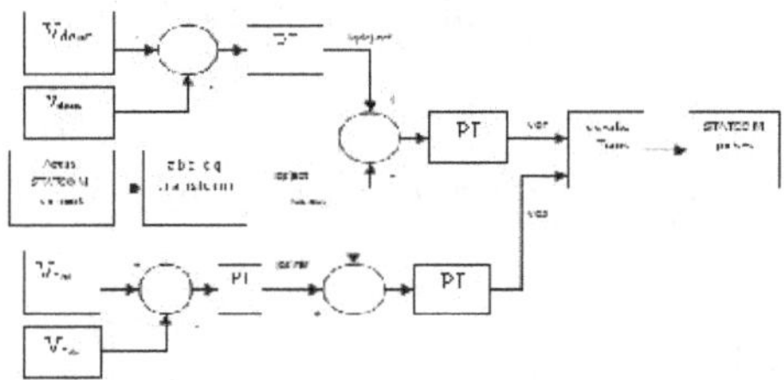

Figure 5. PI Controller in STATCOM

4. Fuzzy Logic Controller and ANFIS Controllers

The Voltage controller determines the value of the leading voltage should be injected to the grid bus to compensate the inverter voltage. The compensated voltage can be adjusted by changing the firing angle of the STATCOM compensator IGBTS. The output signal of the fuzzy controller, Vc, is used to determine the suitable firing angle of the compensator IGBTS. To compensate high voltage to the grid bus the controlling voltage, Vc, should be increased, and Vic versa.

The error of the Voltage of the inverter and the rate of change of voltage error are calculated as following:

$$Ve = Vref - Vinv \tag{6}$$

In this paper these error criteria will be minimized by applying an existing tuning algorithm through the application of a fuzzy logic controller, as will presently be elucidated for a comprehensive and introduction to fuzzy [17]. Considering the fuzzy controller shown in figure (4.12) there are several scaling gains are introduced to the proportional plus integral, and at the same time gain go between the fuzzy controller and the STATCOM. The dynamic behavior of the fuzzy controller (Figure 6) is highly dependent on these scaling factors. These factors have to be selected carefully in order to achieve good performance.

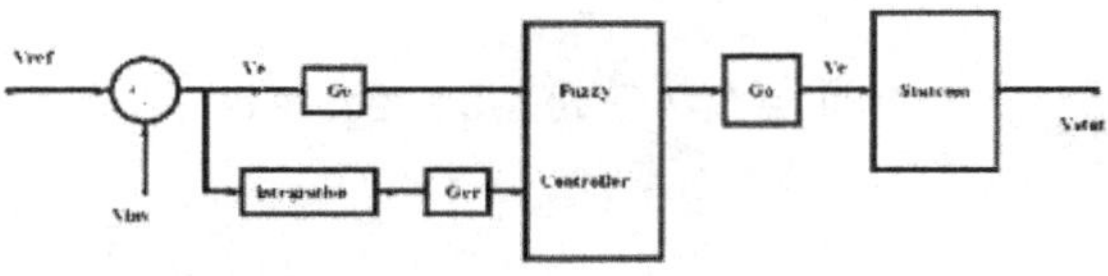

Figure 6. Fuzzy Controller for Voltage Stabilization

Voltage source converter (VSC) based STATCOM is developed with Adaptive Neuro Fuzzy Inference System (ANFIS) controllers. ANFIS controller requires less computation time and characteristics of both fuzzy and neuron controllers. The ANFIS has the ability to generalize and can interpolate in between the training data as well as the membership function has been decided according to the training data.

The difference between AC voltage measured and reference AC voltage reference is input to ANFIS1 gives an error signal one e1(t), also the difference between DC voltage measured and reference DC voltage reference is input to ANFIS2 gives an error signal two e2(t) as shown in Figure 7. First of two input signals of ANFIS is the error signal e(t), the second one is the changing of error signal depending on time de(t)/dt and is expressed with correlations as:

$$e_1(t) = V_{AC} - V_{ACref} \tag{7}$$

$$e_2(t) = V_{DC} - V_{DCref} \tag{8}$$

The output signal of controller element is the value of voltage (v_{2d}, v_{2q}). The output of ANFIS is entering to current regulation of STATCOM.

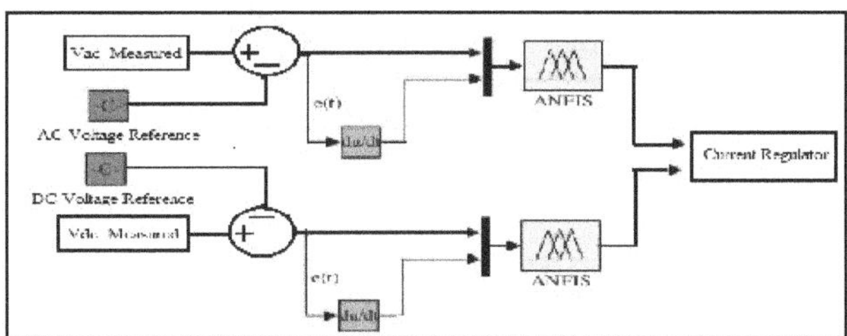

Figure 7. Simulation Block Diagram of STATCOM Controller with ANFIS

5. Simulation Results

5.1. Simulation of The proposed System without STATCOM

In this case, the dynamic response of the wind will be studied at a different step change in reference wind speed.

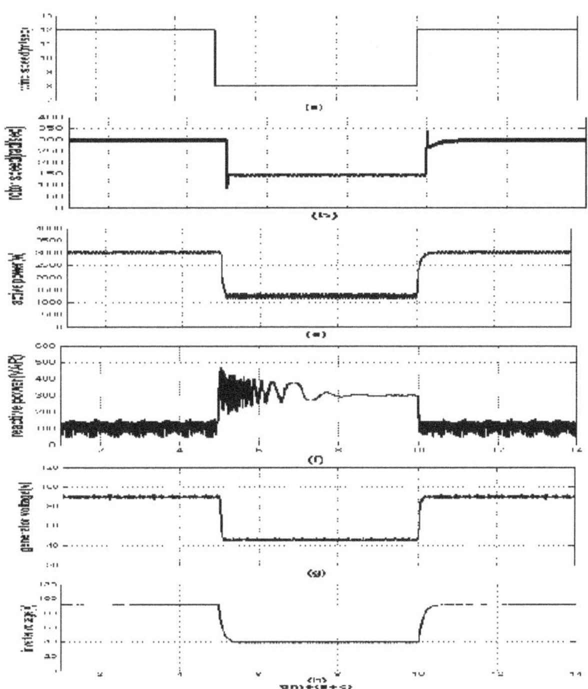

Figure 8. Open loop wind system for a small step change in wind speed

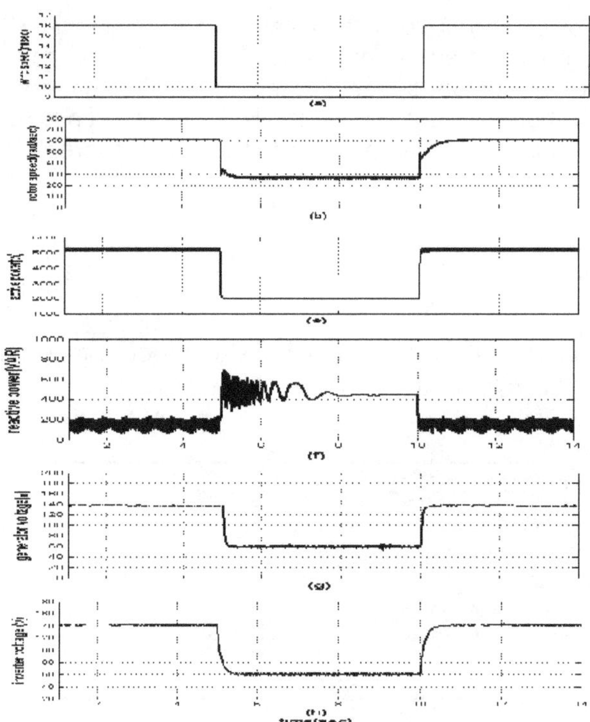

Figure 9. Open loop for large step change in wind speed

From Figure 8 and Figure 9 it can be noticed that increase in the wind turbine speed from small step change to large step change, the rotor speed increasing with increase the wind speed. The active power increases with the wind speed. As the same the inverter voltage increases with large step.

5.2. Simulation Results of the Proposed System with STATCOM and PI Controller

The voltage at the generator terminals is monitored to check whether the system has recovered. The generator speed is monitored to detect if the generator goes into over speed.

The dynamic response of the wind will be studied at a step change in reference wind speed. The proportional plus integral controller is the simple type of controllers that could be applied to the system. The controller gains are fixed and designed using trial and error at a certain operating condition.

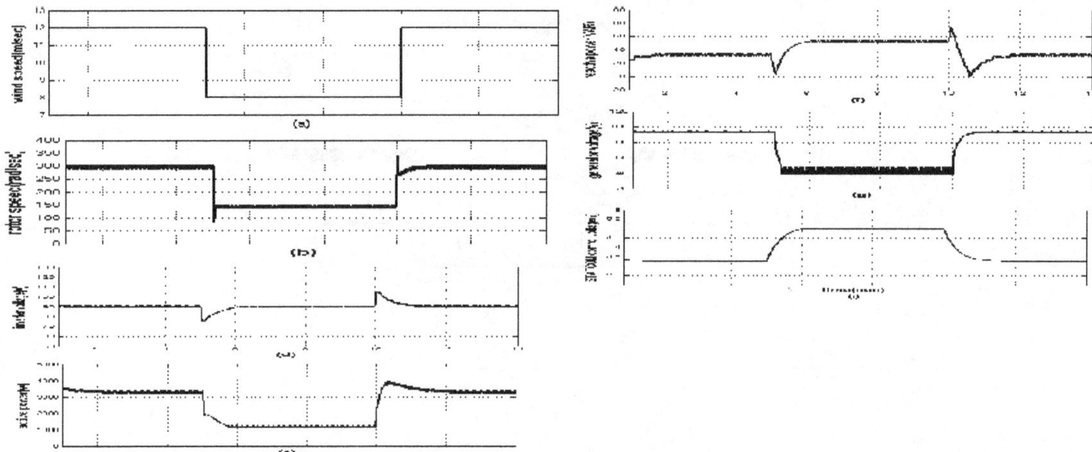

Figure 10. Simulation results of the proposed system with PI controller at small step change in wind speed

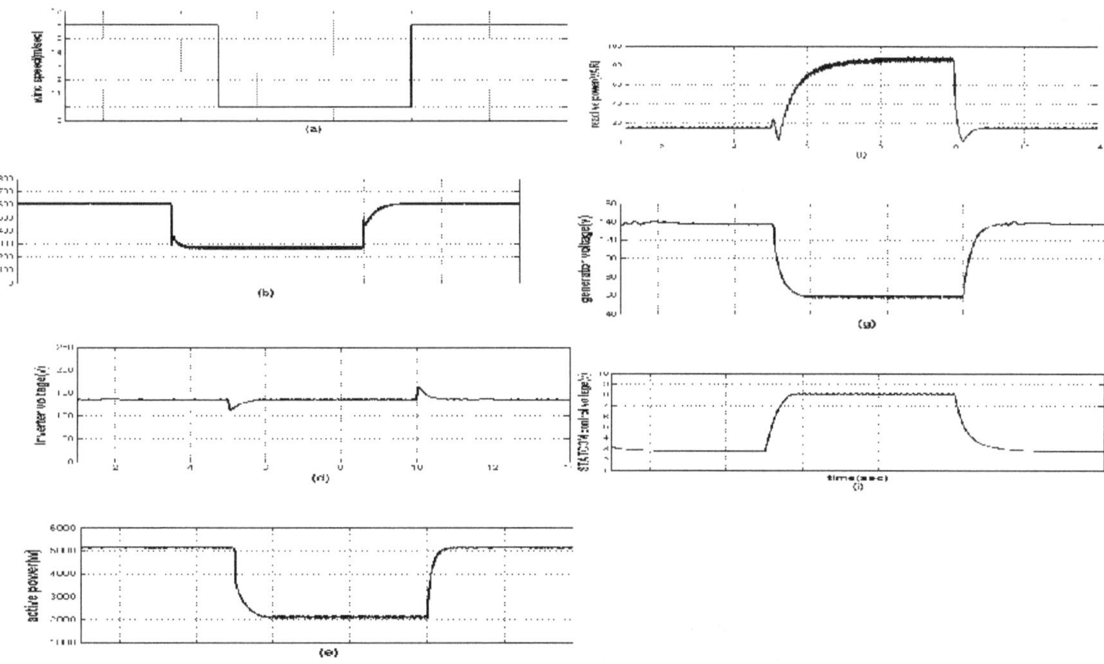

Figure 11. Simulation results of the proposed system at large step change in wind speed using PI controller

5.2. Modelling of the Proposed System with STATCOM and FLC Controller

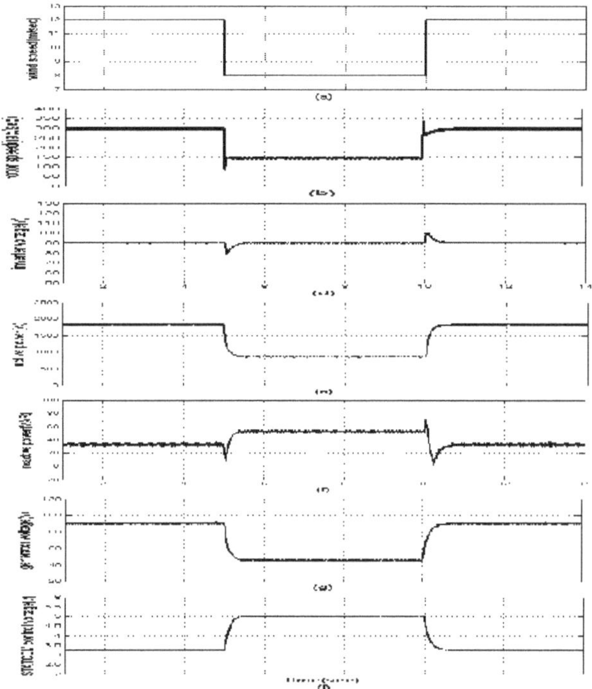

Figure 13. Simulation results of the proposed system at small step change in wind speed using FLC

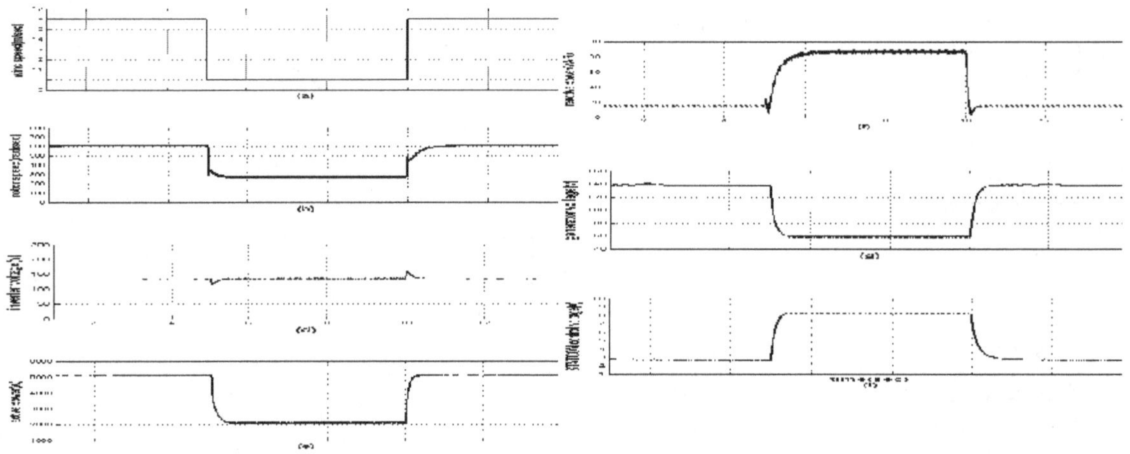

Figure 13. Simulation results of the proposed system at high step change in wind speed using FLC

The dynamic behaviour of the fuzzy controller is highly dependent on these scaling factors. These factors had selected carefully in order to achieve good performance for both steady state and transient conditions. In the above two figures small and large wind speed are more stable than proportional plus integral testes

5.3. Modelling of the Proposed System with STATCOM and ANFIS Controller

ANFIS is an adaptive network. An adaptive network is network of nodes and directional links. Associated with the network is a learning rule - for example back propagation. It's called adaptive because some, or all, of the nodes have parameters which affect the output of the node. [100] these networks are learning a relationship between inputs and outputs. Adaptive networks cover a number of different approaches but for our purposes we will investigate in some detail the method proposed by Jang known as ANFIS.

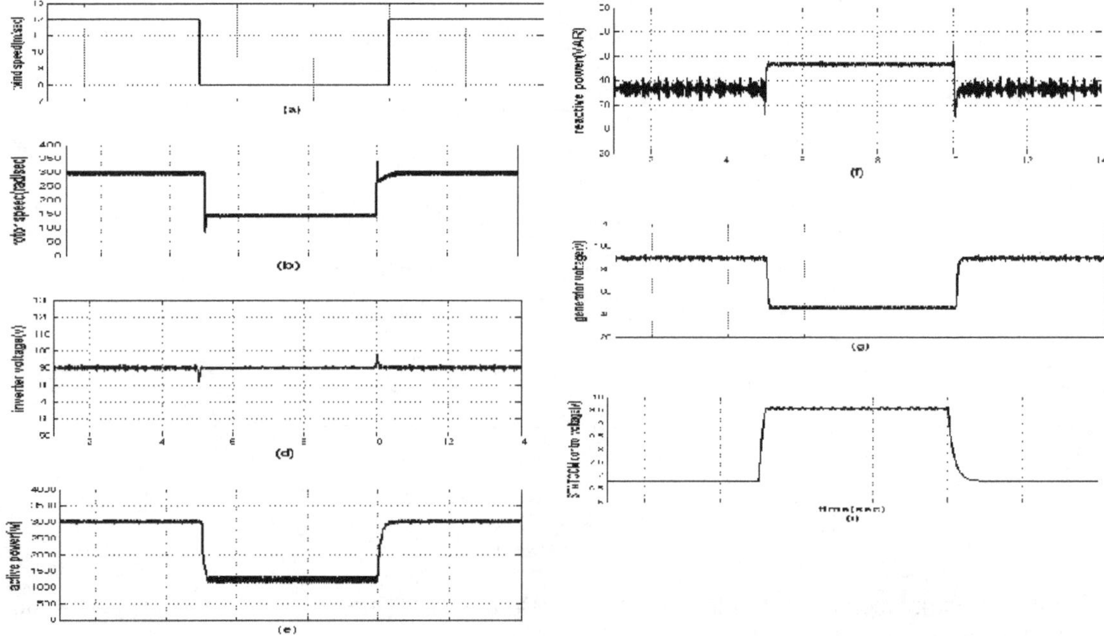

Figure 14. Simulation results of the proposed system at small step change in wind speed using ANFIS

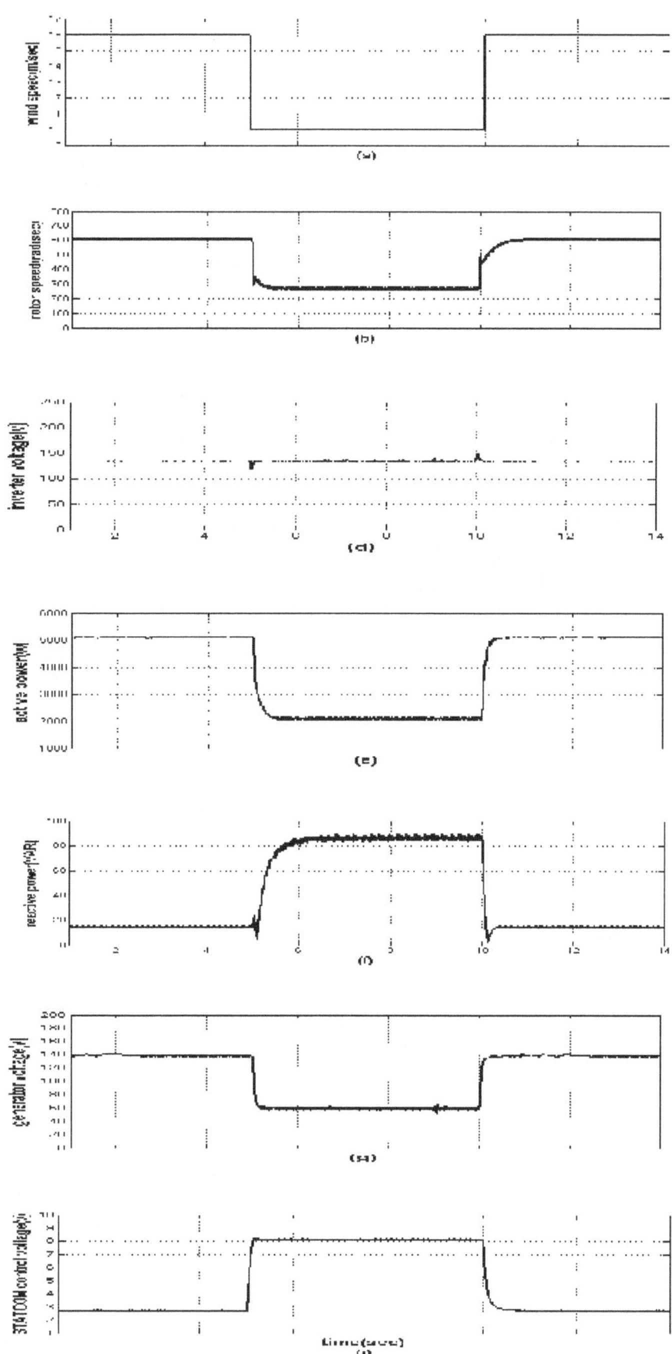

Figure 15. Simulation results of the proposed system at high step change in wind speed using ANFIS

6. Comparison between Conventional and Intelligent Control

Figure 16 shows the comparison between the ANFIS controller, fuzzy logic controller and proportional plus integral controller. The gains of the proportional plus integral are Kp = 5, and Ki = 0.00851 give the best performance of the regulate system with PI controller. Both controllers are tested with the same constraints and the operating conditions.

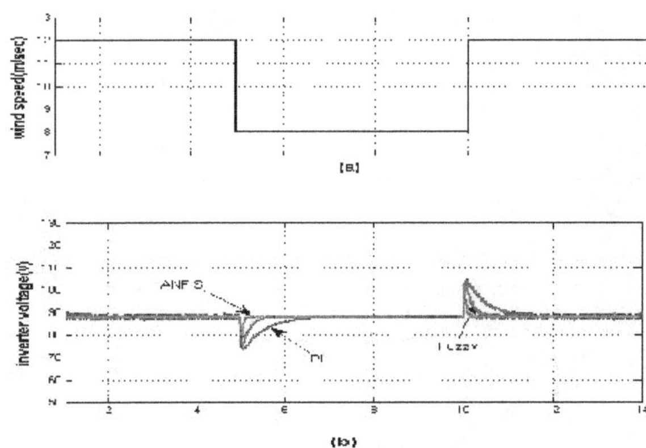

Figure 16. Comparison of PI, fuzzy, and ANFIS controllers at small step change in wind speed

For a large step in wind speed, Figure 17 shows the system performance by with PI controller and intelligent controllers. ANFIS controller is the best controller where the voltage response of the wind system is fastest and least steady-state error.

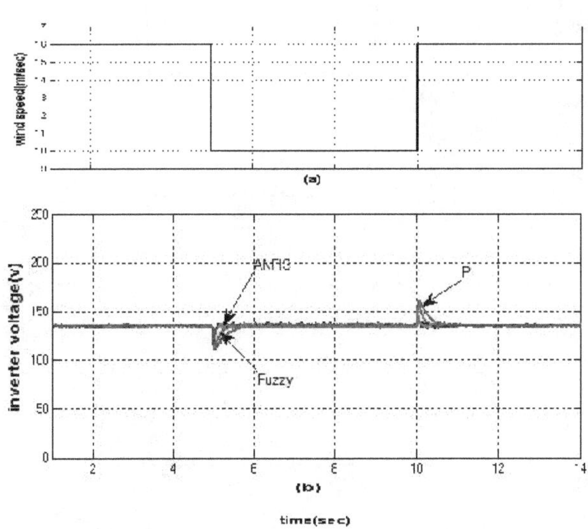

Figure 17. Comparison of PI, fuzzy, and ANFIS controllers at large step change in wind speed

7. Conclusion

This paper concerned with solution of the voltage problems of the wind energy system. Several techniques are proposed to regulate the output voltage of the wind system. Digital controller is developed such as proportional plus integral, ANFIS and fuzzy logic to achieve the best performance of the STATCOM. Wide range of operating conditions of wind speed was considered in this paper to achieve the most suitable controlling technique.

Mathematical models were used to achieve linear models of all the system, as wind turbine model, rectifier model, DC-link model, inverter model, grid model, and STATCOM model

With STATCOM controller the voltage of the inverter was independent of the winds speed. That means, the STATCOM was fully regulated to achieve a constant inverter voltage independent the wind speed variations.

The intelligent controllers (FLC, ANFIS) are better than the conventional PI controller, where the wind system performance is faster and less steady-state error.

References

[1] Ackermann T. "*Wind Power in Power Systems*". John Wiley & Sons. 2005.

[2] Hunter REG. "*Wind-diesel systems a guide to technology and its implementation*". Cambridge University Press. 1994.

[3] Lipman NH. "*Wind-diesel and autonomous energy systems*". Elservier Science Publishers Ltd. 1989.

[4] Bhatti TS, Al-Ademi AAF & Bansal NK. "Load frequency control of isolated wind diesel hybrid power systems". *International Journal of Energy Conversion and Management*. 1997; 39: 829-837.

[5] Das D, Aditya SK & Kothari DP. "Dynamics of diesel and wind turbine generators on an isolated power system". *International Journal of Elect Power & Energy Syst*. 1999; 21: 183-189.

[6] NG Hingorani and L Gyugyi. "Understanding FACTS: Concepts and Technology of Flexible AC Transmission Systems". *IEEE, New York*, ISBN 0-7803. 2000: 3455-3458.

[7] JB Park, KS Lee, JR Shin and KY Lee. "A particle swarm optimization for economic dispatch with nonsmooth cost functions". *IEEE Trans. on Power Systems*. 2005; 20(1): 34-42.

[8] Hamesh babu Nanvala, Gajanan K Awari. "Review on use of Swarm Intelligence Meta heuristics in Scheduling of FMS". *International Journal of Engineering and Technology (IJET)*. 2011; 3(2): 80 – 86.

[9] H Mori and Y Goto. "*A parallel tabu search based method for determining optimal allocation of FACTS in power systems*". Proc. Of the International Conference on Power System Technology (PowerCon 2000). 2000; 2: 1077-1082.

[10] W Ongsakul and P Jirapong. "*Optimal allocation of FACTS devices to enhance total transfer capability using evolutionary programming*". Proc. of the IEEE International Symposium on Circuits and Systems (ISCAS 2005). 2005; 5: 4175-4178.

[11] LJ Cai, I Erlich and G Stamtsis. "*Optimal choice and allocation of FACTS devices in deregulated electricity market using genetic algorithms*". Proc. of the IEEE PES Power Systems Conference and Exposition. 2004; 1: 201-207.

[12] S Gerbex, R Cherkaoui and AJ Germond. "Optimal location of multitype FACTS devices in a power system by means of genetic algorithms". *IEEE Trans. on Power Systems*. 2001: 16(3): 537-544.

[13] Wind/Diesel Systems Architecture Guidebook. *AWEA*. 1991.

[14] Siegfried Heier. "*Grid integration of wind energy conversion systems*". Jon Wiley & Sons Ltd. 1998.

[15] E Uzunovic. "Transient Stability and Power Flow Models of VSC FACTS controllers". Ph.D. dissertation, University of Waterloo, Waterloo, ON, Canada. 2001.

[16] Sim Power Systems User guide. Available http://www.mathworks.com.

[17] Davis and Lawrence. "*Handbook of Genetic Algorithms*". Van Nostrand Reinhold, 115 Fifth Avenue, New York, NY 10003. 1991.

Site Diversity Technique Application on Rain Attenuation for Lagos

Abayomi Isiaka O. Yussuff[1,*], Nana Hamzat[1], Nor Hisham Haji Khamis[2]
[1]Lagos State University, Lagos, Nigeria
[2]Universiti Teknologi Malaysia, Skudai, Malaysia
e-mail: ayussuff@yahoo.com*, hisham@fke.utm.my

Abstract

This paper studied the impact of site diversity (SD) as a fade mitigation technique on rain attenuation at 12 GHz for Lagos. SD is one of the most effective methods to overcome such large fades due to rain attenuation that takes advantage of the usually localized nature of intense rainfall by receiving the satellite downlink signal at two or more earth stations to minimize the prospect of potential diversity stations being simultaneously subjected to significant rain attenuation. One year (January to December 2011) hourly rain gauge data was sourced from the Nigerian Meteorological Agency (NIMET) for three sites (Ikeja, Ikorodu and Marina) in Lagos, Nigeria. Significant improvement in both performance and availability was observed with the application of SD technique; again, separation distance was seen to be responsible for this observed performance improvements.

***Keywords**: site diversity, rain rate, attenuation, slant path, FMT*

1. Introduction

Satellite Communication requires the provisioning and deployment of two or more interconnected earth stations at spatially separated locations. Also, the necessity of operating at millimeter wavebands (Ku, Ka or Q/V bands) was informed by the eventual congestion of lower frequency bands resulting from commercial activities. It has been established that rain produces significant attenuation on radio waves of frequencies beyond 10 GHz. With respect to Ku band (14/12GHz. Diversity technique is one of the common fade mitigation techniques (FMTs) used in attempts to overcome signal fades resulting from convective rain events that typifies tropical stations. Frequency diversity (FD) employs the method of adaptively changing from frequency of propagation to a lower frequency to circumvent impending signal outages due to rain precipitations, and then switch back to the primary frequency after the disruptive events. Other diversity schemes include: Time diversity (TD), which is used to re-send the information at that time that propagation channel has been configured to allow it. Orbital diversity (OD), in which information is switched from one satellite to another. A major disadvantage of this technique is cost. Also, there is site diversity (SD), which is one of the most effective methods to overcome such large fades due to rain attenuation [1].

Site diversity is anchored on the proposition that the probability of attenuation being exceeded simultaneously at two sites is less than the probability of the same attenuation being exceeded at one of the sites by a factor which decreases with increasing distance between the sites and with increasing attenuation [2]. Intense rain cells cause large attenuation values on an earth-space link and often have horizontal dimensions of no more than a few kilometers. SD systems can re-route traffic to alternate earth stations with consequent considerable improvements in the system reliability. A balanced SD system (with attenuation thresholds on the two links equal) uses a prediction method that computes the joint probability of exceeding attenuation thresholds and is considered the most accurate and is preferred by ITU [3,4]. Site diversity takes advantage of the usually localized nature of intense rainfall by receiving the satellite downlink signal at two or more earth stations designed to minimize the prospect of potential diversity stations being simultaneously subjected to significant rain attenuation. Conceptually, a site diverse system comprises two or more spatially separated ground stations arranged in such a way to exploit the fact that the probability of attenuation due to rain will occur simultaneously on a typical slant path is significantly unlikely than the relative probability of attenuation occurrence on either individual paths. The effects of rain attenuation can then be consequently diminished or eliminated altogether. It has been suggested that the ground station with the higher received signal strength at any instant in time should be selected in order to significantly reduce the effect of rain attenuation [5].

Furthermore, rain has been identified as the major culprit for slant path signal impairment at frequencies above 10 GHz. It is also one of the most variable elements of weather. It varies in intensity, duration, frequency, and spatial pattern. The convective precipitations are of the highest rainfall intensities, localized, and also of short duration. While the medium or low rainfall intensities of stratiform precipitations have longer durations, and widespread. Rainfall intensity is an inverse function of its duration and can thus vary considerably with duration from one region to another, and for varying geographical features, such as the presence of mountains, hills or water masses, vegetation. The models predicting rain attenuation and those concerning the performance of site diversity systems can be The Hodge model [6], which is a regression model based on available attenuation statistics valid only for a few specific locations and Physical models based on the understanding of the rain process and the rainfall medium exhibiting a good performance globally [7-9]. The EXCELL [8], Matricciani [9] and Paraboni-Barbaliscia [10] models are well known physical prediction models for enhancement of site diversity performance. The Paraboni-Barbaliscia model [10] have been reported to submit that single and joint sites' rain attenuation indicate log-normal distributions, such that the site diversity gain can be derived from the measured single site rainfall rate and attenuation.

Ikeja with a geographical coordinates of 6.35oN (Lat.) and 3.23oE (Long.), Ikorodu (6.600N, 3.50oE), and Marina (6.45oN, 3.42oE) are all stations located in Lagos, a coastal region located in the rain forest area in the southwestern tropical Nigeria. Lagos is bordered on the south by the Atlantic Ocean, and with mean annual rainfall of 1425 mm and altitude of 38 m above sea level.

1.1. Site Diversity Gain

Site diversity is one of the advanced techniques that employ a master and a remote station configuration separated several kilometres apart in order to take advantage of the inhomogeneity of convective rainfall, which occurs within localized rain cells with a diameter of a few tens of kilometres [11]. This has however been found to be much smaller than this [12] in tropical stations because of the peculiar nature of such regions. The inhomogeneity of rainfalls results in a decorrelation (depolarization) of the rain attenuation on the paths. Therefore, the prospect of transmitting and/ or receiving the signal via alternate paths reduces or eliminates the possibility of experiencing deep fades on both channels simultaneously. It has been observed that the deployment of multiple stations with the site separations in excess of the average horizontal dimensions of individual intense rain cells remarkably improved system availability because the joint path outages are presumed to be random and infrequent for diversity schemes [13]. Figures 1 and 2 depict the conceptual model and geometrical configuration of a typical two-terminal site diverse system. Hence, the jointly received signals are sent to the master station where they are further processed based on signal selection, switching, or a combination of both [8]. A diversity control unit coordinating the signal flow and a signal processing unit is incorporated at the master and the other earth stations respectively.

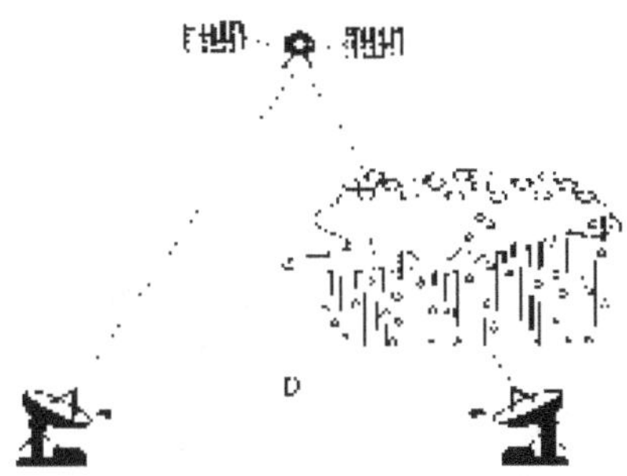

Figure 1. Conceptual model for site diversity

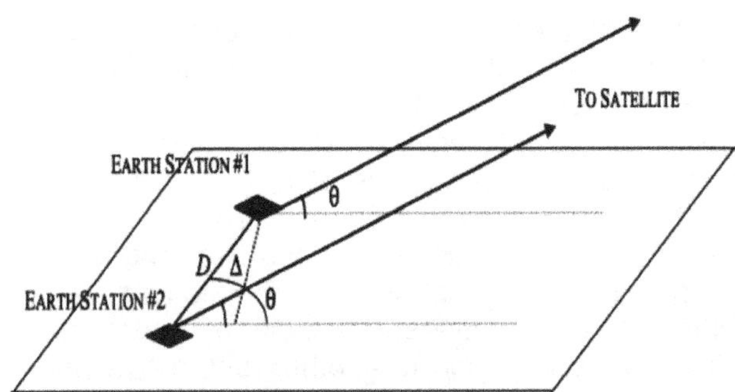

Figure 2. Geometrical configuration for 2-terminal diversity scheme 5[2]

1.2. Models description

Concerted efforts have been made over the past decade to develop reliable techniques for the prediction of path rain attenuation for a given location and frequency, and the availability of satellite beacon measurements has provided a database for the validation and refinement of the predictions models [14,15]. However, majority of these rain attenuation models were developed from data acquired from stations located in temperate regions [16]; with climatological characteristics far distinct from that experienced in tropical regions. Hence, the need for deliberate efforts to formulate rain attenuation prediction models specifically for the tropics. The ITU-R. Rec. P. 618-12 [3] is the globally accepted rain attenuation model for design and testing of any proposed model.

1.2.1. Hodge model

Hodge [6] proposed the first empirical prediction model for site diversity gain as a product function of Individual gains contributed by single site attenuation, site separation, baseline orientation, link frequency and path elevation angle. This model has been adopted by ITU-R. Rec. P. 618-12 [3] for the estimation of gain up to a site separation of 20 km; and it is represented mathematically as:

$$G_{SD} = G_D * G_\theta * G_f * G_\Delta \tag{1}$$

In the above relationship, $G_D, G_\theta, G_f, G_\Delta$,$G_D$ are factors expressing the dependence of the site diversity gain on the the site separation distance, d (km), the common elevation angle of both slant paths, θ (degrees), the frequency of operation, f (GHz), and the orientation of the baseline between the two earth stations , Δ (degrees). Each dependence factor is given by the expression:

$$G_D = a(1 - e^{bd}) \tag{2}$$

where:

$a = 0.78A - 1.94(1 - e^{-0.11A})$,$b = 0.59(1 - e^{-0.1A})$, $G_\theta = 1 + 0.006\theta$, $G_f = \exp(-0.025f)$, and $G_\Delta = 1 + 0.002\Delta$ (3)

Again, the diversity gain, G_{SD} is the difference between the single site antenna attenuation and the joint diversity systems for a specified percentage of probability, p %. Hodge's model is particularly suitable for Ku band although the diversity gain here is observed to be smaller than what is achievable on Ka band. Also, it has been noted that the diversity gain sharply reduces for elevation angles that are below 30 degrees. This is premised on the fact that for lower elevation angles, the slant path tends to be longer and hence, the higher probabilities of experiencing signal propagation impairments due to rainfall. This is especially true for stations located in tropical regions. However, for separation distances in excess of 20 km, the Paraboni-Barbaliscia model [10] has been co-opted into the ITU-R. Rec. P. 618-12 model [3].

2. Research Method

The slant path rain attenuation values were computed in line with the procedure outlined in ITU-R. Rec. P. 618-12 [3], The rain rate (mm/hr) recorded at time t for each pixel location indicating the reference and diversity earth stations were used to compute the specific attenuation $\gamma_{0.01}$ (dB/km) given by Equation (4), where a and b are the link frequency dependent coefficients such as frequency, rain drop size, polarization, antenna's elevation angle and temperature, are defined in [3].

$$\gamma_t = aR_t^{\ b} \tag{4}$$

More so, ITU-R Rec. P. 618-12 [3] rain attenuation prediction model is used to predicts the long-term statistics of the slant-path rain attenuation at a given location for frequencies up to 55 GHz, rainfall rate at 0.01% of the time ($A_{0.01}$) as the basic input and with subsequent resulting attenuation at 0.01% ($A_{0.01}$) applied as the basis for estimating the attenuation exceeded at other percentages ($A_{\%p}$). The step-by-step procedures for calculating rain attenuation cumulative distribution function over the satellite link are as follow:

The effective path length L_{eff} (km) through rain is obtained by multiplying the horizontally adjusted slant path by the vertical reduction factor:

$$L_{eff} = L_{h0.01} r_{v0.01} \ (km) \tag{5}$$

Therefore predicted slant path attenuation exceeded for 0.01% of an average year is

$$A_{0.01} = \gamma_{0.01} L_{eff} \ (dB/km) \tag{6}$$

Hence, the predicted attenuation exceeded for other percentages $\%p$ of an average year may be obtained from the value of $A_{0.01}$ by using the following extrapolation according to ITU-R P. 618-12 [3]. That is:

$$A_{\%p} = A_{0.01}\left(\frac{p}{0.01}\right)^{-\left[0.655+0.033\ln p-0.045\ln A_{0.01}-z\sin\theta(1-p)\right]} (dB) \tag{7}$$

Where p is the percentage probability of interest and z is given by

$$For\ p \geq 1.0\%, z = 0 \tag{8}$$

$$For\ p<1.0\%,\ z=\begin{cases} 0;\ for\ /\phi/\geq 36^0 \\ z=-0.005(/\phi/-36)\ for\ \theta\geq 25^0\ and/\phi/<36^0 \\ z=-0.005(/\phi-36/)+1.8-4.25\sin\theta,\ for\ \theta<25^0\ and/\phi/<36^0 \end{cases} \tag{9}$$

2.1 Data Collection

One year (January to December 2011) hourly rain gauge data was sourced from the Nigerian Meteorological Agency (NIMET). Measurement setup comprise buck-type rain gauge installed at the measurement at the three sites (Ikeja, Ikorodu and Marina) to record the rain rate exceedances. One-minute rainfall rate was subsequently obtained from these data using the empirical model developed by Chebil and Rahman [17,18]. The 1-minute rain rate for arbitrary percentage of time, $p\%$; and R_p can be estimated by using only the average annual total rainfall. The model is given by Equations (10) through (12).

Furthermore, the rain rate, R (mm/h) were recorded for different % probability, p indicating the reference (Ikeja) and diverse earth stations (Marina and Ikorodu) are shown in Table 1 below. Chebil and Rahman's [15,16] proposed rain rate conversion model for converting these hourly data to the equivalent one-minute rainfall rate values was applied as follow:

$$CF_{60} = \mathrm{R}_{1\,(\mathrm{P})}/\mathrm{R}_{60\,(\mathrm{P})} \tag{10}$$

$$CF_{60} = 0.772 * p^{-0.041} + 1.141 * \exp(-2.57 * p) \tag{11}$$

where CF_{60} is the rain rate conversion factor, defined as the ratio of rain rates R_1 at a given percentage of time p with an integration time of 1 and 60 minutes, respectively. The constraint of this model is: $0.001\% \leq p \leq 1.0\%$; hence $\mathrm{R}_{1\,(\mathrm{P})}$ the 1-minute rain rate is:

$$\mathrm{R}_{1\,(\mathrm{P})}/= \mathrm{R}_{60\,(\mathrm{P})} \; * CF60 \tag{12}$$

Table 1. One-minute rain rate for selected locations

PROBABILTY	0.001	0.002	0.003	0.005	0.01	0.02	0.03	0.1	0.5	1.0
IKEJA (mm/h)	119.8	118.0	116.9	115.5	113.2	110.2	107.8	95.8	61.5	47.6
MARINA (mm/h)	126.0	124.1	123.0	121.5	119.1	115.9	113.5	100.8	64.7	50.1
IKORODU (mm/h)	129.4	127.5	126.3	124.8	122.3	119.1	116.5	103.5	66.4	51.4

Ikorodu and Marina stations were set at 16.67 km and 16.69 km apart, as shown in Figure 3. Figure 3 displays the images of the selected locations using Google Map Ruler to determine the site line-of-sight (LOS) separation distances between the reference station and the two other stations.

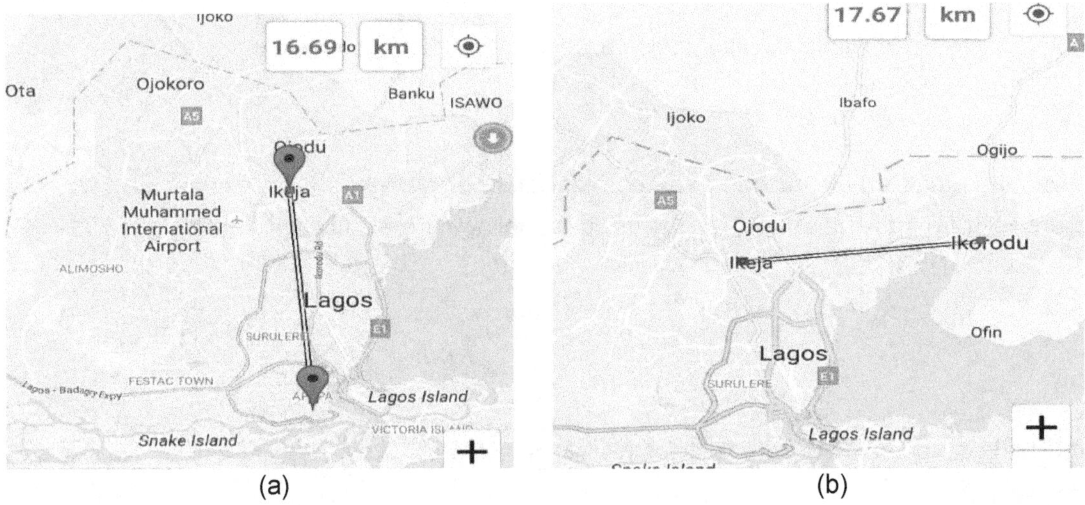

Figure 3. Site separation distances between (a) Ikeja and Marina, and (b) Ikeja and Ikorodu

In Table 2, the parameters used in determining the site diversity gains for the respective earth locations are shown.

Table 2. Climatological parameters for selected stations

Parameters	Values
Elevation angle (degree)	44.5
Frequency (gigahertz)	12
Base line orientation (degree)	45
Site separation distance (km) (IKEJA - MARINA)	16.69
Site separation distance (km) (IKEJA – IKORODU)	17.67

Furthermore, the rain rate values at different percentage of time were plotted and analyzed using some programming codes in MATLAB. The diversity improvement factor, I which is the ratio of the probabilities for a specific attenuation exceeded, is subsequently derived. According to ITU-R. Rec. P.618-12 [3], the diversity improvement factor model is given as:

$$\mathrm{I} = \mathrm{P1}/\mathrm{P2} = 1/\left(1+\beta^2\right) * \left(1+\frac{100\,\beta^2}{\mathrm{P1}}\right) \approx 1 + 100\,\beta^2/\mathrm{P1} \qquad (13)$$

where $\beta^2 = d^{1.33} * 10^{-4}$, d is the seation distance while $\mathrm{P1}$ and $\mathrm{P2}$ refer to the percentages of time for the observed single and joint's rain attenuation exceeded respectively.

3. Results and Analysis

Figure 4.0 shows the cumulative distribution function (CDF) of path attenuation and rain rates for Ikeja, Ikorodu and Marina, Lagos. The plot of the individual site's slant path attenuation exceeded against the corresponding attenuation gain are shown in Figure 5, while Figure 6 displays the evident improvement in performance in terms of reduction in rain attenuation effects accrued to implementing joint diversity system. It is clearly seen that there is significant reduction in attenuation exceeded in Figure 6.

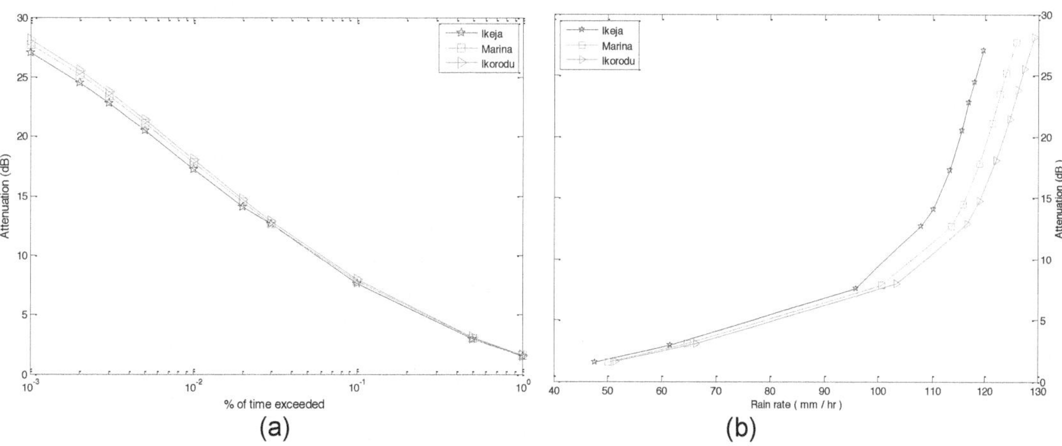

Figure 4. (a) CDFs of attenuation for and (b) Attenuation against. rain rate for Ikeja, Marina and and Ikorodu.

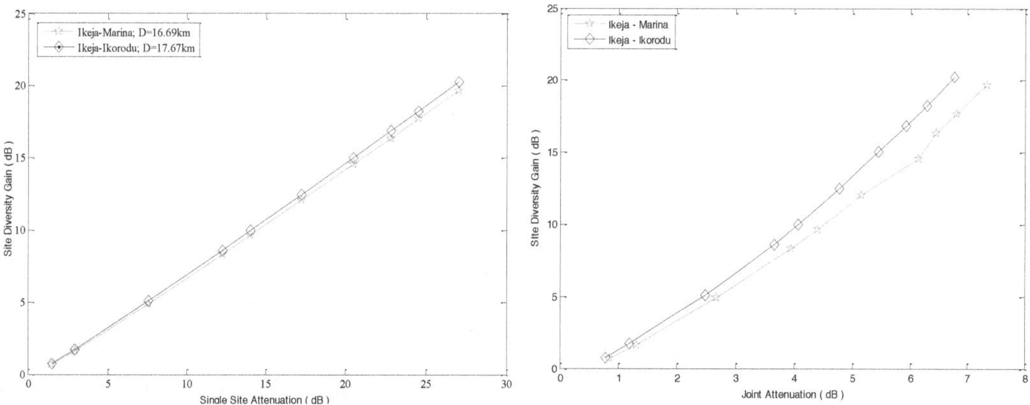

Figure 5. Site diversity gain vs. single site attenuation

Figure 6. Site diversity gain versus joint attenuation

Moreover, site separation distance is the major factor that affects the amount of diversity gain for any two diverse systems. Shown in Figure 7 is the plot of diversity gains (G_{SD}, which consists of G_{SD1} and G_{SD2}) improvement factors I_1 and I_2) and comparison of attenuation exceedances for single and joint diversity systems.

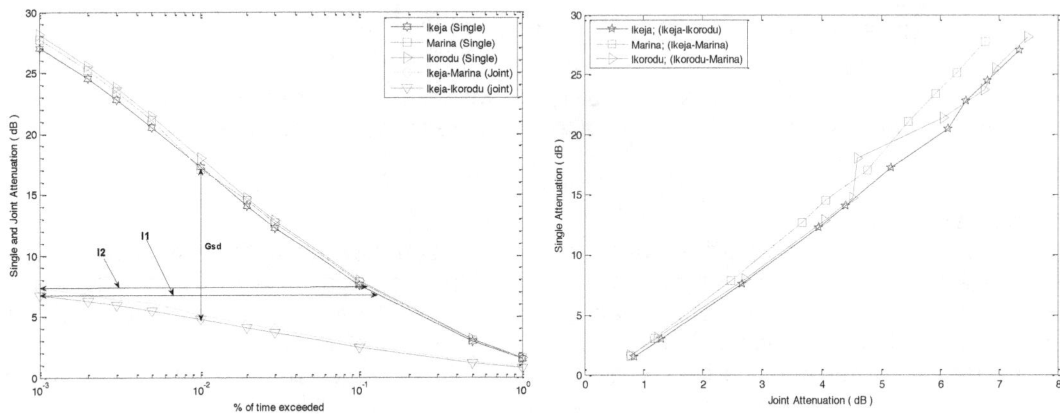

Figure 7. Comparison of single and joint attenuation

Figure 8. CDFs of single site attenuation site against joint attenuation

Here, G_{SD1} and $G_{SD2}$are found to be12.07 dB and 12.48 dB respectively, while $I_1 \approx I_2 \approx 43.23$.

In summary, the improved link availability for distances 16.69 km (Ikeja to Marina) and 17.67 km (Ikeja to Ikorodu) for 0.01% of the time exceeded are 99.9998% and 99.9999% respectively. Hence, site diversity provides significant improvement in both performance and availability of the system. Figure 8 depicts the effect of applying Hodge model to a separation distance exceeding the recommended maximum distance of 20 km (in this instance, 22.70 km). It can be observed that there is no significant increment in joint attenuation between Ikeja to Ikorodu (17.67km) and Ikorodu to Marina (22.70 km). Also, the Ikorodu-Marina system largely deviate from the Hodge model's trend between $p=0.02\%$ and $p=0.003\%$. This effect is more pronounced at $p=0.01\%$ with a corresponding attenuation of 4.06 dB.

4. Conclusion

This effect of site separation distance with keeping other factors such as propagation frequency, antenna elevation angle, polarization, and baseline orientation angle constant was investigated in this work. Significant improvement in both performance and availability of the site diversity system was observed; and it is safe to conclude that separation distance is a key factor to be considered when opting for site diversity as propagation impairment mitigation technique. Furthermore, efficiency starts to diminish and performance begins to suffer when the 20 km maximum separation distance threshold as recommended by ITU-R. Rec. P. 618-12 is exceeded.

References

[1] Ippolito LJ. Satellite Communications Systems Engineering: Atmospheric Effects. *Satellite Link Design and System Performance*. Hoboken, NJ: Wiley Publication. 2008.

[2] ITU-R BO. 791. Choice of Polarization for the Broadcasting-Satellite Service. *Geneva*. 2002.

[3] ITU-R. P. 618-12, Propagation Data and Prediction Methods Required for the Design of Earth-Space Telecommunications Systems, in Recommendation ITU-R P Series. 2015.

[4] Yussuff AI, Khamis NH. Rain Attenuation Modelling and Mitigation in the Tropics: Brief Review. *International Journal of Electrical and Computer Engineering*. 2012; 2(6): 748-757.

[5] Athanasios D. et al. Long-Term Rain Attenuation Probability and Site Diversity Gain Prediction Formulas. *IEEE Transactions on Antennas and Propagation*. 2005; 53(7): 2307-2313.

[6] Hodge DB. An improved model for diversity gain in earth-space propagation paths. *Radio Sci.* 1982; 17(6): 1393–1399.

[7] Kanellopouloset JD. al. Rain attenuation problems affecting the performance of microwave communication systems. *Ann. Telecommun*. 1990; 45(7–8): 437–451.

[8] Bosisio AV, Riva C. A novel method for the statistical prediction of rain attenuation in site diversity systems: Theory and comparative testing against experimental data. *Int. J. Satellite. Communication.* 1998; 16: 47–52.

[9] Matricciani E. Prediction of site diversity performance in satellite communications systems affected by rain attenuation: Extension of the two layer rain model. *European Trans. Telecommunication.* 1994; 5(3): 27–36.

[10] Luglio M. et al. Large-scale site diversity for satellite communication networks. *Int. J. Satellite Commun*. 2002; 20: 251–260.

[11] Lin SH. Method for calculating rain attenuation distribution on microwave paths. *Bell Syst. Tech. J.* 1975; 54(6): 1051–1086.

[12] Khamis NH. H. Path Reduction Factor for Microwave Terrestrial Links Derived from the Malaysian Meteorological Radar Data. PhD Thesis. Universiti Teknologi Malaysia, Malaysia; 2005.

[13] Allnutt J. E. Satellite-to-Ground Radiowave Propagation. Second Edition. *IET Electromagnetic wave series 54*. 2011: 293.

[14] Tseng C. H. et al. Prediction of Ka-band terrestrial rain attenuation using 2-year rain drop size distribution measurements in Northern Taiwan. *Journal of Electromagnetic Waves and Applications.* 2005; 19(13): 1833-1841.

[15] Li Y, Yang P. The permittivity based on electromagnetic wave attenuation for rain medium and its applications. *Journal of Electromagnetic Waves and Applications*. 2006; 20(15): 2231-2238.

[16] Singh M. S. et al. Rain attenuation model for S. E. Asia countries. IET Electronic Letters. 2007; 43(2): 75-77.

[17] Chebil J, Rahman T.A. Rain rate statistical conversion for the prediction of rain attenuation in Malaysia. *Electronic Letts*. 1999; 35: 1019-1021.

[18] Chebil J, Rahman T.A. Development of 1-min rain rate contour map for microwave application in Malaysia Peninsula. *Electronics Letts*.1999; 35: 1712-1774.

17

Birefringence Control in Silicon Wire Waveguide by using Over-etch

Ika Puspita*, A. M. Hatta
Photonics Engineering Laboratory, Department of Engineering Physics, Institut Teknologi Sepuluh Nopember
e-mail: ika.tf10@gmail.com

Abstract

Silicon wire waveguide technology becomes great issue in optical communication system. The high index contrast of the silicon wire waveguide induced the birefringence. It played important role in silicon wire waveguide loss since it caused polarization dependent loss (PDL), polarization mode dispersion (PMD) and wavelength shifting. Hence, controlling birefringence in silicon wire waveguide become very important. The current birefringence controlling techniques by using cladding stress and geometrical variation in bulk silicon waveguide was presented. Unfortunately, it could not obtain zero birefringence when applied to silicon wire waveguide. The over-etching technique was employed in this paper to obtain zero birefringence. The tall silicon wire waveguide obtained minimum birefringence

***Keywords**: Birefringence control, silicon wire waveguide, over-etching*

1. Introduction

Silicon wire waveguide structure was fabricated by depositing the silica layer onto the silicon using plasma-enhanced chemical vapor deposition (PECVD). The deposition process involves temperature of 300o C. Silicon wire waveguide has high refractive index contrast ($\Delta \sim$ 40%). It may lead to birefringence [1]. The presence of birefringence is an important issue in silicon wire waveguide because it can induce polarization mode dispersion (PMD), polarization dependent loss (PDL) and wavelength shifting [2-4]. In silicon wire waveguide, the birefringence is caused by the anisotropic distribution of the stresses induced by thermal coefficient mismatch and lattice defect between core and cladding material during deposition process [5-7].

The birefringence controlling in large cross-section silicon on insulator (SOI) has been reported [8-10]. In [8,9], stress engineering was employed to control birefringence in an Array Waveguide Grating (AWG) by varying upper cladding thickness. The birefringence was also controlled by changing the silicon waveguide geometry, especially the ridge width. Under stress condition, silicon waveguide would have zero birefringence in a wider ridge width [10].

The birefringence in highly-contrast waveguide such as silicon-on-insulator (SOI) waveguide and silicon wire waveguide is largerly dominated by geometrical birefringence [11]. In the other hand, the birefringence in waveguides with over-etch has been studied in [11,12]. The purpose of over-etching process is to create a pedestal. The pedestal could modify the stress distribution in the core [12]. However, the silicon wire waveguide with over-etching has not been studied yet. In this paper, the effect of over-etching to control birefringence in silicon wire waveguide has been studied theoretically and numerically. Since, the dominant birefringence in silicon wire waveguide is geometrical birefringence, the stress birefringence from thermal stress can be neglected.

2. Research Method

Figure 1(a) shows the silicon wire waveguide structure. The waveguide core, typically of 250 nm or 300 nm on thickness and formed from a silicon layer, through photolitography process and etched down to the lower cladding. However, the core dimension was varied to know the each effect of the dimension. The upper cladding is silicon nitride (Si_3N_4) with 1 μm in thickness. While the lower cladding is silicon dioxide (SiO_2) with 3 μm in thickness. The refractive index of silicon, silicon nitride, and silicon dioxide in wavelength of 1550 nm are 3.444, 2.4629, and 1.444 respectively. The numerical approach used Finite Difference Method (FDM) that was performed by using GUI Octave software. The method in this paper divided in four main following steps.

a. Step 1: The core thickness was set fixed. The core width and The over-etched depth would be varied. This step as means to know the effect of the core thickness and over-etched depth to the guided mode in the core region.
b. Step 2: The core width was set fixed. The core thickness and the over-etched depth would be varied. This step as means to know the effect of core thickness and over-etched depth to the guided mode in the core region.
c. Step 3: As the result of the step 1 and step 2, the possible core dimension for guiding the mode was used to do numerical approach of birefringence. This step is used to obtain the smallest birefringence. Figure 1(b) shows the over-etched waveguide.
d. Step 4: The final step is validation by comparing the result in step 3 to the published data. This step is used to know the validity of the over-etched silicon wire waveguide numerical approach.

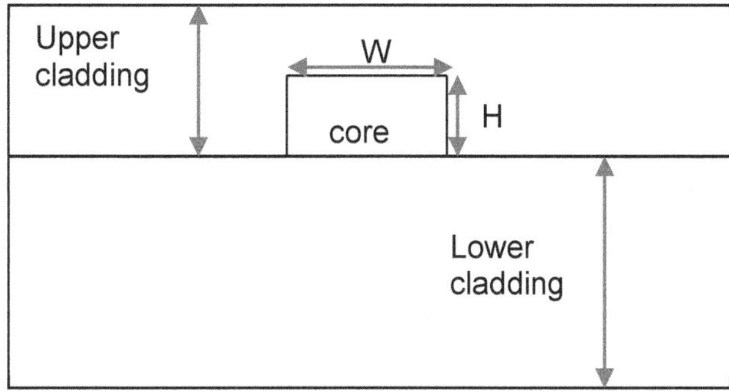

Figure 1(a). Silicon wire waveguide cross-section

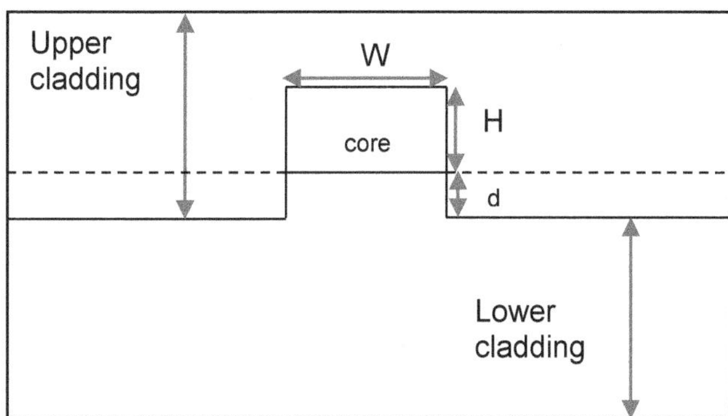

Figure 1(b). Over-etched silicon wire waveguide

3. Birefringence in over-etched silicon wire waveguide

In silicon wire waveguide without over-etch, the core aspect ratio is defined as,

$$a \quad \frac{H}{W} \tag{1}$$

while, the over-etching process in a silicon wire waveguide creates a pedestal that wil modify the aspect ratio of the core region. The over-etching can be assumed as the change of the aspect ratio of the core accompanied by the a change in material composition (as the core material differs from the pedestal material). The aspect ratio of core is redefined as,

$$b \quad \frac{(H+d)}{W} \tag{2}$$

As change in core aspect ratio and also the compition, the effective index of x and y-direction could be change. It induced the change in number of guide mode and birefringence. The birefringence is defined as the difference between effective index of TM and TE polarized fundamental modes [13].

$$B \quad n_{eff}TM - n_{eff}TE \tag{3}$$

The total birefringence in over-etched silicon wire waveguide can be approximated as the sum of the geometrical birefringence and the over-etched birefringence:

$$B_{total} \quad B_{geo} + B_{over} \tag{4}$$

where the geometrical birefringence, defined as the difference in effective indexes corresponding to solutions of the scalar wave equation. The geometrical birefringence is stress-free birefringence. Since the thermal stress was neglected, hence the over-etching birefringence defined as the difference in effective indexes of fundamental mode corresponding to solutions of the wave equation, with the boundary condition as a function of over-etched depth [14].

$$B_{over} \quad 2n_{core} \sin(\theta_2 - \theta_1)\, cos(\theta_1 - \theta_2) \tag{5}$$

where the $\theta_1$and $\theta_2$can be obtained by using eigen value equation of TE and TM modes as follows,

TE Eigen value equation

$$[\mathrm{k}_o \mathrm{n}_{core}(h + d)\cos\theta_1 - m\pi] \quad \tan^{-1}\left[\frac{\sqrt{\sin^2\theta_1 - \left(\frac{n_{lc}}{n_{core}}\right)^2}}{\cos\theta_1}\right] + \tan^{-1}\left[\frac{\sqrt{\sin^2\theta_1 - \left(\frac{n_{uc}}{n_{core}}\right)^2}}{\cos\theta_1}\right] \tag{6}$$

TM Eigen value equation

$$\mathrm{k}_o \mathrm{n}_{effx} w \cos\theta_2 \quad 2\tan^{-1}\left[\frac{\sqrt{\left(\frac{n_{effx}}{n_{uc}}\right)^2 \sin^2\theta_2 - 1}}{\left(\frac{n_{effx}}{n_{uc}}\right)\cos\theta_2}\right] \tag{7}$$

4. Result and Analysis

Silicon wire waveguide typically fabricated in thickness of 250 nm or 300 nm. Table 1 and 2 show the numerical result of the effective index calculation with the width variation.

Table 1. Effective index calculation for thickness of 250 nm

width	TE	TM
300	2.530162	2.448033
350	2.590297	2.470601
400	2.644973	2.492485
450	2.69178	2.512629

Table 2. Effective index calculation for thickness of 300 nm

width	TE	TM
150	2.419512	2.416048
200	2.457573	2.446684
250	2.520665	2.486742
300	2.594384	2.52845

From the table above, the silicon wire waveguide with the thickness of 300 nm has lower differences in TE and TM effective indexes. This thickness was used to do step 1. The width of 350 nm was used to do step 2.

Figure 2 shows the numerical result of step 1. The core thickness was fixed at 300 nm and the width was varied from 150 – 350 nm. The over-etched depth effect added to the calculation. The over-etched depth was varied from 0 – 200 nm. The upper and lower cladding thickness were fixed at 1000 nm and 3000 nm respectively.

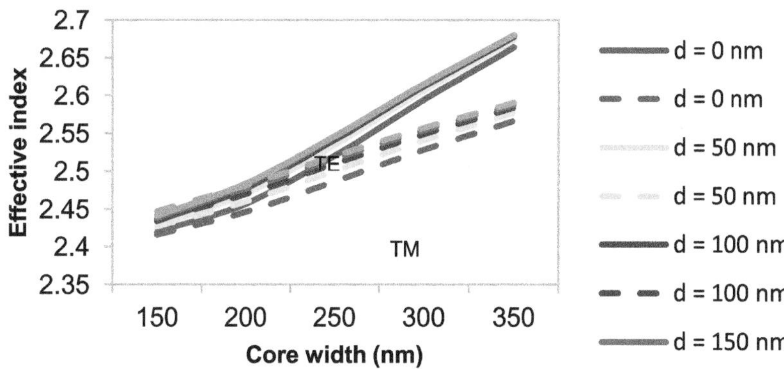

Figure 2. Effective index calculation for step 1

For core thickness range of 150 – 200 nm, no mode guided in the core region. While, for core thickness range of 200 – 350 nm, both TE and TM mode guided in the core region. This result shows that the waveguide with the core width 200 nm and over-etched depth of 200 nm has birefringence of -00253.

Figure 3 shows the numerical approach result of over-etched silicon wire waveguide. The core width was fixed at 350 nm and the thickness was varied from 220 – 300 nm. The over-etched depth was varied from 0 – 200 nm. The upper and lower cladding thickness were fixed at 1000 nm and 3000 nm respectively

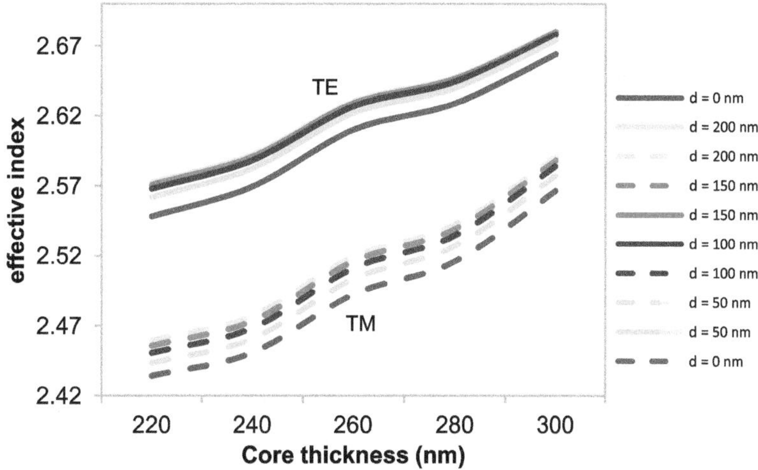

Figure 3. Effective index calculation for step 2

. For core thickness range of 220 – 240 nm, only TE mode guided in the core region. After over-etching process, both TE and TM mode could be guided in the core region. While, for core thickness range of 240 – 300 nm, both TE and TM mode guided in the core. The result shows that over-etching process pulls the mode to the confined part.

From table1, without over-etching, the core width and thickness that possible to guide modes are at 350 nm and 250 nm, respectively. Then the effect of over-etched depth added to the numerical calculation of birefringence. The over-etched depth was varied from 0 – 225 nm. The upper and lower cladding thickness were fixed at 1000 nm and 3000 nm respectively. The result shows in Figure 4.

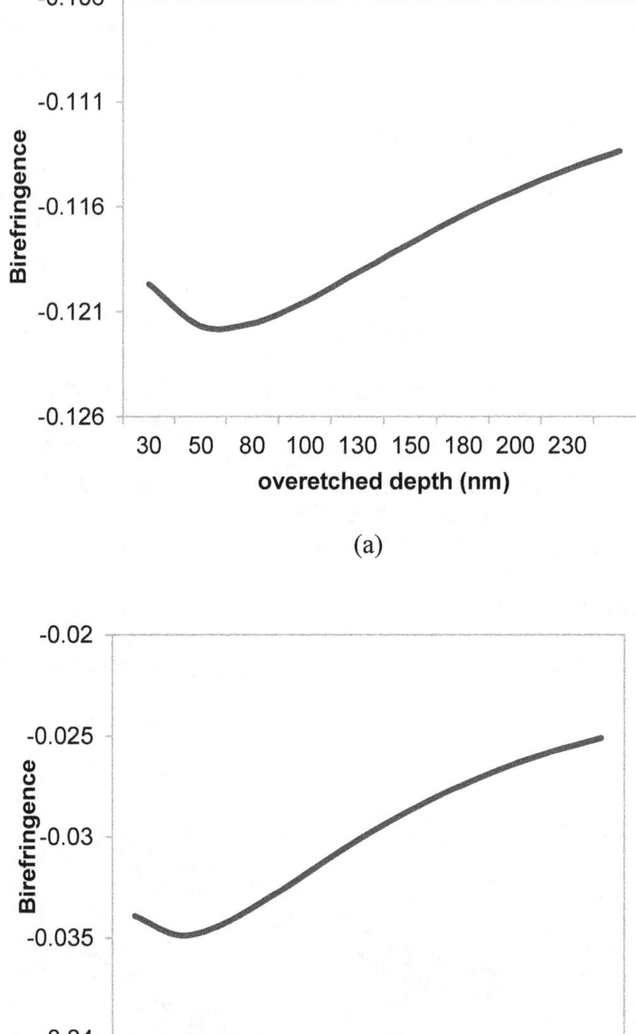

Figure 4. birefringence calculation of (a) 350 nm in width and 250 in thickness (b) 250 in width and 300 nm in thickness

Figure 4(b) shows the numerical result of the over-etched silicon wire waveguide with the core width and thickness was fixed at 250 nm and 300 nm, respectively. The over-etched depth was varied from 0 – 225 nm. The upper and lower cladding thickness were fixed at 1000 nm and 3000 nm respectively.

The small birefringence value could be obtained by changing the core width to be narrower. The adding of over-etched part would pull the non-guided mode to the confined part. The tall rectangular core dimension has lower birefringence than the wide one.

5. Conclusion

The effect of over-etching in silicon wire waveguide has been studied theoretically and numerically. The thermal stress effect to the waveguide was neglected since the dominant birefringence is geometrical birefringence. Without over-etched the tall waveguide core has lower birefringence than the wider one. The ove-etching pull the unguided mode to the confined region and minimize the birefringence in silicon wire waveguide. The smallest birefringence is -0.00253. It could be obtained by the waveguide with the core dimension of 200 nm in width and 300 nm in thickness. The given over-etched depth is 200 nm.

Acknoledgement

This paper work is supported by ITS Fresh graduate Student Scholarship from Indonesia Goverment.

References

[1] H Yamada, T Chu, S Ishida, Y Arakawa. Si photonic wire waveguide devices. *IEEE J. Quantum Elect.* 2006; 12(6): 2006: 1371-1379.

[2] JN Damask. Polarization optics in telecommunication, Chapter 8, New York: Springer. 2005: 297 – 383.

[3] M Huang and X Yan. Thermal-stress effects on the temperature sensitivity of optical waveguides. *J. Opt. Soc. Am. B.* 2003; 20(6): 1326-1333.

[4] S Janz, P Cheben, H Dayan, and R Deakos. Measurement of birefringence in thin-film waveguides by Rayleigh scattering. *Opt. Lett.* 2003; 28(19): 1778-1780.

[5] Huang M. Stresses effect on the performance of optical waveguide. *Int. J. Sol. and Struct.* 2003; 40: 1615 - 1632.

[6] Huang M. Thermal stresses in optical waveguide. *Opt. Lett.* 2003; 28(23): 2327-2329.

[7] WD Callister. Material Science and Engineering: An Introduction Seventh Ed, Chapter 6, pp. 137 – 139, USA: John Wiley and Sons Inc, 2007.

[8] WN Ye, et.al. Stress-induced birefringence in silicon-on-insulator (SOI) waveguides. *Opt.elect Integrat on Si. proc. SPIE.* 2004; 5357: 57-66.

[9] J Lockwood and L Pavesi, Silicon photonics II component an7d integration. Springer: Heidelberg. 2010; 119.

[10] DX Xu, et.al. Eliminating the birefringence in silicon-on-insulator ridge waveguide by use of cladding stress. *Opt. Lett.* 2004; 29(20): 2384-2386.

[11] Dumais P. Thermal stress birefringence in buried-core wavguide with over-etch. *IEEE J. Quant. Elect.* 2011; 47(7): 989 – 996.

[12] Dumais P. Modal birefringence analysis of strained buried-core waveguides. J Lightwave Tech. 2015; 30(6): 906-912.

[13] K Okamoto, Fundamental of Optical Waveguides, Academic Press, USA, 2000.

[14] Schriemer HP. Modal birefringence amd power density distribution in strained buried-core square waveguides. *EEE J. Quant. Elect.* 2004; 40(8): 1131 – 1139.

18

Modeling of Six Pulse Voltage Source Inverter based STATCOM with PWM and Conventional Triggering

Sachin Sharma, Alok Pandey, Nitin Kumar Saxena*
MIT Moradabad, UP India
e-mail: nitinsaxena.iitd@gmail.com

Abstract

In the present study, a six pulse inverter is triggered using conventional sequential and pulse width modulation (PWM) technique simultaneously for comparing their results. FACTS devices such as STATCOM are the most emerging areas in today's power system. This STATCOM is a power electronic based converter circuit in which inverter circuit is fabricated for the purpose of voltage control and reactive power compensation through its current flow. A six pulse inverter is developed that can be generalized for developing any number of multiple of six pulse inverters by cascading them in parallel using transformers connection. Voltage source Inverter (VSI) is the most commonly used inverter that can be modeled for electrical system based studies. A methodology is presented to trigger the STATCOM inverter circuit using conventional and PWM techniques, and then the results are compared using MATLAB simulink model.

***Keywords**: Six Pulse Voltage Source Inverter, IGBT, DC source, PWM, Active and Reactive Power*

1. Introduction

In today's scenario the different kind of industrial machines, household electronic gadgets design to operate with AC but their operating performances go down due to the various power quality issues. The use of power electronic devices are the major cause for poor output waveform. In industry the required AC is actually converted by rectified DC which contains the harmonics which leads electrical power and temperature losses [1, 2]. Thus a suitable converter is used after filtering the rectified DC to operate AC operated gadgets [3]. In industry mostly inverter takes input from rectifiers and they have further convert into AC at usable places. Also FACT DEVICES which are playing an important role in power quality improvement in power transmission and renewable energy transmission systems are also uses the six pulse voltage source inverters for transmission of power [4]. And in dc power transmission system the ac power first rectified and then transmitted over the long distances but for use of ac power at domestic and industrial level a circuit is required to convert it into ac so voltage source inverter is also uses for the purpose of converting the power in power transmission system. A six pulse voltage source inverter gives an ac wave nearly equal to the sine wave which can further be used in many applications [5-8]. A six pulse voltage source inverter is can be consist thyristors, GTOs, IGBTs such devices. Here we used the fully controlled devices i.e. IGBT. The circuit consist six IGBTs (Insulator Gate Bipolar Transistor), a dc source and triggering circuits. A single phase inverter are also exists but three phase is familiar in industrial applications. Six pulse voltage source inverter is also be used in power grid devices like in STATCOM as converter circuit to provide six pulse ac to the flexible transmission lines. And this six pulse ac power can be increase by means of pulse by cascading the inverter [9, 10].

2. Three Phase Six Pulse Voltage Source Inverter (VSI)

The three phase six pulse voltage source inverter is used to convert the DC power into AC power at adjustable frequency. And the inverter can be used for providing ac supply to 3 phase load or gadgets and it can be used as the converter in STATCOM, a FACTS device. Figure 1 shows the basic block diagram of conversion of power in industry or in power transmission system.

VSI is an important part of STATCOM for conversion of power used in power system. Figure 2 shows the basic block diagram of STATCOM using six pulse VSI connected to power transmission system.

Here we uses the IGBT based VCI, IGBTs are fully control device and they are turned ON only for the duration during which a gate pulse is given to it. The Figure 3 shows the basic diagram of a six pulse three- phase inverter. In this model, six IGBTs are connected in such a way that each phase of 3-Phase balance load is connected in middle of two IGBTs and they are triggered properly in a particular sequence as the name with number given to each IGBT. And another side is connected to the DC voltage source. A six pulse VSI has less harmonic content than that of directly obtained single pulse ac [11].

Conversion of firing angle in time scale from radian or degree is required for MATLAB simulink model and therefore, for 50 Hz frequency system, firing angle can be expressed in time scale using the expression given in Equation (1). Table 1 represents the firing angle conversion in time scale using this expression.

$$t = \frac{0.02}{360} d \tag{1}$$

Where,

t: Firing angle in time scale

d: Firing angle in degree

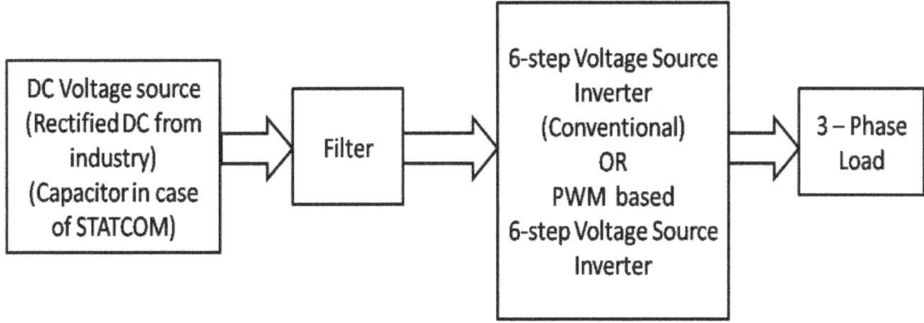

Figure 1. Block diagram of VSI connected to 3- Phase load

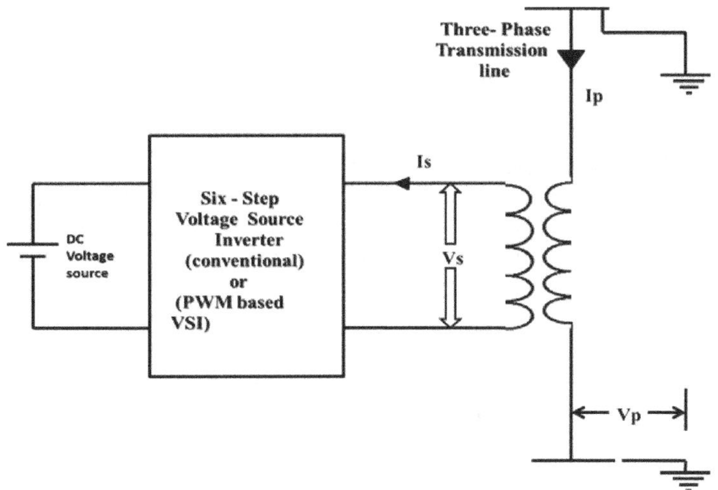

Figure 2. Block diagram of VCI connected as a element of STATCOM.

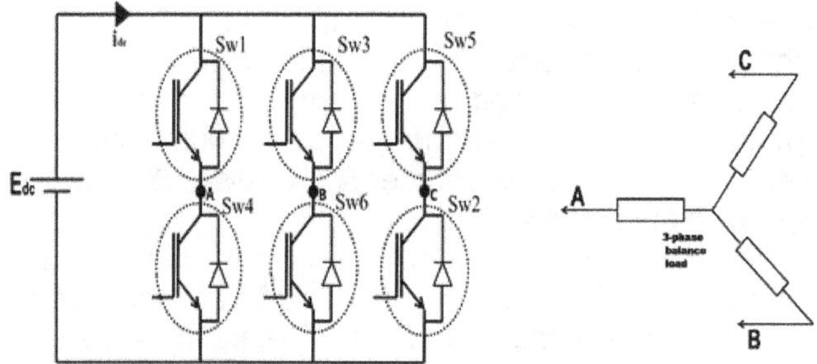

Figure 3. Basic structure of three-phase six-pulse voltage source inverter.

Table 1. Angles of triggering and their corresponding time values.

No.	Firing angle (Angle)	Correspond value in time scale
1.	30o	0.00167
2.	60o	0.00333
3.	90o	0.005
4.	120o	0.0067
5.	150o	0.0083
6.	180o	0.01
7.	210o	0.0116
8.	240o	0.0133
9.	270o	0.015
10.	300o	0.0167
11.	330o	0.0183
12.	360o	0.02

The voltage source inverter may work in two different possible modes. In one mode, each IGBT conducts for 180o and in another mode each IGBT conducts for 120o. In 180o mode the gate pulses are given continuously for 180o to each IGBT and in 120o mode the gate pulses are given continuously for 120o to each IGBT. A triggering sequence is the important factor that must be in consideration for smooth conduction of six pulse output [12, 13].

3. Working Methodology

3.1. Operation of VSI for 180° mode

In 180o mode of operation three IGBTs are worked together at the same time. The pulses are given for the 50% duration of a complete cycle. Each works for every 60o of a complete cycle with the help of specific triggering operation, in which gating signals are applied and removed at 60o intervals. The triggering operation on which IGBTs work sequentially is shown on the following Table 2. In 180o mode of operation, Table 3 shows the sequence of triggered IGBTs in each cycle of 60o. Their corresponding output phase voltages are also mentioned in this Table.

Table 2. The angles duration for which corresponding IGBTs are in working

Conducting Switch	Sw1 Sw6 Sw5	Sw1 Sw6 Sw2	Sw1 Sw3 Sw2	Sw4 Sw3 Sw2	Sw4 Sw3 Sw5	Sw4 Sw6 Sw5
Angle Duration	0° to 60°	60° to 120°	120° to 180°	180° to 240°	240° to 300°	300° To 360°

Table 3. Values of three phase voltages corresponding to input dc voltage and the working switches

Phase Voltages			Conducting Switch
AN	BN	CN	
Vs/3	-2Vs/3	Vs/3	[1.6.5]
2Vs /3	-Vs/3	-Vs/3	[1,6,2]
Vs/3	Vs/3	-2Vs/3	[1,3,2]
-Vs/3	2Vs/3	-Vs/3	[4,3,2]
-2Vs/3	Vs/3	Vs/3	[4,3,5]
-Vs/3	-Vs/3	2Vs/3	[4,6,5]

3.2. Operation of VSI for 120° mode

In 120o mode of operation two IGBTs are worked together at the same time. The pulses are given for 33.33% of the full cycle. Each works for every 60o of a complete cycle with the help of specific triggering operation, in which gating signals are applied and removed at 60o intervals. The triggering operation on which the IGBTs works sequentially is shown on the following Table 4. In 120o mode of operation, Table 5 shows the sequence of triggered IGBTs in each cycle of 60o. Their corresponding output phase voltages are also mentioned in this Table.

Table 4. The angles duration for which corresponding IGBTs are in working

Conducting Switch	Sw1 Sw6	Sw1 Sw2	Sw2 Sw3	Sw4 Sw3	Sw4 Sw5	Sw5 Sw6
Angle Duration	0° to 60°	60° to 120°	120° to 180°	180° to 240°	240° to 300°	300° to 360°

Table 5. Values of three phase voltages corresponding to input dc voltage and the working switches

Phase Voltages			Conducting Switch
AN	BN	CN	
Vs/3	-2Vs/3	Vs/3	[1,6]
2Vs /3	-Vs/3	-Vs/3	[1,2]
Vs/3	Vs/3	-2Vs/3	[3,2]
-Vs/3	2Vs/3	-Vs/3	[4,3]
-2Vs/3	Vs/3	Vs/3	[4,5]
-Vs/3	-Vs/3	2Vs/3	[6,5]

3.3. VSI triggering using PWM

In this method a PWM generator is modeled for triggering purpose. This is achieved by comparing sine wave with repeated sequence signal. VSI give the output voltages in the form of modulated voltages [14]. Figure 4 gives a basic block diagram of VSI triggering circuit using PWM.

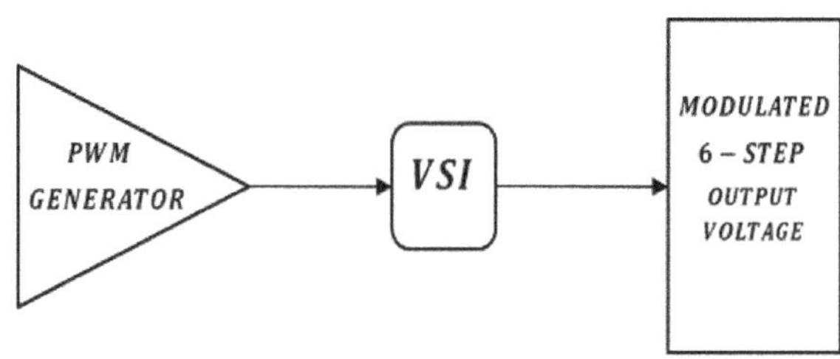

Figure 4. Block Diagram of triggering of VSI by PWM generator.

4. Developement Of Simulink Models

For the study of six pulses VSI two methods as discussed above are modeled. In first model conventional triggering method is used in which six IGBTs are triggered individually by pulse generator in a right sequence of 60o phase difference. A complete simulink model is represented in Figure 5. A 100 volt dc source is connected to convert into ac through this inverter. A three phase resistive load is connected to output side. The diagram also a three vase measurement block which measures the output three phase (phase to ground) voltages and current whose result is shown on scope. There is also an active and reactive power measurement block which measures the active and reactive powers which further measures individually by the use of DMUX.

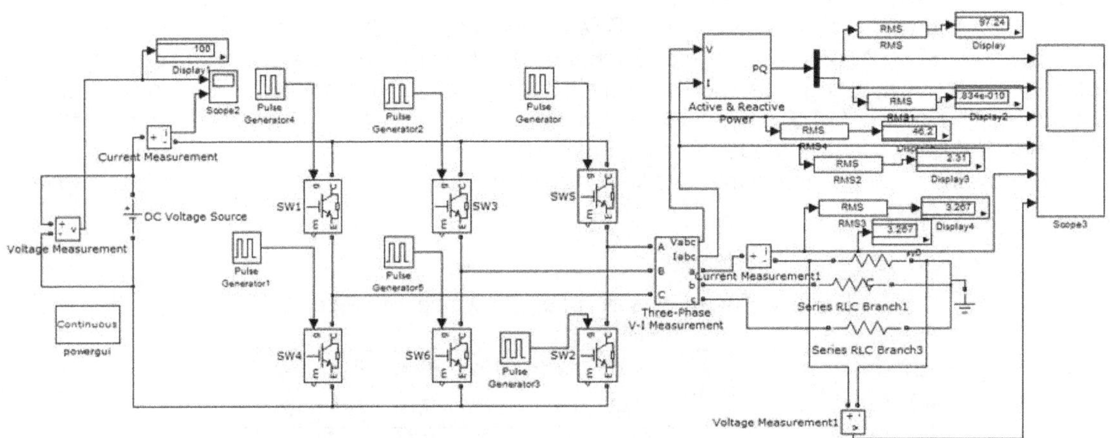

Figure 5. Simulink model of six pulse VSI with resistive load

In second model, VSI is triggered using PWM technique. A complete simulink model is represented in Figure 6. In this triggering technique, IGBTs' switches SW1, SW2 and SW3 are triggered by PWM generator while the IGBTs' switches SW2, SW4 and SW6 are triggered through PWM generator after passing through NOT gate to provide complement signals of PWM generated signals. In Figure 6 PWM generator is shown using its subsystem while detail PWM generator simulink model is further elaborated in Figure 7. Two sine wave signals are added first in which one is working as harmonic signals whose frequency is 3 times of the main sine signal and then compare with the repeating sequence signal after delaying through transport delay.

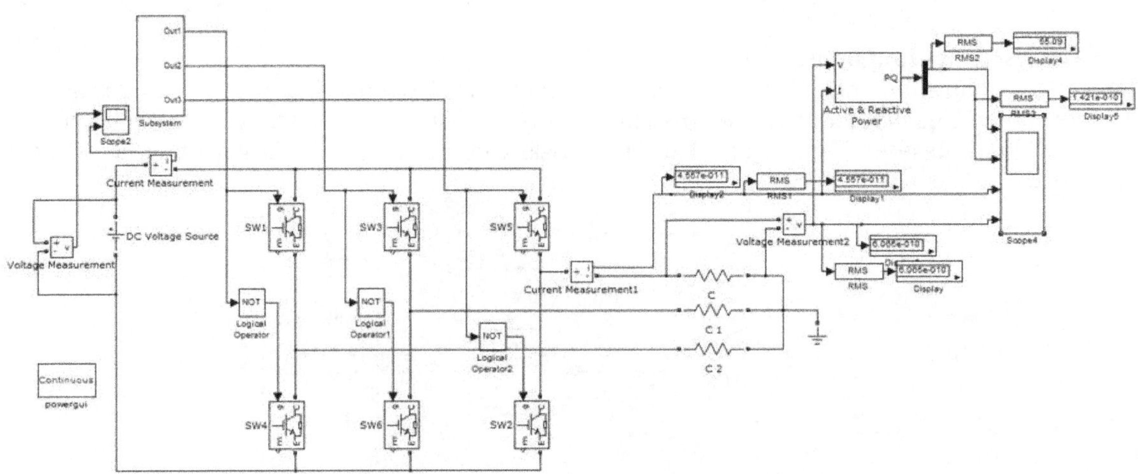

Figure 6. Simulation Diagram of 6-pulse PWM based VSI with R load

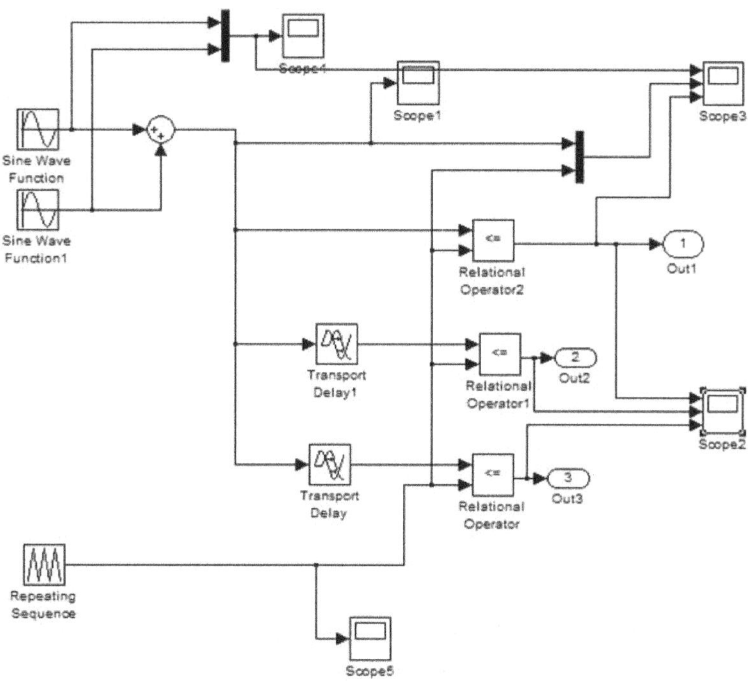

Figure 7. Simulation Diagram of PWM generator

5. Results and Discussion

In this paper, STATCOM internal configuration is represented and modeled using MATLAB simulink blockset. This STATCOM is developed with six pulse voltage source inverter circuit. Results for two triggering procedures, conventional and PWM techniques are elaborated for resistive load. A generalized approach is presented to develop STATCOM simulink model that can be expanded for any number of pulses keeping the fact in mind that more pulses can improve the waveform of alternating output. The neccessacity of output performance helps to decide the exact number of pulses of VSI because more pulses includes the power quality problems along with their complex structure.

For six pulse VSI based STATCOM, triggering results by both techniques are compared. Table 6 compares the output voltage, current, real power and reactive power for both the techniques for resistive load of 20Ω. Results are comparable from both the techniques. The results are also verified by their response waveforms. Figure 8 and 9 represent the output voltage and current for conventional triggering and Figure 10 and 11 represent the output voltage and current for PWM technique.

Table 6. Result comparison for conventional and PWM techniques

Method of triggering	DC input voltage (V)	Peak value of Single phase output voltage (V)	Peak value of Single phase output current (A)
Conventional	100	65.33	3.267
PWM	100	65.33	3.267

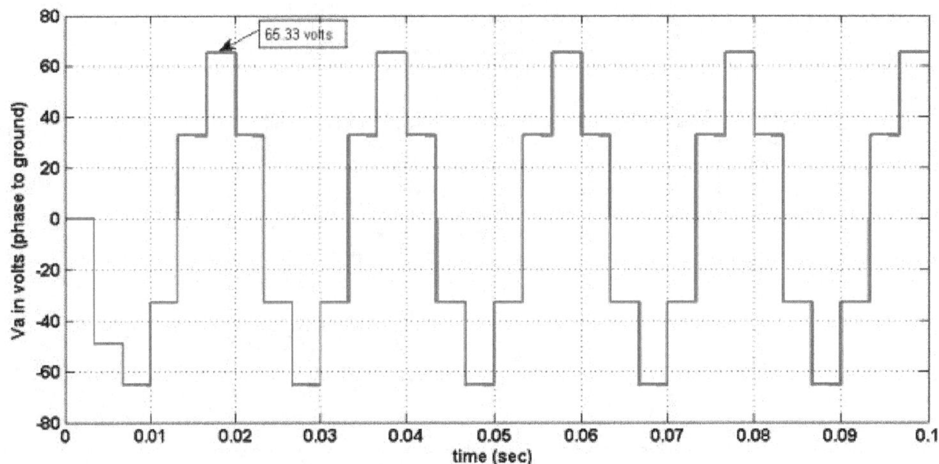

Figure 8. Voltage response of six pulse VSI based STATCOM using conventional triggering

Figure 8 shows that the single phase voltage obtained from conventional triggering and its shape is approaching towards sinusoidal wave in steps with maximum voltage of 65.33 volt (peak value). Figure 9 shows the similar nature of wave with peak value of 3.267amp.

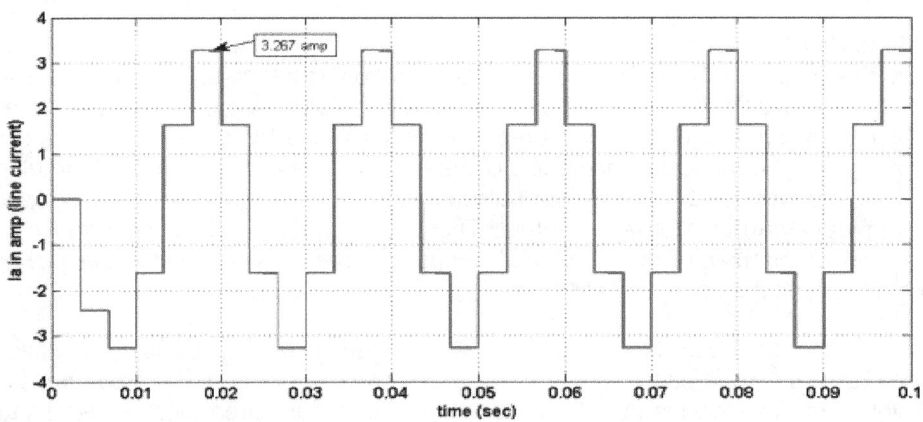

Figure 9. Current response of six pulse VSI based STATCOM using conventional triggering

For triggering through PWM technique, Figure 10 and 11 represent voltage and current responses at output respectively. It must be noted that response is similar to convenmtional triggering exceopt the width of the pulses that make a waveform approaching towards sinusoidal nature.

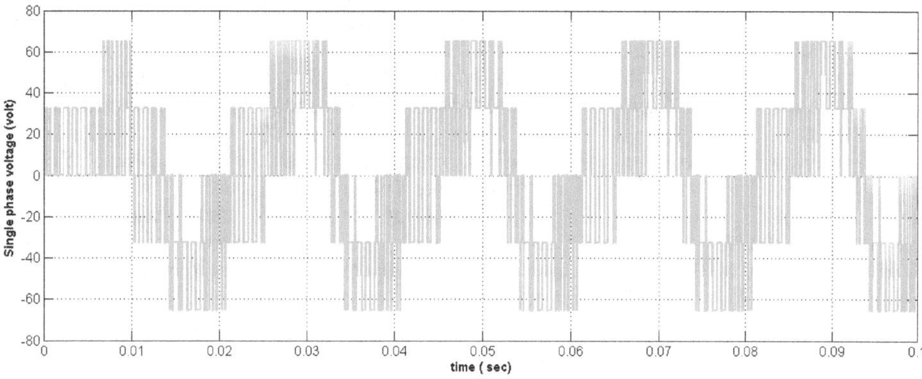

Figure 10. Waveform of phase voltage of six- pulse VSI with Resistive load with PWM triggering.

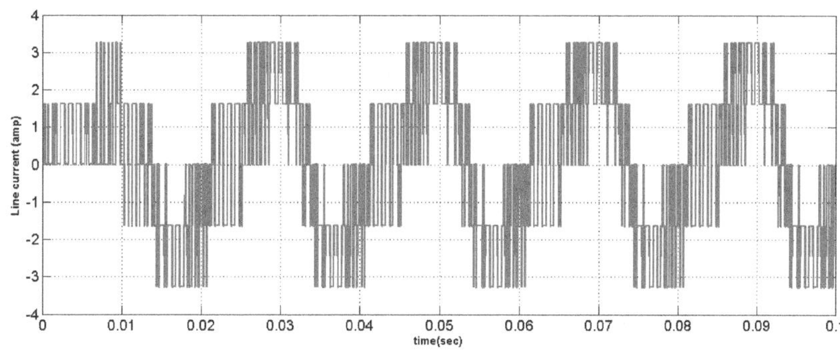

Figure 11. Waveform of phase output line current of six- pulse VSI with Resistive load with PWM triggering.

6. Conclusion

Triggering procedures play an important role in modeling and hence a comparative study for two triggering procedures are modeled for six pulse voltage source inverter based STATCOM. Results are compared in terms of output voltage, current, and power. A generalized procedure for developing any number of pulses STATCOM is depicted in this paper. Since in advance power system based studies, STATCOM is one of the frequently used device that helps in voltage control, reactive power compensation and power transfer capability studies, therefore a systematic studies to analyses its model and its expansion is mainly elaborated in this work.

References

[1] B Suechoey, S Tadsuan and S Bunjongjit. A Comparison of Power Losses and Winding Temperature of Three-phase Induction Motor with PWM Inverter and Six-Step Inverter Supply.

[2] A Boglietti, P Ferraris, M Lazzari and F Profumo. Effect of Different Modulation Index on the Iron Losses in soft Magnetic Materials Supplied by PWM Inverter. *IEEE Transaction on Magnetics*. 1993; 29(6): 3234-3236.

[3] Mihaela Popescu, Alexandru Bitoleanu, and Mircea Dobriceanu. On the AC-Side Interface Filter in Three-Phase Shunt Active Power Filter Systems. *World Academy of Science, Engineering and Technology International Journal of Electrical, Computer, Energetic, Electronic and Communication Engineering*. 2010; 4(10).

[4] Yun Xu, Yunping Zou, Wei Chen, Chengzhi Wang, Xiong Liu and Feng Li. A novel STATCOM based on hybrid cascade multilevel inverter. *IEEE International Conference on Industrial Technology*. 2008.

[5] Sungho Jung and Jung-Ik Ha. *Low Voltage Modulation Method in Six-step Operation of Three Phase Inverter*. 9th International Conference on Power Electronics-ECCE Asia June 1 – 5. 2015 / 63 Convention Center, Seoul, Korea.

[6] Mohamed H Saied, MZ Mostafa, TM Abdel- Moneim, and HA Yousef. *New Three-Level Voltage Source Inverter With Different 25 Space Voltage Vectors*. 2006 2nd International Conference on Power Electronics Systems and Applications.
[7] Wang Jinyu, Ye Xitai, Liu Junli. *The Research of A Novel Single-Phase Hybrid Asymmetric 6-level Inverter Based on Epwm Control*. 2nd International Conference on Measurement, Information and Control. 2013.
[8] Kenta Kawasugi, Shigeo Masukawa. Development of Quasi 24-step Voltage Source Inverter. Department of Electrical and Electronic Engineering, Tokyo Denki University, Japan.
[9] Cheng-Han Hsieh, Tsorng-Juu Liang, Shih-Ming Chen, and Shih-Wen Tsai. Design and Implementation of a Novel Multilevel DC-AC Inverter. *IEEE Transactions on Industry Applications*.
[10] NNV Surendra Babu, BG Fernandes. Cascaded two-level inverter-based multilevel static VAr compensator using 12-sided polygonal voltage space vector modulation. *IET Power Electronics*.
[11] A Mahechwari, KDT Ngo. Synthesis of six-step pulsewidth-modulated waveforms with selective harmonic elimination. *IEEE Trans. Power Elec.*1993; 8(4): 554-561.
[12] KB Khanchandani, MD Singh. Power Electronics Second Edition, Tata McGraw Hill Education Private Limited. 2006: 565-579.
[13] PS Bhimbhra. Power Electronics. Third Edition, Khanna Publishers, 2005: 488-497.
[14] Yanlei Zhao. Design and implementation of inverter in dynamic voltage restorer based on selective harmonic elimination PWM. in DRPT. 2008: 2239 – 2244.

19

Universal Data Logger System for Environmental Monitoring Applications

Osman Abd Allah[1]*, Mohammed Abdalla[2] , Suliman Abdalla[3] , Amin Babiker[4], and Alaa Awad Allah[5]
[1,2,5]Sudan Academy of Science, Khartoum, Sudan
[3]Sudan Atomic Energy Commission, Khartoum, Sudan
[4]Faculty of Engineering, Alneelain University, Khartoum, Sudan
e-mail: abudahdah@live.com

Abstract

Collecting huge amount of data in long time acquisition systems like in environmental monitoring, there is a need to collect and save data over time for further use or analysis. A data logger is an electronic device that records data over time or in a relation to location either with a built-in instrument, sensor or via an external instruments and sensor. In this paper, a data logger system is designed to use as a stand-alone or computer based device. When used as a standalone system, all data acquired are saved in SD memory card, which must be copied and erased periodically depending on the memory size. When used as computer based device, all the data sent to the computer via the serial port and stored automatically in achieved files. The limit of those files size only restricted by the capacity of the disk. The data logger is designed using an Arduino UNO board and LabView software, and it has the flexibility to set it up for different user options. With this system, the user could be able to record and read back sensory data to or from existing files, or in automatically generated files and plot these readings in a graph. Also, the user have the ability to choose the periodic time at which a sample record in a file in term of seconds, minuits or hours. The system designed to monitor and record a single channel data, but it could be adapted to monitor more than one channel.

***Keywords**: Data Logger, Monitoring, SD memory, Arduino, LabView*

1. Introduction

A data logger is an electronic instrument that records environmental parameters such as temperature, relative humidity, wind speed and direction, light intensity, water level and water quality over time[1,2]. Typically, data loggers are compact, battery-powered devices that are equipped with microprocessor input channels and data storage. Most data loggers utilize turnkey software on a personal computer to initiate the logger and view the collected data. The use of data loggers for environmental monitoring became common during the 1980's; coinciding with the explosion in personal computers (PC's), since a data logger system consists of many of the same, or similar, components used to manufacture a PC. Before then, chart recorders were commonly used as well as manual measurements. Both of these methods were labor intensive and time consuming so the advent of stand-alone data loggers was welcomed. [3]

A data logger used to collect readings, or output, from sensors. These sensors could be measuring industrial parameters such as pressure, flow and temperature or environmental parameters such as water level, wind speed or solar radiation. Today there are sensors available, which can virtually measure any physical parameter. Sensors developed to measure gas pressure within human cells to cloud height and density[4].

Data loggers in general needs eight basic components to achieve a proper function[5]:

a. Input channel: The output from a sensor connected to a data logger channel. A channel consists of circuitry designed to 'channel' a sensor signal (typically a voltage or current) from the sensor to the data logger processor. A single data logger can have a variety of channel types and from one to many channels (multi-channel data logger) – one channel is required for every sensor signal output.
b. Analog to Digital Converter: All sensor signals must be in binary format for the data logger system to recognize and record. Since most of the sensors output is analog, they must be digitized using an ADC before feeding them to the system.

c. Microprocessor: A processor is the logic circuitry that responds to and processes the basic instructions that drive a computer or data logger.
d. Memory: for storing the data
e. Power supply to power up the data logger system
f. Data output port: Most data loggers communicate with a PC via a serial port, which allows data to be transmitted in a series (one after the other). The RS-232 interface has been a standard for decades as an electrical interface between data terminal equipment, such as a PC, and data communications equipment employing serial binary data interchange, such as a data logger or modem.
g. Whether proof enclosure: Since environmental data loggers are usually installed in remote and harsh locations, a weatherproof case is an obvious requirement. Most loggers are packaged in weather resistant plastic cases.
h. Software: Proprietary software is usually required to program and download data from a data logger. Data logging functions such as sensor scan rate and scaling, log interval, communication protocol and output format (Excel, ASCII, plot, etc.) are programmed using software loaded on a PC.

Most often, data logger systems are portable stand-alone devices, which allocated to measure and record sensory data, then connected to a PC to analyze data. In this work, the system could be used as a stand-alone –as in normal cases- or use any general PC with the software installed to act as a data logger system.

2. Materials and Method

The designed data logger system consists mainly from Arduino UNO board as the main controller and data acquisition system. The system represents a single channel that record the temperature over time, which acquired from an LM35 temperature sensor. The system equipped with an SD memory card adaptor, which loaded with a micro SD card for storing files in stand-alone mode. Figure 1 below shows the system components used in the design.

Figure 1. the system components

The adaptor connected to the board and exchange data through the standard SPI interface, and it has the "CS" pin which determined according to the adaptor type or brand and in this case its "8"[6] . In this mode, the card has to be formatted first in FAT/16 or FAT/32 system before using it with the board. The SDFormatter is a suitable program used to format the card [7]. Then the card tested and initialized from the Arduino software and a text file is created in the memory card to store data. Of course, there could be many files in the memory card to save data, but only one file could be opened and if another file is targeted, the first one must be closed before opening the new one. The size of the memory eliminates the files size, and the acquisition time determine the duration at which the memory came to an overflow point. For these reasons, the memory must be copied and cleared periodically.

In the computer-based mode, the board send the data to the PC through the serial port and the system designed using the LabView software from National Instruments, which offers huge graphical programming capabilities makes data logging more flexible and powerful.

LabView block diagram enables dealing with files operations like saving, clearing or even creating files from the program, as well as monitoring and plotting data in a single user interface. The system has the ability to save data in an existing text file, or in a file that created from the program automatically. It also has the option to clear or read any saved file manually through control buttons.

The system could be developed by adding many features such as printing a specific file, determine a critical reading, searching for specific measures in a specific period of time, compare file contents with another, sending data to e-mail or even plotting files data online using IoT applications.

3. The System

The temperature sensor LM35 connected to the board in the analog pin A0 to read ambient temperature and many other sensors could be added to the design in the same way as desired. In stand-alone mode, the card module connected to the SPI interface, which consists of the standard signals MOSI, MISO, SCK and additional chip select pin (CS) which determined by the type and brand of the adaptor. The board itself connected to the PC through a USB connector used to program the board and send data to the LabView software.

The computer-based system consists of many user-defined options described below.

The read/clear option enables the user to read any file and display its contents in an indicator. It also enables the user to clear any existing file to start over saving.

Figure 2. File read/clear options

Saving files option offers two options for saving data, either saving data in a an existing file which called "One File" option, or saving data daily in separate files in the option " Daily". In the first option, the user creates a text file in any directory in the PC and choose that file for saving. In the second option, the system automatically creates a file and name it with the current date, when a new day exists, a different file generated with the new day name.

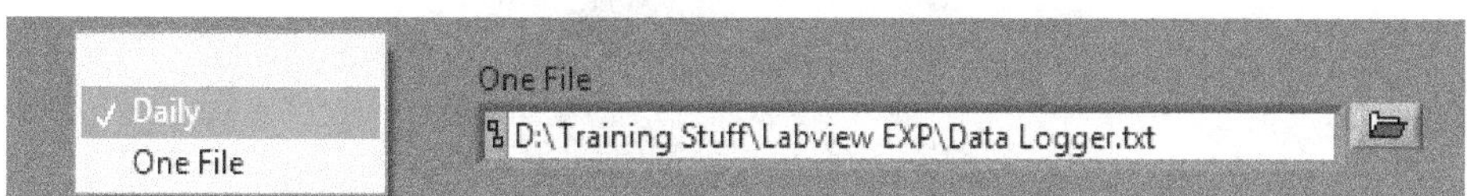

Figure 3. Saving files options

Interval option enables the user to choose an interval scale for saving data either in seconds, minuits or hours and an interval period to determine the specific period for saving data. For example, if the user choose "minuits" in the "interval scale" control and "5" in the "interval period" control, the system will save a sample data every five minuits.

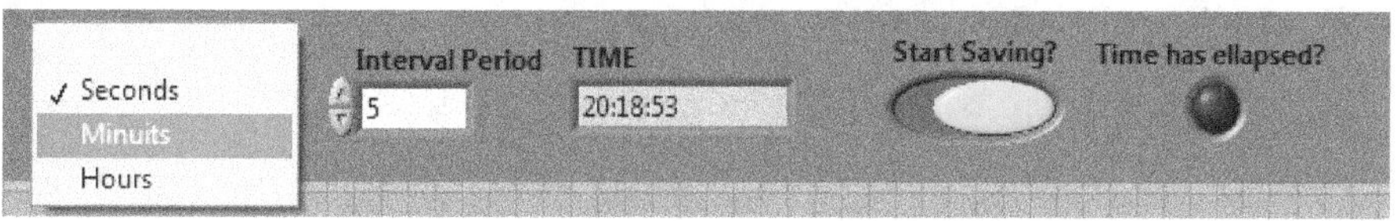

Figure 4. Sample saving intervals options

In addition, there is an indicator for the local time and a control for start saving data. The LED indicator blinks once when the predefined period for saving data has elapsed.

Monitoring option shows a live reading for the temperature and a graph for plotting the readings instantly.

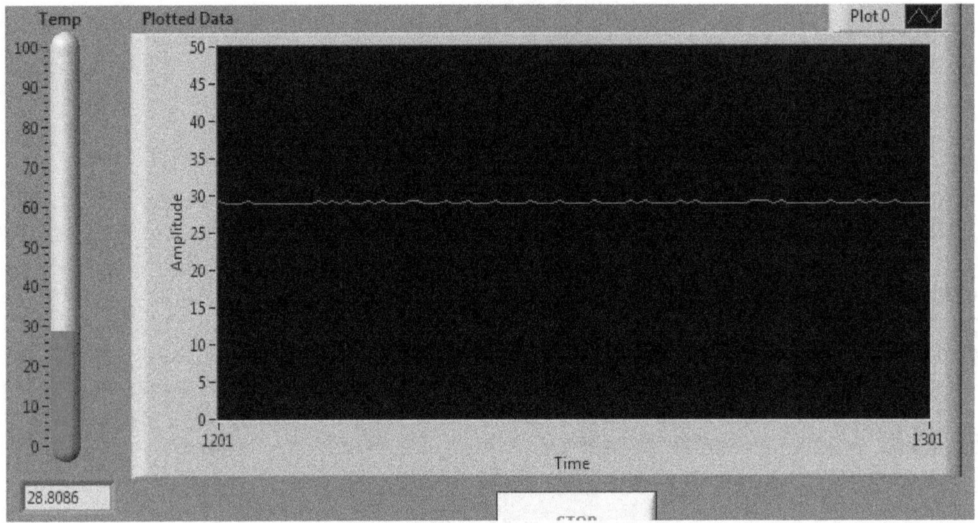

Figure 5. The temperature reading and data plotting

Figure 6 shows the designed block diagram in LabView program, and figure 6 shows the circuit.

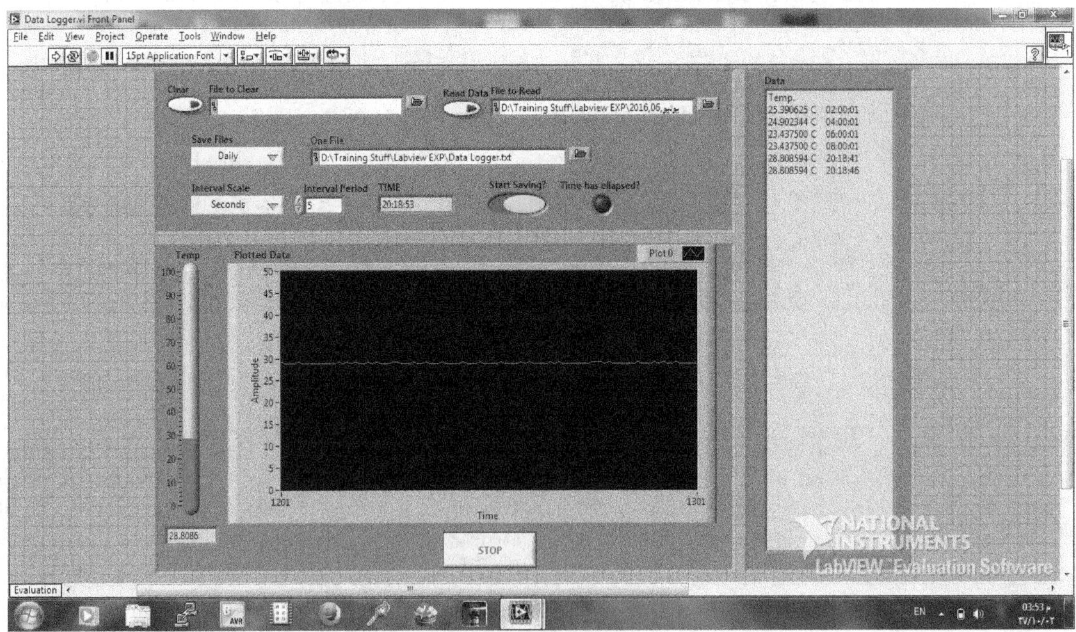

Figure 6. The Data logger system in LabView

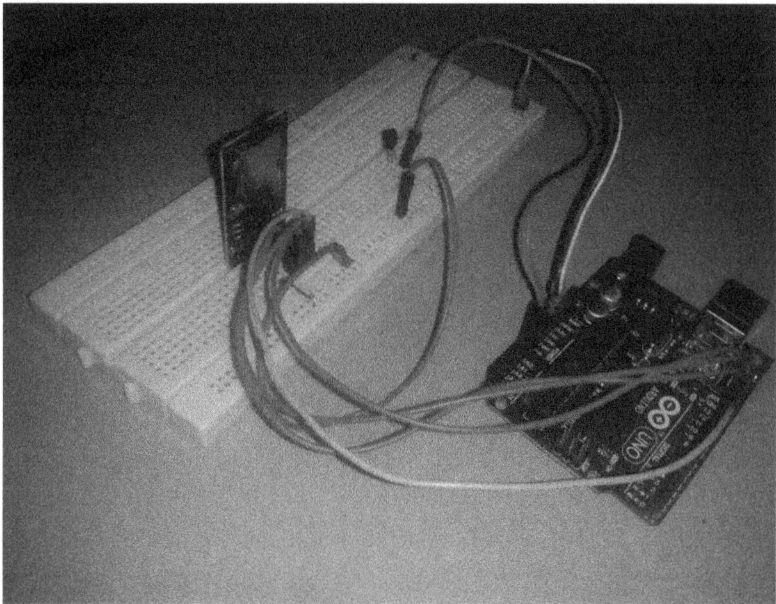

Figure 7. the Data Logger circuit

The figure 8a and 8b shows the files that used to store data while measuring temperature.

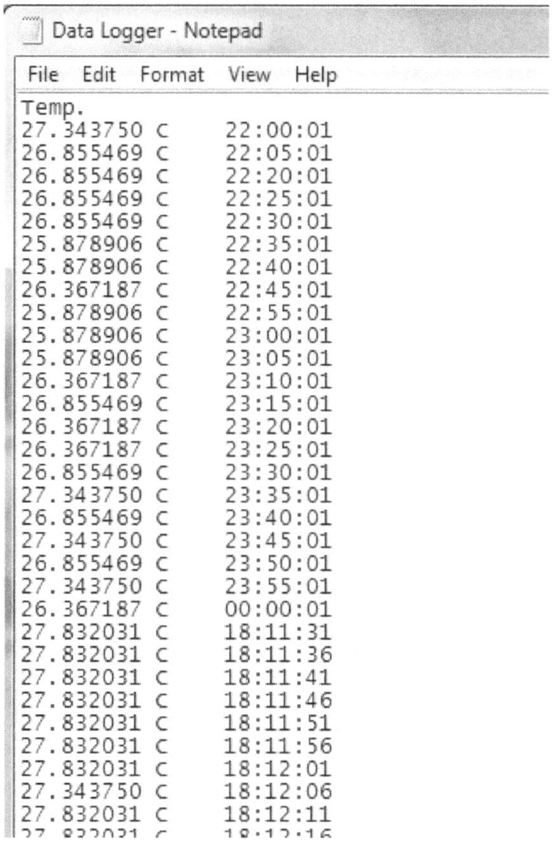

```
Data Logger - Notepad
File  Edit  Format  View  Help
Temp.
27.343750 C     22:00:01
26.855469 C     22:05:01
26.855469 C     22:20:01
26.855469 C     22:25:01
26.855469 C     22:30:01
25.878906 C     22:35:01
25.878906 C     22:40:01
26.367187 C     22:45:01
25.878906 C     22:55:01
25.878906 C     23:00:01
25.878906 C     23:05:01
26.367187 C     23:10:01
26.855469 C     23:15:01
26.367187 C     23:20:01
26.367187 C     23:25:01
26.855469 C     23:30:01
27.343750 C     23:35:01
26.855469 C     23:40:01
27.343750 C     23:45:01
26.855469 C     23:50:01
27.343750 C     23:55:01
26.367187 C     00:00:01
27.832031 C     18:11:31
27.832031 C     18:11:36
27.832031 C     18:11:41
27.832031 C     18:11:46
27.832031 C     18:11:51
27.832031 C     18:11:56
27.832031 C     18:12:01
27.343750 C     18:12:06
27.832031 C     18:12:11
```

Figure 8a. Saving data in "One File" option

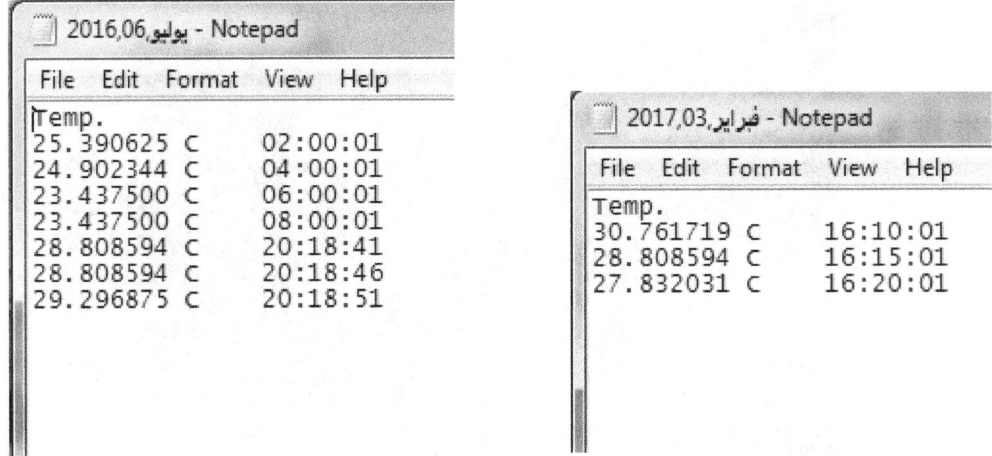

Figure 8b. Saving data in "Daily" option

Figure 8a shows the data stored in a file named as "Data Logger" which is an existing file used to store data in the "One File" option. From the time column, it is clear that the periodic time chosen by the user to save samples is "five Minuits". This option tested and data acquired and saved in different time during the day, which the system can plotted and analyzed.

Figure 8b shows two files generated automatically by the system in the "Daily" option, and the file name takes the current date when the data saved during this day. In the daily option, the system automatically generates a new file daily with the appropriate date. The figure shows a file generated in the day 6, July2017 with "Tow Hours" interval for saving data, and the other shows a file generated in the date 3, February 2017 with "Five Minuits" intervals.

4. Conclusion

As a result, for testing and evaluating the system, the data logger shows a very good results and it could be developed to adapt any user requirements. The system flexibility enables easy upgrade for multi-channel data logger system for monitoring more environmental changes like humidity, air pressure, wind speed or the percentage of CO2 gas. Additional features could be also added to the system concerned the files such as print a hard copy via a printer or send a file to a particular e-mail.

References

[1] Geo scientific ltd. Watershed management instrumentation

[2] Kang J, Park S. Integrated comfort sensing system on indoor climate. *Sensors and Actuators*. 2000: 302-307.

[3] Ravishanker A, Pandian R. Embedded system based sensor failure detection and industrial environment control over wireless network. 2014.

[4] Riva Marco, Piergiovanni, Schiraldi, Luciano Schiraldi, Alberto. Performances of time-temperature indicators in the study of temperature exposure of packaged fresh foods. *Packaging Technology and Science.* 2001.

[5] GS Nhivekar, RR Mudholker. Data Logger and Remote Monitoring System for Multiple Parameter Measurement Application, submitted to e -Journal of Science & Technology (e-JST) (2011)

[6] Information on http://www. www.campbellsci.com

[7] SanDisk Corporation. SanDisk SD card product manual. Version-2.2.2004. www.sandisk.com.

Real Power Loss Reduction by Enhanced Imperialist Competitive Algorithm

K. Lenin
Department of EEE, Prasad V.Potluri Siddhartha Institute of Technology,
Kanuru, Vijayawada, Andhra Pradesh -520007.
e-mail: gklenin@gmail.com

Abstract

In this paper, an Enhanced Imperialist Competitive (EIC) Algorithm is proposed for solving reactive power problem. Imperialist Competitive Algorithm (ICA) which was recently introduced has shown its decent performance in optimization problems. This innovative optimization algorithm is inspired by socio-political progression of imperialistic competition in the real world. In the proposed EIC algorithm, the chaotic maps are used to adapt the angle of colonies movement towards imperialist's position to augment the evading capability from a local optima trap. The ICA is candidly stuck into a local optimum when solving numerical optimization problems. To overcome this insufficiency, we use four different chaotic maps combined into ICA to augment the search ability.Proposed Enhanced Imperialist Competitive (EIC) algorithm has been tested on standard IEEE 30 bus test system and simulation results show clearly the decent performance of the proposed algorithm in reducing the real power loss.

***Keywords**: Optimal Reactive Power, transmission loss, Enhanced Imperialist Competitive Algorithm.*

1. Introduction

Optimal reactive power dispatch (ORPD) problem is a multi-objective optimization problem that minimizes the real power loss and bus voltage deviation. Various mathematical techniques like the gradient method [1-2], Newton method [3] and linear programming [4-7] have been adopted to solve the optimal reactive power dispatch problem. Both the gradient and Newton methods have the complexity in managing inequality constraints. If linear programming is applied then the input- output function has to be uttered as a set of linear functions which mostly lead to loss of accuracy. The problem of voltage stability and collapse play a major role in power system planning and operation [8]. Global optimization has received extensive research awareness, and a great number of methods have been applied to solve this problem. Evolutionary algorithms such as genetic algorithm have been already proposed to solve the reactive power flow problem [9, 10]. Evolutionary algorithm is a heuristic approach used for minimization problems by utilizing nonlinear and non-differentiable continuous space functions. In [11], Genetic algorithm has been used to solve optimal reactive power flow problem. In [12], Hybrid differential evolution algorithm is proposed to improve the voltage stability index. In [13] Biogeography Based algorithm is projected to solve the reactive power dispatch problem. In [14], a fuzzy based method is used to solve the optimal reactive power scheduling method. In [15], an improved evolutionary programming is used to solve the optimal reactive power dispatch problem. In [16], the optimal reactive power flow problem is solved by integrating a genetic algorithm with a nonlinear interior point method. In [17], a pattern algorithm is used to solve ac-dc optimal reactive power flow model with the generator capability limits. In [18], F. Capitanescu proposes a two-step approach to evaluate Reactive power reserves with respect to operating constraints and voltage stability. In [19], a programming based approach is used to solve the optimal reactive power dispatch problem. In [20], A. Kargarian et al present a probabilistic algorithm for optimal reactive power provision in hybrid electricity markets with uncertain loads. This paper proposes a new Enhanced Imperialist Competitive Algorithm (EIC) to solve the optimal reactive power problem. Recently, a new algorithm known as ICA has been proposed by Atashpaz-Gargari and Lucas [21], which is inspired from a socio-human phenomenon [22]. In this paper, a new-fangled method using the chaos theory is proposed to regulate the angle of colonies movement in the direction of the imperialist's positions. The projected method uses some chaotic maps to produce chaotic movement angle. This chaotic movement angle, will act like the mutation operator in genetic algorithm. With the chaos theory the semi-random variation of movement angle causes the planned algorithm escape from the local optimums during the search process. ProposedEIC algorithm has been evaluated in standard IEEE 30 bus test system and the simulation results show that the proposed approach outperforms all the entitled reported algorithms in minimization of real power loss.

2. Problem Formulation

The optimal power flow problem is treated as a general minimization problem with constraints, and can be mathematically written in the following form:

Minimize f(x, u) (1)
subject to g(x,u)=0 (2)
and
$(x, u) \leq 0$ (3)

where f(x,u) is the objective function. g(x.u) and h(x,u) are respectively the set of equality and inequality constraints. x is the vector of state variables, and u is the vector of control variables.

The state variables are the load buses (PQ buses) voltages, angles, the generator reactive powers and the slack active generator power:

$$x = \left(P_{g1}, \theta_2, .., \theta_N, V_{L1}, ., V_{LNL}, Q_{g1}, .., Q_{gng}\right)^T \quad (4)$$

The control variables are the generator bus voltages, the shunt capacitors/reactors and the transformers tap-settings:

$$u = \left(V_g, T, Q_c\right)^T \quad (5)$$

or

$$u = \left(V_{g1}, ..., V_{gng}, T_1, .., T_{Nt}, Q_{c1}, .., Q_{cNc}\right)^T \quad (6)$$

where ng, nt and nc are the number of generators, number of tap transformers and the number of shunt compensators respectively.

3. Objective Function

3.1. Active Power Loss

The objective of the reactive power dispatch is to minimize the active power loss in the transmission network, which can be described as follows:

$$F = PL = \sum_{k \in Nbr} g_k \left(V_i^2 + V_j^2 - 2V_iV_j\cos\theta_{ij}\right) \quad (7)$$

or

$$F = PL = \sum_{i \in Ng} P_{gi} - P_d = P_{gslack} + \sum_{i \neq slack}^{Ng} P_{gi} - P_d \quad (8)$$

where gk: is the conductance of branch between nodes i and j, Nbr: is the total number of transmission lines in power systems. Pd: is the total active power demand, Pgi: is the generator active power of unit i, and Pgsalck: is the generator active power of slack bus.

3.2. Voltage profile improvement

For minimizing the voltage deviation in PQ buses, the objective function becomes:

$$F = PL + \omega_v \times VD \quad (9)$$

where ωv: is a weighting factor of voltage deviation.

VD is the voltage deviation given by:

$$VD = \sum_{i=1}^{Npq} |V_i - 1| \quad (10)$$

yy

3.3. Equality Constraint

The equality constraint g(x,u) of the Optimal reactive power problem is represented by the power balance equation, where the total power generation must cover the total power demand and the power losses:

$$P_G = P_D + P_L \quad (11)$$

This equation is solved by running Newton Raphson load flow method, by calculating the active power of slack bus to determine active power loss.

3.4. Inequality Constraints

The inequality constraints h(x,u) reflect the limits on components in the power system as well as the limits created to ensure system security. Upper and lower bounds on the active power of slack bus, and reactive power of generators:

$$P_{gslack}^{min} \leq P_{gslack} \leq P_{gslack}^{max} \quad (12)$$

$$Q_{gi}^{min} \leq Q_{gi} \leq Q_{gi}^{max}, i \in N_g \quad (13)$$

Upper and lower bounds on the bus voltage magnitudes:

$$V_i^{min} \leq V_i \leq V_i^{max}, i \in N \quad (14)$$

Upper and lower bounds on the transformers tap ratios:

$$T_i^{min} \leq T_i \leq T_i^{max}, i \in N_T \quad (15)$$

Upper and lower bounds on the compensators reactive powers:

$$Q_c^{min} \leq Q_c \leq Q_C^{max}, i \in N_C \quad (16)$$

where N is the total number of buses, NT is the total number of Transformers; Ncis the total number of shunt reactive compensators.

4. Imperialist Competitive Algorithm

Imperialist Competitive Algorithm (ICA) is a new-fangled evolutionary algorithm in the Evolutionary Computation ground based on the human's socio-political progression. The algorithm starts with a primary random population called countries. Some of the paramount countries in the population selected to be the imperialists and the rest form the colonies of these imperialists. In an N dimensional optimization problem, a country is $1 \times N$array. This array defined as below

$$\text{country} = [p_1, p_2, .., p_N] \quad (17)$$

The cost of a country is found by calculating the cost function f at the variables $(p_1, p_2, .., p_N)$. Then

$$c_i = f(\text{country}_i) = f(p_{i1}, p_{i2}, .., p_{iN}) \quad (18)$$

The algorithm begins with N initial countries and the Nimpbest of them (countries with minimum cost) selected as theimperialists. The left over countries are colonies that belong to a kingdom. The primary colonies belong toimperialists in convenience with their authority. To allocatethe colonies among imperialists proportionally, thestandardized cost of an imperialist is defined as follow

$$c_n = \max_i c_i - c_n \quad (19)$$

where, costnis the cost of nth imperialist and $c_n$is its standardized cost. Every imperialist that has more cost value, will have less standardized cost value. Having the standardized cost, the authority of each imperialist is computed as below and based on that the colonies spread among the imperialist countries.

$$p_n = \left| \frac{c_n}{\sum_{i=1}^{N_{imp}} c_i} \right| \tag{20}$$

On the other hand, the standardized power of an imperialist is weighed up by its colonies. Then, the primary number of colonies of an empire will be

$$Nc_n = \text{rand}\{p_n \cdot (N_{col})\} \tag{21}$$

where, $Nc_n$is initial number of colonies of nth empire andN_{col} is the number of all colonies.

To allocate the colonies among imperialist, $Nc_n$of the colonies is selected arbitrarily and allocated to their imperialist. The imperialist countries absorb the colonies towards themselves using the absorption policy. The imperialists take in these colonies towards themselves with respect to their power that described in (22). The entire power of each imperialist is determined by the power of its both parts, the empire power with addition of its average colonies power.

$$TC_n = \text{cost}(\text{imperialist}_n) + \xi \text{mean}\{\text{cost}(\text{colonies of Empire}_n)\} \tag{22}$$

where $TC_n$is the total cost of the nth empire and$\xi$is a positive number which is considered to be less than one.

$$x \sim U(0, \beta \times d) \tag{23}$$

In the amalgamation strategy, the colony moves in the direction of the imperialist by x unit. The direction of movement is the vector from colony to imperialist. The distance between the imperialist and colony shown by d and x is a random variable with uniform distribution. Where $\beta$is greater than 1 and is near to 2. So, a suitable option can be β =2. In our execution $\gamma$is $\frac{\pi}{4}$(rad)respectively.

$$\theta \sim U(-\gamma, \gamma) \tag{24}$$

In ICA to investigate different points in the region of the imperialist, an arbitrary amount of deviation is added to the way of colony movement in the direction of the imperialist. While moving in the direction of the imperialist countries, a colony may reach to a superior position, so the colony position alters according to position of the imperialist. In this algorithm, the imperialistic competition has a significant role. During the imperialistic competition, the weak empires will lose their authority and their colonies. To model this competition, firstly we compute the probability of possessing all the colonies by each empire taking into consideration with the total cost of empire.

$$NTc_n = \max_i\{TC_i\} - TC_n \tag{25}$$

where, $TC_n$is the total cost of nth empire and $NTc_n$is the normalized total cost of nth empire. Having the normalized total cost, the possession probability of each empire is calculated as below

$$p_{pn} = \left| \frac{NTC_n}{\sum_{i=1}^{N_{imp}} NTC_i} \right| \tag{26}$$

After a while all the empires except the most powerful one will fall down and all the colonies will be under the control of this unique kingdom. Figure 1 shows the flow chart of the ICA.

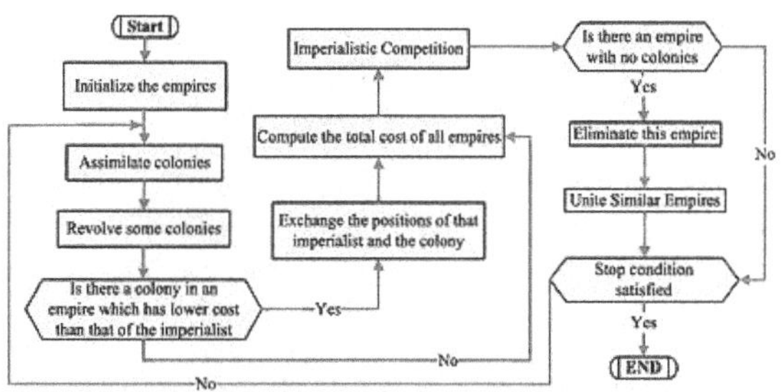

Figure 1. Flowchart of Imperialist Competitive Algorithm (ICA)

Steps of Imperialist Competitive Algorithm
Define objective function

1) Initialization of the algorithm. Generate some random solution in the search space and create initial empires.
2) Assimilation: Colonies move towards imperialist states in different in directions.
3) Revolution: Random changes occur in the characteristics of some countries.
4) Position exchange between a colony and Imperialist. A colony with a better position than the imperialist, has the chance to take the control of empire by replacing the existing imperialist.
5) Imperialistic competition: All imperialists compete to take possession of colonies of each other.
6) Eliminate the powerless empires. Weak empires lose their power gradually and they will finally be eliminated.
7) If the stop condition is satisfied, stop, if not go to 2.
8) End

5. Chaos Theory

In the chaos theory behaviour between unbending regularity and arbitrariness based on unadulterated possibility is called a chaotic system, or chaos. Chaos emerges to be stochastic but it arises in a deterministic non-linear system under deterministic conditions [23]. Chaotic map is very significant for optimization problem. In addition, it has a very susceptible dependence upon its primary condition and parameter. A chaotic map is a discrete-time dynamical system. Since the ICA algorithm experience from being trapped at local optima, chaotic local search has been initiated to conquer the local optima and speed up the convergence.

6. Enhanced Imperialist Competitive (EIC)Algorithm

The primary Imperialist Competitive Algorithm (ICA) utilizes a local search mechanism as like as many evolutionary algorithms. Therefore, the chief ICA may fall into local minimum trap during the search procedure and it is likely to get far from the global optimum. To solve this problem, we augmented the exploration aptitude of the ICA algorithm, using a chaotic behaviour in the colony movement in the direction of the imperialist's position. So it is planned to improve the global convergence of the ICA and to avoid it to stick on a local solution.

In this paper, to augment the universal exploration capability, the chaotic maps are integrated into ICA to augment the ability of absconding from a local optimum. The angle of movement is altered in a chaotic way during the search procedure. Adding this chaotic behaviour in the imperialist algorithm amalgamation policy we make the conditions proper for the algorithm to run away from local peaks. Chaos variables are usually generated by the some well-known chaotic maps [23, 24]. Table 1shows the mentioned chaotic maps for adjusting θparameter (Angle of colonies movement in the direction of the imperialist's position) in the planned algorithm.

Table 1.Chaotic Maps

	Chaotic maps
CM1	$\theta_{n+1} = a\theta_n(1-\theta_n)$
CM2	$\theta_{n+1} = a\theta_n^2 sin(\pi\theta_n)$
CM3	$\theta_{n+1} = \theta_n + b - (a/2\pi)sin(2\pi\theta_n)mod(1)$
CM4	$\theta_{n+1} = \begin{cases} 0 & \theta_n = 0 \\ 1/X_n mod(1) \theta_n \in (0,1) \end{cases}$

In table I, ais a control parameter. θ is a chaotic variable in kth iteration which belongs to interval of (0,1). During the search process, no value of θis repeated.

Enhanced Imperialist Competitive (EIC) Algorithm to solve reactive power problem

(i) Initialize the kingdom and their colonies location arbitrarily.

(ii) Calculate the chaotic θ (colonies movement angle in the direction of the imperialist's location) using the chaotic maps.

(iii) Calculate the sum cost of all kingdom.

(iv) Choose the weakest colony from the weakest kingdom and give it to the empire that has the most likelihood to own it.

(v) Eradicate the powerless empires.

(vi) If there is just one empire then stop else continue.

(vii) Check the stop conditions

7. Simulation Results

Validity of proposed Enhanced Imperialist Competitive (EIC) Algorithm has been verified by testing in IEEE 30-bus, 41 branch system and it has 6 generator-bus voltage magnitudes, 4 transformer-tap settings, and 2 bus shunt reactive compensators. Bus 1 is taken as slack bus and 2, 5, 8, 11 and 13 are considered as PV generator buses and others are PQ load buses. Control variables limits are given in Table 2.

Table 2 Primary Variable Limits (Pu)

Variables	Min.	Max.	category
Generator Bus	0.95	1.1	Continuous
Load Bus	0.95	1.05	Continuous
Transformer-Tap	0.9	1.1	Discrete
Shunt Reactive Compensator	-0.11	0.31	Discrete

In Table 3 the power limits of generators buses are listed.

Table 3. Generators Power Limits

Bus	Pg	Pgmin	Pgmax	Qgmin	Qmax
1	96.00	49	200	0	10
2	79.00	18	79	-40	50
5	49.00	14	49	-40	40
8	21.00	11	31	-10	40
11	21.00	11	28	-6	24
13	21.00	11	39	-6	24

Table 4 shows the proposed EIC approach successfully kept the control variables within limits.Table 5 narrates about the performance of the proposed EIC algorithm. Fig 2 shows about the voltage deviations during the iterations and Table 6 list out the overall comparison of the results of optimal solution obtained by various methods.

Table 4. After optimization values of control variables

Control Variables	EIC
V1	1.0481
V2	1.0401
V5	1.0205
V8	1.0314
V11	1.0701
V13	1.0500
T4,12	0.00
T6,9	0.01
T6,10	0.90
T28,27	0.91
Q10	0.10
Q24	0.10
Real power loss	4.2762
Voltage deviation	0.9092

Table 5. Performance of EIC algorithm

Iterations	30
Time taken (secs)	8.83
Real power loss	4.2762

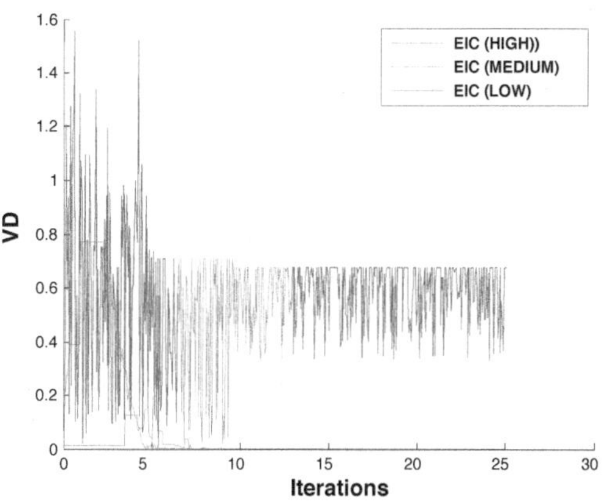

Figure 2. Voltage deviation (VD) characteristics

Table 6 Comparison of results

Techniques	Real power loss (MW)
SGA(Wu et al., 1998) [25]	4.98
PSO(Zhao et al., 2005) [26]	4.9262
LP(Mahadevan et al., 2010) [27]	5.988
EP(Mahadevan et al., 2010) [27]	4.963
CGA(Mahadevan et al., 2010) [27]	4.980
AGA(Mahadevan et al., 2010) [27]	4.926
CLPSO(Mahadevan et al., 2010) [27]	4.7208
HSA (Khazali et al., 2011) [28]	4.7624
BB-BC (Sakthivel et al., 2013) [29]	4.690
MCS(Tejaswini sharma et al.,2016) [30]	4.87231
Proposed EIC	4.2762

8. Conclusion

Enhanced Imperialist Competitive (EIC) Algorithm has been successfully applied for reactive power problem.In the proposed EIC algorithm, the chaotic maps are used to adapt the angle of colonies movement towards imperialist's position to augment the evading capability from a local optima trap. Proposed EIC algorithm has been tested in standard IEEE 30 bus system and simulation results reveal about the better performance of the proposed EIC algorithm in reducing the real power loss when compared to other stated standard algorithms.

References

[1] O Alsac, B Scott. Optimal load flow with steady state security. *IEEE Transaction*. PAS -1973: 745-751.
[2] Lee KY, Paru YM, Oritz JL. A united approach to optimal real and reactive power dispatch. *IEEE Transactions on power Apparatus and systems.* 1985: 1147-1153
[3] A Monticelli, MVF Pereira, S Granville. Security constrained optimal power flow with post contingency corrective rescheduling. *IEEE Transactions on Power Systems*. 1987; 2(1): 175-182.
[4] Deeb N, Shahidehpur SM. Linear reactive power optimization in a large power network using the decomposition approach. *IEEE Transactions on power system*. 1990: 5(2) : 428-435
[5] E Hobson. Network consrained reactive power control using linear programming. *IEEE Transactions on power systems PAS*. 1980; 99(4): 868-877.
[6] KY Lee, YM Park, JL Oritz. *Fuel –cost optimization for both real and reactive power dispatches.* IEE Proc; 131(3): 85-93.
[7] MK Mangoli, KY Lee. Optimal real and reactive power control using linear programming. *Electr.Power Syst.Res.*1993; 26: 1-10.
[8] CA Canizares, ACZ de Souza, VH Quintana. Comparison of performance indices for detection of proximity to voltage collapse. 1996; 11(3): 1441-1450.
[9] K Anburaja. Optimal power flow using refined genetic algorithm. *Electr.Power Compon.Syst*. 2002; 30: 1055-1063.
[10] D Devaraj, B Yeganarayana. *Genetic algorithm based optimal power flow for security enhancement.* IEE proc-Generation.Transmission and. Distribution. 2005; 152.
[11] A Berizzi, C Bovo, M Merlo, M Delfanti. A ga approach to compare orpf objective functions including secondary voltage regulation. *Electric Power Systems Research*. 2012; 84(1): 187 – 194.
[12] CF Yang, GG Lai, CH Lee, CT Su, GW Chang. Optimal setting of reactive compensation devices with an improved voltage stability index for voltage stability enhancement. *International Journal of Electrical Power and Energy Systems.* 2012; 37(1): 2012.
[13] P Roy, S Ghoshal, S Thakur. Optimal var control for improvements in voltage profiles and for real power loss minimization using biogeography based optimization. *International Journal of Electrical Power and Energy Systems.* 2012; 43(1): 830 – 838.
[14] B Venkatesh, G Sadasivam, M Khan. A new optimal reactive power scheduling method for loss minimization and voltage stability margin maximization using successive multi-objective fuzzy lp technique. *IEEE Transactions on Power Systems*. 2000; 15(2): 844 – 851.
[15] W Yan, S Lu, D Yu. A novel optimal reactive power dispatch method based on an improved hybrid evolutionary programming technique. *IEEE Transactions on Power Systems*. 2004; 19(2): 913 – 918.
[16] W Yan, F Liu, C Chung, K Wong. A hybrid genetic algorithminterior point method for optimal reactive power flow. *IEEE Transactions on Power Systems*. 2006; 21(3): 1163 –1169.
[17] J Yu, W Yan, W Li, C Chung, K. Wong. An unfixed piecewiseoptimal reactive power-flow model and its algorithm for ac-dc system. *IEEE Transactions on Power Systems*. 2008; 23(1): 170 –176.
[18] F Capitanescu. Assessing reactive power reserves with respect to operating constraints and voltage stability. *IEEE Transactions on Power Systems*. 2011; 26(4): 2224–2234.
[19] Z Hu, X Wang, G Taylor. Stochastic optimal reactive power dispatch: Formulation and solution method. *International Journal of Electrical Power and Energy Systems*. 2010; 32(6): 615 – 621.
[20] A Kargarian, M Raoofat, M Mohammadi. Probabilistic reactive power procurement in hybrid electricity markets with uncertain loads. *Electric Power Systems Research*. 2012; 82(1): 68 – 80.
[21] E Atashpaz-Gargari, C Lucas. Imperialist Competitive Algorithm: An Algorithm for Optimization Inspired by Imperialistic Competition. *IEEE Congress on Evolutionary Computation*. 2007: 4661- 4667.
[22] Helena Bahrami, Karim Faez, Marjan Abdechiri. *Imperialist Competitive Algorithm using Chaos Theory for Optimization.* 12th International Conference on Computer Modelling and Simulation. 2010.
[23] Zheng WM. Kneading plane of the circle map. Chaos, Solitons & Fractals. 1994;4:1221.
[24] Schuster HG. Deterministic chaos: an introduction. 2nd revised. Weinheim, Federal Republic of Germany: Physick-Verlag GmnH. 1988.
[25] Wu QH, YJ Cao, JY Wen. Optimal reactive power dispatch using an adaptive genetic algorithm. *Int.J.Elect.Power Energy Syst*. 1998; 20: 563-569.
[26] Zhao B, CX Guo, YJ CAO. Multiagent-based particle swarm optimization approach for optimal reactive power dispatch. *IEEE Trans. Power Syst*. 2005; 20(2): 1070-1078.
[27] Mahadevan K, Kannan PS. Comprehensive Learning Particle Swarm Optimization for Reactive Power Dispatch. *Applied Soft Computing*. 2010; 10(2): 641–52.
[28] Khazali AH, M Kalantar. Optimal Reactive Power Dispatch based on Harmony Search Algorithm. *Electrical Power and Energy Systems.* 2011; 33(3): 684–692.
[29] Sakthivel S, M Gayathri, V Manimozhi. A Nature Inspired Optimization Algorithm for Reactive Power Control in a Power System. *International Journal of Recent Technology and Engineering*. 2013; 2(1): 29-33.
[30] Tejaswini Sharma, Laxmi Srivastava, Shishir Dixit. *Modified Cuckoo Search Algorithm For Optimal Reactive Power Dispatch*. Proceedings of 38 th IRF International Conference. 2016; 4-8. 20th March, 2016, Chennai, India, ISBN: 978-93-85973-76-5.

21

A Real-Time Implementation of Moving Object Action Recognition System based on Motion Analysis

Kamal Sehairi[1*], Cherrad Benbouchama[2], Kobzili El Houari[2], Chouireb Fatima[1]
[1] Laboratoire LTSS, Departement of Electrical Engineering, Université Amar Telidji Laghouat, Route de Ghardaia, Laghouat 03000, Algeria
[2] Laboratoire LMR, École Militaire Polytechnique, Bordj El Bahri, Algeria
e-mail: k.sehairi@lagh-univ.dz, sehairikamel@yahoo.fr

Abstract

This paper proposes a PixelStreams-based FPGA implementation of a real-time system that can detect and recognize human activity using Handel-C. In the first part of our work, we propose a GUI programmed using Visual C++ to facilitate the implementation for novice users. Using this GUI, the user can program/erase the FPGA or change the parameters of different algorithms and filters. The second part of this work details the hardware implementation of a real-time video surveillance system on an FPGA, including all the stages, i.e., capture, processing, and display, using DK IDE. The targeted circuit is an XC2V1000 FPGA embedded on Agility's RC200E board. The PixelStreams-based implementation was successfully realized and validated for real-time motion detection and recognition.

***Keywords**: from papermoving object, recognition system, FPGA, real-time system, motion detection*

1. Introduction

In modern society, there is a growing need for technologies such as video surveillance and access control to detect and identify human and vehicle motion in various situations. Intelligent video surveillance attempts to assist human operators when the number of cameras exceeds the operators' capability to monitor them and alerts the operators when abnormal activity is detected. Most intelligent video surveillance systems are designed to detect and recognize human activity. It is difficult to define abnormal activity because there are many behaviors that can represent such activity. Examples include a person entering a subway channel, abandonment of a package, a car running in the opposite direction, and people fighting or rioting. However, it is possible not only to set criteria to detect abnormal activity but also to zoom in on the relevant area to facilitate the work of the operator.

In general, an intelligent video surveillance system has three major stages: detection, classification, and activity recognition [1]. Over the years, various methods have been developed to deal with issues in each stage.

2. Related Work

Many methods for motion detection have already been proposed. They have been classified [1]–[3] into three major categories: background subtraction, [4],[5] temporal differencing [6] , [7] and optical flow[8],[9]. Further, motion detection methods have been recently classified into matching methods, energy-based methods, and gradient methods. The aim of the motion detection stage is to detect regions corresponding to moving objects such as vehicles and human beings. It is usually linked to the classification stage in order to identify moving objects. There are two main types of approaches for moving object classification:[1],[2],[10] shape-based identification and motion-based classification. Different descriptions of shape information of motion regions such as representations of points, boxes, silhouettes, and blobs are available for classifying moving objects. For example, Lipton et al.[11] used the dispersedness and area of image blobs as classification metrics to classify all moving object blobs into human beings, vehicles, and clutter. Further, Ekinci et al.[12] used silhouette-based shape representation to distinguish humans from other moving objects, and the skeletonization method to recognize actions. In motion-based identification, we are more interested in detecting periodic, non-rigid, articulated human motion . For example, Ran et al.[13] examined the periodic gait of pedestrians in order to track and classify it. The final stage of surveillance involves behavior understanding and activity recognition. Various techniques for this purpose have been categorized into seven types: dynamic time warping algorithms, finite state machines, hidden Markov models, time-delay neural networks, syntactic techniques, non-deterministic finite automata, and self-organizing neural networks. Such a wide variety of techniques is attributable to the complexity of the problems and the extensive research conducted in this field. The computational complexity of these methods and the massive amount of information obtained from video streams makes it difficult to achieve real-time performance on a general-purpose CPU or DSP. There are four main architectural approaches for overcoming this challenge: application-specific integrated circuits (ASICs) and field-programmable gate arrays (FPGAs), parallel computing, GPUs, and multiprocessor architectures. Evolving high-density FPGA architectures, such as those with embedded multipliers, memory blocks, and high I/O (input/output) pin counts, are ideal solutions for video processing applications [14]. In the field of image and video processing, there are many FPGA implementations for motion segmentation and tracking. For example, Menezes et al. [5] used background subtraction to detect vehicles in motion, targeting Altera's Cyclone II FPGA with Quartus II software. Another similar study on road traffic detection [15] adopted the sum of absolute differences (SAD) algorithm, implemented on Agility's RC300E board using an XC2V6000 FPGA with Handel-C and the PixelStreams library of Agility's DK Design Suite. Other methods for motion detection such as optical flow have been successfully implemented [8],[9] on an FPGA. For example, Ishii et al.[8] optimized an optical flow algorithm to process 1000 frames per second. The algorithm was implemented on a Virtex-II Pro FPGA.

Many video surveillance systems have been developed for behavior change detection. For example, in the framework of ADVISOR, a video surveillance system for metro stations, a finite state machine (with scenarios) [16] is used to define suspicious behavior (jumping over a barrier, overcrowding, fighting, etc.). The W4 system [17] is a system for human activity recognition that has been implemented on parallel processors with a resolution of 320×240. This system can detect objects carried by people and track body parts using background detection and silhouettes. Bremond and Morioni [18] extracted the features of moving vehicles to detect their behaviors by setting various scenario states (toward an endpoint, stop point, change in direction, etc.); the application employs aerial grayscale images.

The objective of this study is to implement different applications of behavior change detection and moving object recognition based on motion analysis and the parameters of moving objects. Such applications include velocity change detection, direction change detection, and posture change detection. The results can be displayed in the RGB format using chains of parallelized sub-blocks. We used Handel-C and the PixelStreams library of Agility's DK Design Suite to simplify the acquisition and display stages. An RC200E board with an embedded Virtex-II XC2V1000 FPGA was employed for the implementation.

3. Mixed Software-Hardware Design

To make our implementation more flexible, we use the software-hardware platform approach. This approach simplifies not only the use of the hardware but also the change between soft data and hard data, especially for image processing applications that need many parameters to be changed, for example, the parameters of convolution filters and threshold levels. In our implementation, we use Handel-C for the hardware part. Handel-C is a behavior-oriented programming language for FPGA hardware synthesis, and it is adapted to the co-design concept [19].

The software part is developed using Visual C++. After generating the bit file using Agility's DK Design Suite, [20] we use our software interface to load this bit file via the parallel port (with a frequency of 50 MHz) on the RC200E board in order to configure the FPGA. The algorithm parameters are transferred through this port as 8-bit data at the same frequency. For the user, these operations are hidden. The graphical user interface allows the user to configure/erase the FPGA and change the algorithm parameters. For example, in our case, we can change the threshold level according to the brightness of the scene or the velocity level according to the object in motion (human, vehicle).

4. Outline of the Algorithm

4.1. Pixel Streams Library

Before we detail and explain our algorithm and the method used to achieve our goals, we should discuss the tools used for our implementation. We used an RC200E board with an embedded XC2V1000 FPGA [21]. This board has multiple video inputs (S-video, camera video, and composite video), multiple video outputs (VGA, S-video, and composite video), and two ZBT SRAMs, each with a capacity of 2 MB. The language used is Handel-C [22] and the integrated development environment (IDE) is Agility's DK5. This environment is equipped with different platform development kits (PDKs) that include the PixelStreams library [23].

The PixelStreams library is used to develop systems for image and video processing. It includes many blocks (referred to as filters) that perform primary video processing tasks such as acquisition, stream conversion, and filtering. The user has to associate these blocks carefully by indicating the type of the stream (pixel type, coordinate type, and synchronization type). Then, the user can generate the algorithm in Handel-C. Thereafter, the user has to add or modify blocks to program his/her method, and finally, he/she must merge the results. It is worth mentioning that these blocks are parameterizable, i.e., we can modify the image processing parameters, such as the size of the acquired image or the threshold. These blocks are fully optimized and parallelized. Figure 1 shows the GUI of PixelStreams.

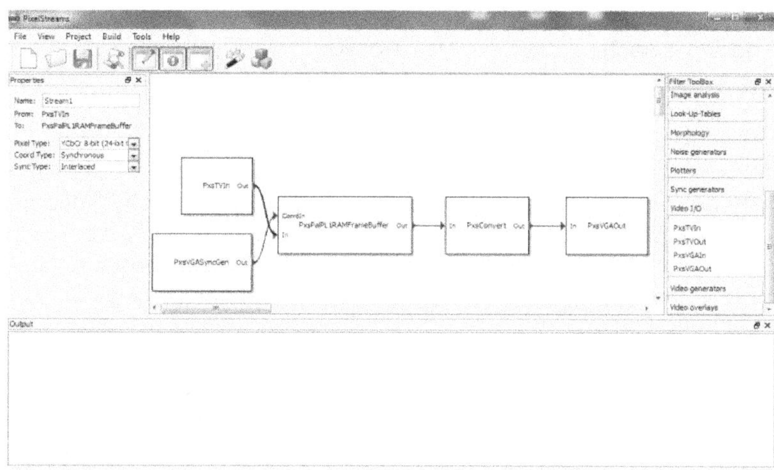

Figure 1. PixelStreams GUI

4.2. Detection Algorithm

We choose to implement the delta frame method for three reasons: its adaptability to changes in luminance, its simplicity, and its low consumption of hardware resources. This method determines the absolute difference between two successive images, and it is executed in two stages: temporal difference and segmentation.

4.2.1. Temporal difference

In this stage, we determine the absolute difference between the previous frame and the current frame as follows.

$$\zeta(x,y) = \left|\frac{dI(x,y)}{dt}\right| = \left|\Delta_{t,t-1}(x,y)\right| = \left|I_t(x,y) - I_{t-1}(x,y)\right| \tag{1}$$

where $\zeta(x,y)$ is the difference between It(x,y) (i.e., the intensity of pixel (x,y) at moment t) and It-1(x,y) (i.e., the intensity of pixel (x,y) at moment t-1).

4.2.2. Segmentation

In this stage, significant temporal changes are detected by means of thresholding:

$$\Psi(x,y)=\begin{cases}0 & \text{if } \zeta(\text{x,y})<\text{Th}\\ 1 & \text{otherwise}\end{cases} \tag{2}$$

This operation yields a binary card that indicates zones of significant variations in brightness from one image to the other.

4.3. Feature Extraction and Behavior Change Detection

In this study, simple behavior change detection refers to motion that can be caused by abrupt movements that might represent suspicious actions. To define these actions, we use the parameters of the objects in motion, such as the center of gravity, width, and length. In general, the actions detected by this method are simple yet useful in video surveillance. For example, velocity change detection is useful for detecting a criminal who is being chased by the police or a car that exceeds the speed limit; direction change detection is useful for detecting a car that is moving in the wrong direction; and posture change detection is useful for detecting a person who bends to place or pick up an object.

Our implementation involves the following stages: acquisition of the video signal, elimination of noise from the input video signal, detection of moving regions, segmentation for separating the moving objects, extraction of the object parameters, classification of the moving objects, and determining whether movements are suspicious.

4.3.1. Velocity change detection

We can detect suspicious behavior of a person from his/her gait as well as his/her change in velocity near sensitive locations such as banks, airports, and shopping centers. In such cases, we can calculate the speed (in pixels/s) or acceleration (in pixels/s2) of the suspect in the image space in real time. There are several ways of representing this anomaly: the most widely adopted method in the literature is the use of a bounding box (a rectangle around the suspect).

It is easy to calculate the speed of a moving object. As soon as the speed or acceleration of the object exceeds a certain threshold of normality (predetermined experimentally or on the basis of statistical studies), a bounding box appears around the suspect. However, the issue that needs to be addressed is the calculation of the speed in real-time circuits owing to the absence of mathematical functions (such as square root), types of data (integer or real values), and the object parameters on which we base our calculation.

In general, the speed and acceleration are calculated as follows:

$$velocity\,(t)=\left(\sqrt{(x_g(t)-x_g(t-dt))^2+(y_g(t)-y_g(t-dt))^2}\right)/\,dt \tag{3}$$

$$acceleration\,(t)=\left(velocity\,(t)-velocity\,(t-dt)\right)/\,dt \tag{4}$$

where $\mathrm{x}_g(t),\mathrm{y}_g(t)$ and $\mathrm{x}_g(t-dt),\mathrm{y}_g(t-dt)$ are the co-ordinates of the center of gravity of the object at moments t and $t-dt$, respectively, dt=40ms in our case, and $velocity\,(t)$ and $velocity\,(t-dt)$ are the velocities of the object at moments t and $t-dt$, respectively.

4.3.2. Direction change detection

Changes in direction or motion in the wrong direction can represent abnormal behavior depending on the situation. For example, roaming around a building or car can be considered as an abrupt cyclic change in direction, possibly indicating the intention of burglary or car theft. Other examples include detection of a car that is moving in the wrong direction or a person who is moving in the opposite direction of a queue at an exit gate or exit corridor in an airport.

To determine the direction, we select parameters that distinguish the object of interest, such as its center of gravity, width, and length. In general, the co-ordinates of the center of gravity can be used to determine whether the object has changed its direction, i.e., whether it has moved rightward or leftward depending on the position of the camera.

The change in direction along the x-axis is given by

$$x_g(t) - x_g(t-dt) \begin{cases} > 0 & \text{The object did not change direction} \\ < 0 & \text{The object changed direction} \end{cases} \quad (5)$$

The change in direction along the y-axis is given by

$$y_g(t) - y_g(t-dt) \begin{cases} > 0 & \text{The object did not change direction} \\ < 0 & \text{The object changed direction} \end{cases} \quad (6)$$

These techniques, which are based on the object parameters, can be improved by integrating them with advanced models such as finite state machines (FSMs).

5. Hardware Implementation

Figure 2 shows the general outline of our FPGA implementation.\

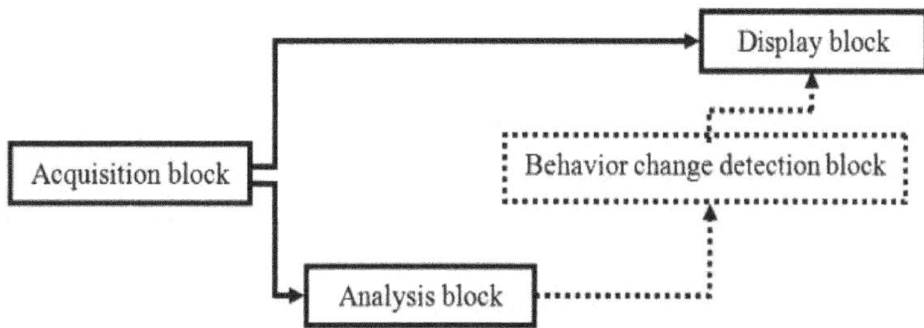

Figure 2. General outline of behavior change detection.

This general outline consists of four blocks: an acquisition block, an analysis block, a display block, and an intermediate block between the display block and the analysis block.

5.1. Acquisition Block

Acquisition is achieved using the standard camera associated with the RC200E board. The video input processor, Philips SAA7113H, acquires the frames in the PAL format at a rate of 25 fps. The pixels are in the YCbCr format. Using the PixelStreams library of Agility's DK Design Suite, we split the input video signal into two identical streams (see Figure 3). The first stream is fed to the display block and it is converted into the RGB format to display the results on a VGA display. The second stream is fed to the analysis block and it is converted into the grayscale format to reduce (by one-third) the amount of data to be processed.

We can choose to perform the conversion into the RGB format before splitting the input signal and then convert the second stream for the analysis block into the grayscale format. However, this method is not preferable because the conversion from the YCbCr format to the RGB format is approximate. Moreover, in the conversion from the YCbCr format to the grayscale format, the brightness is simply represented by the Y component of the YCbCr format. Further, it is not preferable to approximate the input stream that is fed to the analysis block.

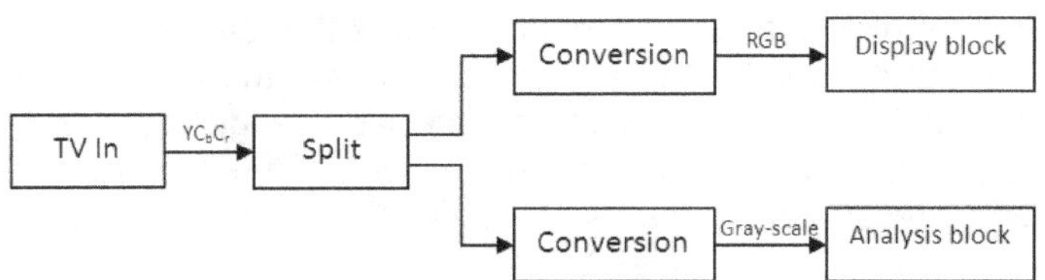

Figure 3. Acquisition block.

5.2. Analysis Block

The analysis block consists of several stages. In the first stage, we use inter-image subtraction (delta frames) and apply thresholding to detect moving regions.

To obtain the delta frames, we start by splitting the video signal in three channels (see Figure 4). The first and second channels are used to save the acquired image, creating a delay cell. The image I(t-1) is recorded in the memory. The third channel is used to acquire the actual frame at moment t. Then, the two image streams are synchronized and fed to the subtraction block. The subtraction block is a modified block that takes the absolute result of subtraction and compares it with a threshold. This function is realized using a macro. The threshold value Th is fixed according to the luminosity of the scene.

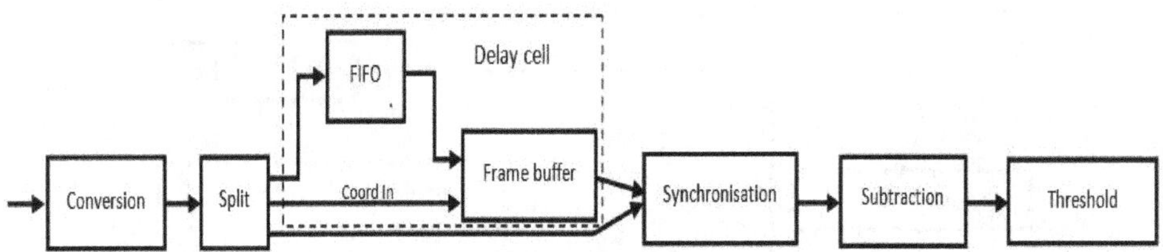

Figure 4. Motion detection block.

The second stage of the analysis block involves statistical analysis. In this stage, we search for the min and max values along the x- and y-axes of the mobile regions (Figure 5). In general, this stage must be preceded by a filter for noise reduction. We employed a morphological filter (e.g., alternating sequential filter, opening/closing filter) using the PixelStreams library.

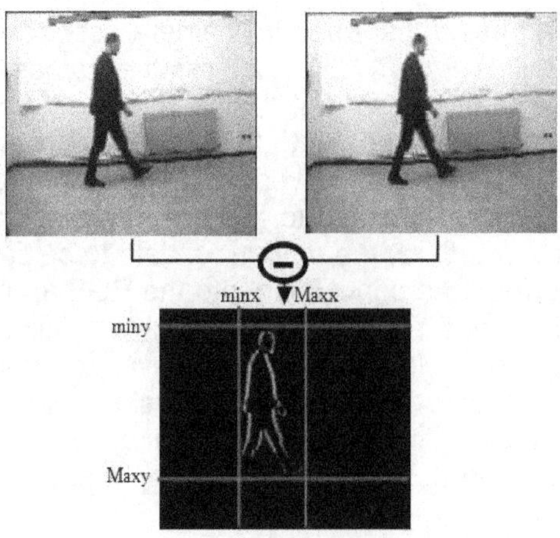

Figure 5. Inter-image difference and calculation of min and max values.

After calculating the min and max values along the two axes, we determine the center of gravity of the detected object. We calculate the sum of the pixel co-ordinates that have non-zero values along the x- and y-axes, and we divide these coordinate values by their sum. However, for our implementation, it is better to avoid this division. Therefore, we use the direct method. We subtract the max from the min and divide the result by 2. Division by 2 is achieved by a simple bit shift (right shift). Once the values minX, MaxX, minY, MaxY, and XG, YG are obtained, we copy these values into the behavior change detection block. Then, we reset these values to zero.

5.3. Behavior Change Detection Block

As stated in the previous section, the analysis block provides the behavior change detection block with the parameters of the moving objects. In this stage, we save the values extracted from the first delta frame (xg(t-1), yg(t-1), minx(t-1), MaxX(t-1), miny(t-1), MaxY(t-1)), and from the second delta frame, we obtain the current values xg(t), yg(t), minx(t), MaxX(t), miny(t), and MaxY(t). From these latter results, we can calculate the width and length of the moving object to classify the object as human, vehicle, or others, as in our previous work [24]. Using the values extracted in two different instants (t-1, t), we define the changes in behavior.

For velocity change detection, the speed and acceleration are calculated using the two equations presented in Sec. 4.4.1. However, we simplify these equations by calculating the absolute differences between two moments (the previous and current values). If the absolute difference exceeds a certain threshold Vth, we assume that the velocity has changed, and we copy the values of the center of gravity in the display block in order to draw a rectangle around the object. Then, the current values are saved as previous values.

Consider a practical problem that involves the values of the center of gravity. In our algorithm, we need to reset all the variables to zero. Consequently, the coordinates of the center of gravity will be zero. If an object enters the scene, the coordinates of the center of gravity change from 0 to xg, yg, and this will cause false detection.

To overcome this problem, we have to ensure that the object has entered the scene entirely. For this purpose, we set a condition on the coordinates of the bounding box for two consecutive instants; if this condition is met (|minXt1 - minXt2 | > S AND |MaxXt1 - MaxXt2 | > S), we can guarantee that the object has entered the scene entirely, either from the right or from the left.

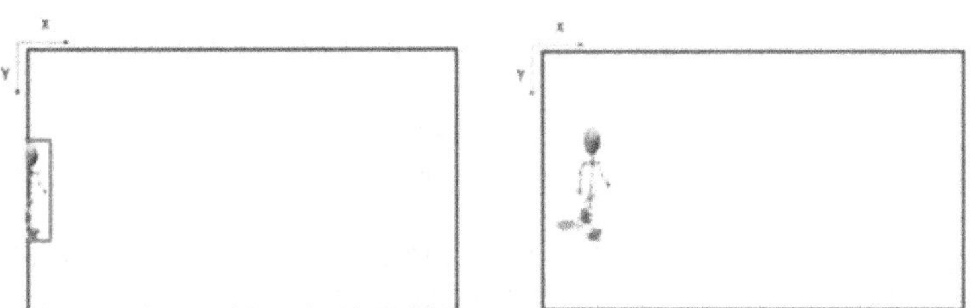

Figure 6. False velocity change detection (the object enters the scene).

Table 1. Solution proposed for false velocity change detection.

$C_1 \Leftrightarrow \|\min x(t-1) - \min x(t)\| > S$	$C_2 \Leftrightarrow \|MaxX(t-1) - MaxX(t)\| > S$	$(C_1 \&\& C_2)$
0	0	0
0	1	0
1	0	0
1	1	1

Figure 7 shows this implementation and represents all the stages realized.

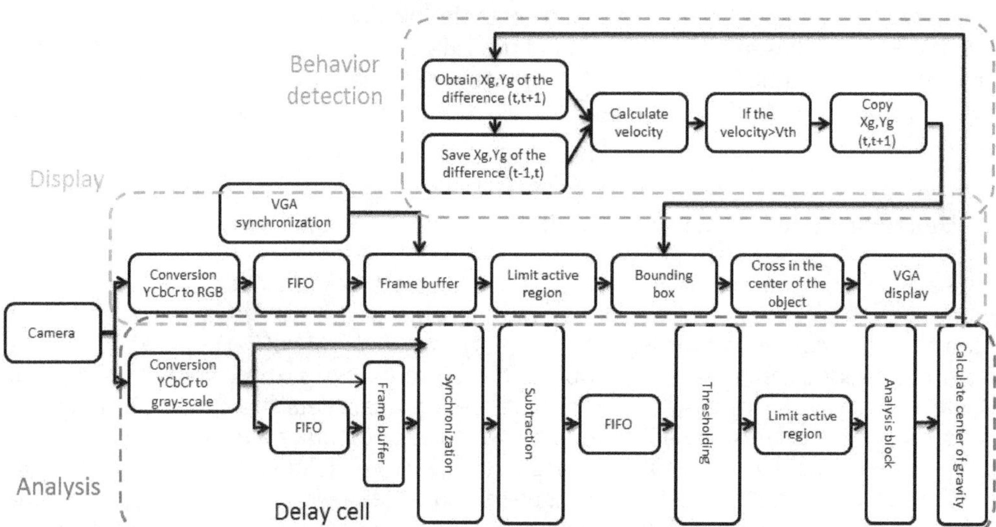

Figure 7. Hardware architecture for velocity change detection.

Direction change detection: To implement this application, we follow the same stages as those used in velocity change detection, except that the condition changes. We use the same parameters, minX, MaxX, minY, and MaxY, in order to avoid the center of gravity problem. We calculate the difference between minX1 and minX2, and MaxX1 and MaxX2. If there is a change in sign, we assume that the object has changed its direction. Otherwise, we assume that the object has not changed its direction.

We can easily determine the direction of motion of an object by applying the same concept as that described above. However, in this case, it is impractical to compare the differences between the previous values and the current values with zero because the presence of a small or non-significant movement (such as that of the arms) can cause false detection. Therefore, to overcome this problem, we compare the difference with a threshold Thd, which should not be very large. Then, the values minX, MaxX, minY, and MaxY are copied to the block that draws the bounding box.

We use two blocks for detection in two directions (a different color for each direction of motion). In order to minimize resource consumption, we used only one block for drawing the bounding box by changing the parameters of entry in our macro. In this macro, we added a parameter that changes the color according to the direction of detected motion (Figure 8).

Posture change detection: We are interested in such an application to detect a person who leans (bends) to place or pick up something, especially in sensitive locations (e.g.,the subways). In this case, we are interested in movements along the y-axis of the image (up/down motion), and we use the same architecture as that used in velocity change detection. We calculate the difference between the previous and current values of miny(t-1), MaxY(t-1), miny(t), MaxY(t).

If the difference between the previous and current values is positive, we assume that the person leans, and we copy the values minX, MaxX, minY, and MaxY to the block that draws the bounding box and fix the color parameter of the rectangle. We can add a warning message using the PxsConsole filter of PixelStreams. In the opposite case, we assume that the person rises, and we copy the values to the block that draws the rectangle, which uses a different color in this case. As in the case of direction change detection, it is better to use a threshold Thp to reduce the occurrence of false detection due to small movements along the y-axis. For such detections, we require a camera whose front sight faces the scene.

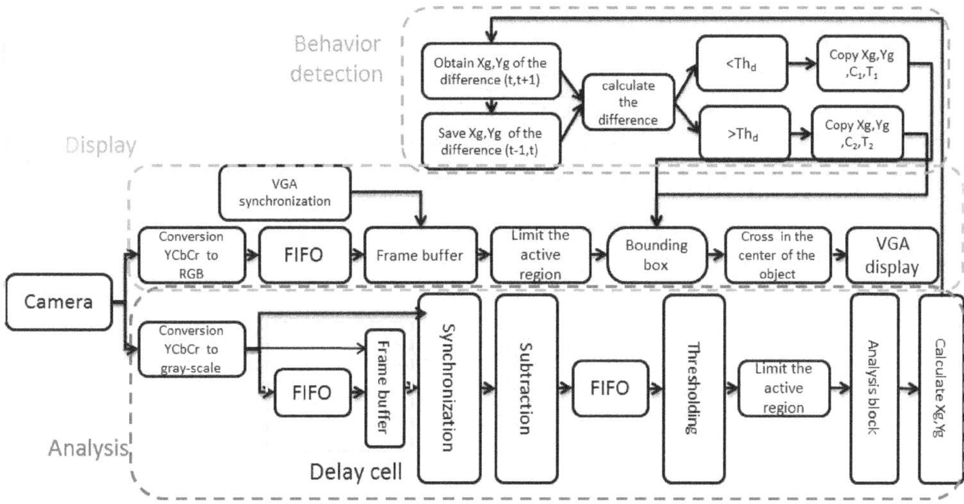

Figure 8. Hardware architecture for direction change detection.

Motion analysis: Here, we tried to collect all the above-mentioned behaviors using a single program in order to practically validate the system. To minimize resource consumption, we considered our problem as a finite state machine with several scenarios. The thresholds of detection for each case were used to define and manage these various scenarios. The differences between the values of minX, MaxX, minY, and MaxY at moments t and t-1 are denoted by ΔminX, ΔMaxX, Δminy, and ΔMaxY, respectively.

In the first state, all the values are initialized (State 0); they represent the initial state of each new inter-image difference. In the second state (State 1), if the absolute values of Δminx and ΔMaxX are higher than VTh, we assume that the velocity changes and we return to the initial state after copying the values of the block to the bounding box filter. In the opposite case, we go to the third state (State 2) and compare Δminx and ΔMaxX with the threshold Thd. According to the result of this comparison, we assume that a leftward or rightward movement has occurred. Then, we return to the initial state. Starting from this state, if the moving object accelerates, we return to the second state of velocity change. For posture change detection, the condition is related to the values of Δminy and ΔMaxY (State 3). We can detect this behavior from any state (e.g., a person runs and leans to collect something). The following figure summarizes these states and the possible scenarios.

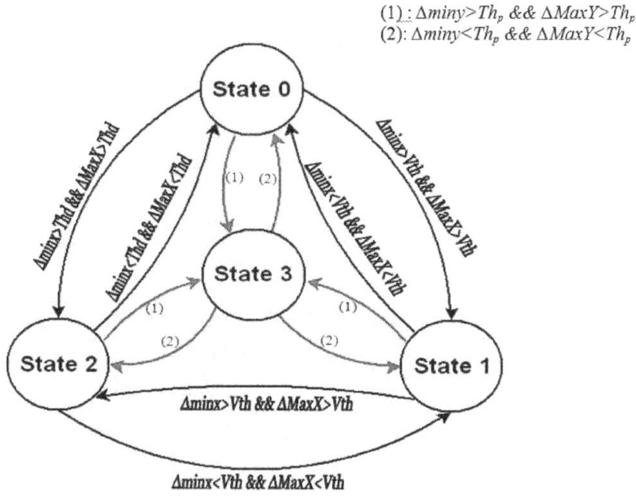

Figure 9. FSM of motion analysis.

5.4. Display Block

In this block, we call the macro PxsAnalyseAwaitUpdate, which allows us to pause the display until an update occurs in the analysis block. We obtain the values minX, MaxX, minY, and MaxY; if there is a motion, we copy these values to the bounding box filter to draw the rectangle. The values of the center of gravity, Xg,Yg, are also copied to the PxsCursor filter in order to draw a cross at the center of the moving object. We can add a warning message, e.g., "Warning: velocity change detection", by using the PxsConsole filter of the PixelStreams library. Finally, the results are displayed in the RGB format on a VGA display.

6. Experimental Results

An RC200E board with an embedded Virtex-II XC2V1000 FPGA was used for our implementation. The language used was Handel-C. The results for each behavior are summarized in Tables 2–5:

Table 2. Resource consumption and maximum frequency of implementation for velocity change detection.

Resources	Total	One object	Two objects
I/O	324	179 (55%)	179 (55%)
LUTs	10240	2286 (22%)	3484 (34%)
Slice Flip/Flops	10240	3046 (29%)	3738 (36%)
CLB slices	5120	3092 (60%)	4040 (78%)
Block RAM	40	9 (22%)	9 (22%)
Frequency	/	67.21 MHz 6.17 ms/image	56.85 MHz 7.29 ms/image

Table 3. Resource consumption and maximum frequency of implementation for direction change detection.

Resources	Total	One object	Two objects
I/O	324	179 (55%)	179 (55%)
LUTs	10240	2052(20%)	2991 (29%)
Slice Flip/Flops	10240	2908(28%)	3491 (34%)
CLB slices	5120	2895(56%)	3698 (72%)
Block RAM	40	9(22%)	9 (22%)
Frequency	/	66.69 MHz 6.22 ms/image	55.12 MHz 7.26 ms/image

Table 4. Resource consumption and maximum frequency of implementation for up/down motion detection.

Resources	Total	One object	Two objects
I/O	324	179 (55%)	179 (55%)
LUTs	10240	2100 (20%)	3233 (31%)
Slice Flip/Flops	10240	2927 (28%)	3595 (35%)
CLB slices	5120	2961 (57%)	3904 (76%)
Block RAM	40	9 (22%)	9 (22%)
Frequency	/	67.14 MHz 6.17 ms/image	58.97 MHz 7.03 ms/image

Table 5. Resource consumption and maximum frequency of implementation for motion analysis.

Resources	Total	Two objects
I/O	324	179 (55%)
LUTs	10240	3640 (35%)
Slice Flip/Flops	10240	4173 (40%)
CLB slices	5120	4537 (88%)
Block RAM	40	9 (22%)
Frequency	/	53.18 MHz 7.80 ms/image

These tables specify the resource consumption and maximal frequency of each implemented detection case for PAL video with a resolution of 720×576.

In all these implementations, the results show that the two main constraints, i.e., the resource limit of our FPGA and the real-time aspect (40 ms/image), are well respected. We note that the consumption of the CLB blocks increases in the case of detection of multiple objects; this is caused by the algorithm used to identify the number of objects in the scene. We also note that the algorithm for motion analysis that collects all the previous behaviors can be implemented on our FPGA in real time, but it consumes nearly all of the CLB resources (88%).

The following figures show the results of all these implementations. Each behavior is represented by a different color, and a warning message is added below the scenes.

Figure 10. Results of velocity change detection in the case of one object.

Figure 10 shows the results of velocity change detection in the case of one object. In Figure 10(a), as soon as the object decreases its speed, the rectangle disappears. In Figure 10(b), as soon as the object starts to run, a rectangle appears around it.

Figure 11. Results of velocity change detection in the case of two objects.

Figure 11 shows the results of velocity change detection in the case of two objects. As soon as the objects start running, a rectangle appears. We note that in the case of occlusion, the algorithm considers both objects as a single object. After the objects separate, two rectangles with different colors appear on them.

Figure 12 shows the results of direction change detection. Figure 12(a) shows direction detection for two directions: right to left movement, represented by the blue rectangle, and left to right movement, represented by the red rectangle. The figures also show warning messages below the images. Figure 12(b) shows the results of direction change detection in one direction for two objects.

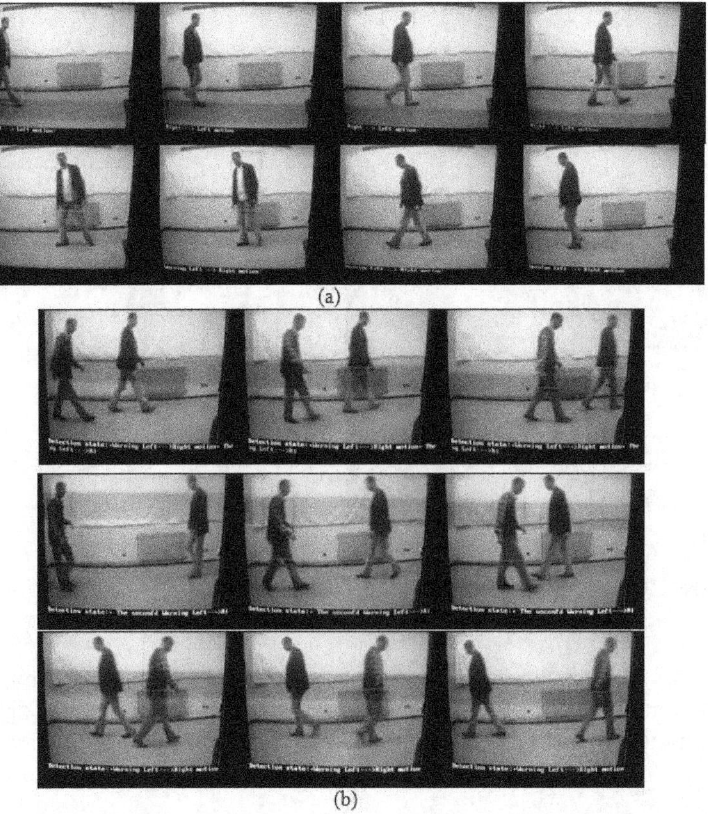

Figure 12. Direction change detection.

Figure 13 shows the results of posture change detection. When the object leans to pick up something, it will be detected. Up/down and down/up motion are represented in different colors. A warning message is added in each case.

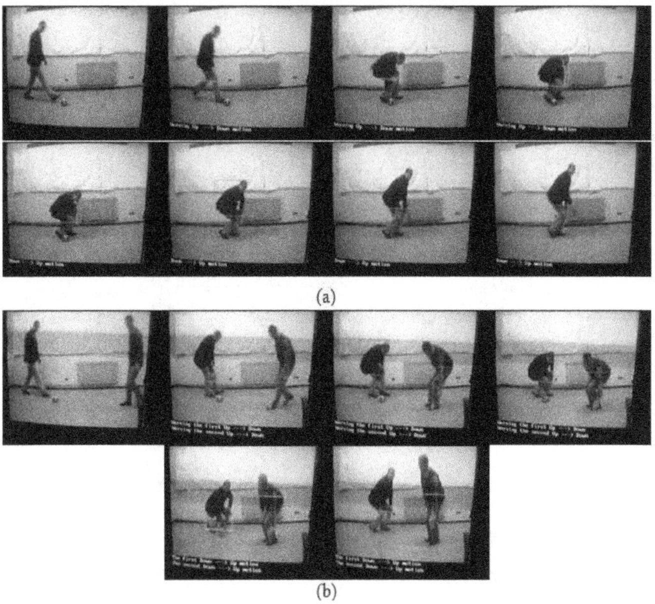

Figure 13. Posture change detection: a) for one object, b) for two objects.

Figure 14 shows the results of collecting all the behaviors using a single program. Motion to the right and left are represented by red and blue rectangles, respectively. Further, up/down and down/up motion are represented by turquoise and yellow rectangles, respectively. Finally, velocity change is represented by a black rectangle. In every case, a warning message is displayed.

Figure 14. Motion analysis.

Figure 15 shows our graphical user interface (GUI), which is divided into four sections. Three of these sections are used to detect just one simple behavior each, whereas the fourth section detects all the behaviors. Using this GUI, we can send the bit-file for configuring or erasing our FPGA, or directly changing the filter parameters without the need to use the IDE.

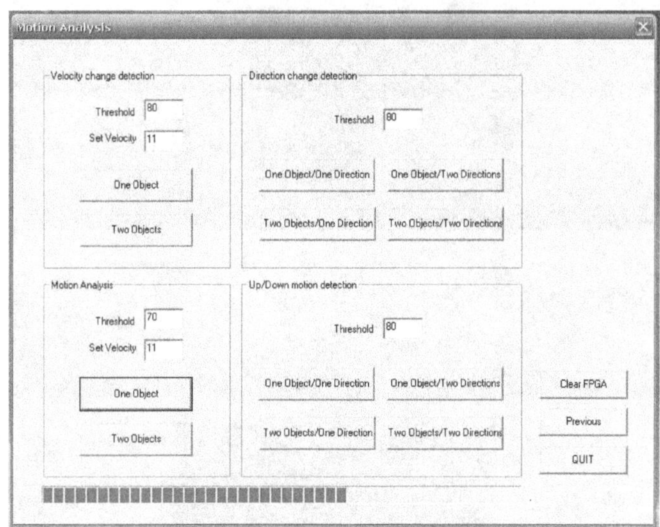

Figure 15. Graphical user interface.

7. Conclusions

We presented a mixed software-hardware approach that simplifies the use of the hardware part by enabling us to communicate with it using the graphical interface. In addition, it simplifies the choice of the algorithm to be implemented and modifies the parameters of this algorithm. We adopted the proposed approach for object detection and behavior recognition based on motion analysis and sudden movements. We exploited the hardware part, which offers the possibility of handling large amounts of data and performing calculations for image processing via parallel processing, guaranteed by the use of the PixelStreams library of Agility's DK Design Suite. Further, we tried to improve our architecture by collecting all the different behaviors using a single program. In addition, we added warning messages using the PxsConsole filter. Thus, we successfully implemented different algorithms that can recognize objects in motion and detect changes in velocity, direction, and posture in real time. The results showed that our approach achieves good recognition and detection of these behaviors, especially in indoor areas. However, in outdoor areas, the results are less promising owing to the simple motion detection algorithm used; this problem is aggravated by occlusion due to overlapping movements of different persons. Therefore, in the future, we will try to use multiple cameras (stereoscopic, Kinect) with improved motion detection and learning methods to detect behavior changes in crowded environments.

Acknowledgements

We would like to thank Pr. Larbes Cherif and Dr. Benkouider Fatiha for their insightful comments.

References

[1] L Wang, W Hu, and T Tan. Recent developments in human motion analysis. *Pattern Recogn.* 2003; 36(3): 585-601.

[2] W Hu, T Tan, L Wang, and S Maybank. A survey on visual surveillance of object motion and behaviors. *IEEE T Syst Man Cyb.* 2004; 34(3).

[3] T Ko. A survey on behavior analysis in video surveillance for homeland security applications. Washington DC: AIPR. 2008.

[4] M Piccardi. Background subtraction techniques: a review. *IEEE SMC*. 2004; 4: 3099-3104.

[5] GGS Menezes and AG Silva-Filho. Motion detection of vehicles based on FPGA. *SPL VI Southern*. 2010: 151-154.

[6] W Shuigen, C Zhen, L Ming, and Z Liang. An improved method of motion detection based on temporal difference. *ISA 2009*. 2009: 1-4.

[7] Widyawan, MI Zul, and LE Nugroho. Adaptive motion detection algorithm using frame differences and dynamic template matching method. *URAI 2012*. 2012: 236-239.

[8] I Ishii, T Taniguchi, K Yamamoto, and T Takaki. *1000 fps real-time optical flow detection system.* Proc. SPIE 7538. 2010. 75380M.

[9] J Diaz, E Ros, F Pelayo, EM Ortigosa, and S Mota. FPGA-based real-time optical-flow system. *IEEE T Circ Syst Vid.* 2006; 16(2): 274-279.

[10] M Paul, S Haque, and S Chakraborty. Human detection in surveillance videos and its applications - a review. EURASIP JASP. Springer International Publishing. 2013.

[11] AJ Lipton, H Fujiyoshi, and RS Patil. Moving target classification and tracking from real-time video. *WACV 98*. 1998: 8-14.

[12] M Ekinci and E Gedikli. Silhouette based human motion detection and analysis for real-time automated video surveillance. *Turk. J. Elec. Eng. & Comp. Sci*. 2005; 13: 199-229.

[13] Y Ran, I Weiss, Q Zheng, and LS Davis. Pedestrian detection via periodic motion analysis. *Int J Comput Vision.* 2007; 71(2): 143-160.

[14] K Ratnayake and A Amer. *An FPGA-based implementation of spatio-temporal object segmentation.* Proc. ICIP. 2006: 3265-3268.

[15] M Gorgon, P Pawlik, M Jablonski, and J Przybylo. FPGA-based road traffic videodetector. DSD 2007.

[16] F Cupillard , A Avanzi , F Bremond, and M Thonnat. Video understanding for metro surveillance. ICNSC 2004.

[17] I Haritaoglu, D Harwood, and LS Davis. W4: Real-time surveillance of people and their activities. *IEEE T Pattern Anal*. 2000; 22 (8): 809-830.

[18] F Bremond and G Medioni. Scenario recognition in airborne video imagery. *IUW 1998*. 1998: 211-216.

[19] M Edwards and B Fozard. Rapid prototyping of mixed hardware and software systems. DSD 2002. 2002: 118-125.

[20] "Agility DK User Manual", Mentor Graphics Agility (2012), http://www.mentor.com/products/fpga/handel-c/dk-design-suite/

[21] Virtex II 1.5v Field-Programmable Gate Arrays. Data sheet, Xilinx Corporation, 2001.

[22] DK5 Handel-C language reference manual. Agility 2007.

[23] "PixelStreams Manual", Mentor Graphics Agility (2012), http://www.mentor.com/products/fpga/handel-c/pixelstreams/

[24] K Sehairi, C Benbouchama, and F Chouireb. Real Time Implementation on FPGA of Moving Objects Detection and Classification. *International Journal of Circuits, Systems and Signal Processing*. 2015: 9; 160-167.

22

A New Hybrid Wavelet Neural Network and Interactive Honey Bee Matting Optimization based on Islanding Detection

Nasser Yousefi
Young Researchers and Elite club, Ardabil Branch, Islamic Azad University, Ardabil, Iran
email: nasseryousefi2472@gmail.com

Abstract

In this paper a passive Neuro-wavelet on the basis of islanding detection procedure for grid-connected inverter-based distributed generation has been developed. Moreover, the weight parameters of neural network are optimized by Interactive Honey Bee Matting optimization (IHBMO) to increase the efficiency of the capability of suggested procedure in tendered problem. Islanding is the situation where the distribution system including both distributed generator and loads is disconnected from the major grid as a consequence of lots of reasons such as electrical faults and their subsequent switching incidents, equipment failure, or pre-planned switching events like maintenance. The suggested method uses and combines wavelet analysis and artificial neural network together to detect islanding. It can be used in removing discriminative characteristics from the acquired voltage signals. In passive schemes have a large Non Detection Zone (NDZ), concern has been raised on active method because of its lowering power quality impact. The main focus of the proposed scheme is to decrease the NDZ to as close as possible and to retain the output power quality fixed. The simulations results, performed by MATLAB/Simulink, demonstrate that the mentioned procedure has a small non-detection zone. What is more, this method is capable of detecting islanding precisely within the least possible amount of standard time.

Keywords: *IHBMO, islanding detection, neuro-wavelet, non-detection zone, distributed generation*

1. Introduction

The traditional distributed generation systems are wind power generation, photovoltaic power generation, fuel cell power generation, and micro-turbine power generation. A Distributed Generator (DG) is usually related to each other with the existing utility grid at one point so that they are dividing the local load. Islanding detection is an important issue when a DG works in connection with the power grid [1]-[3].

Classical view of power system is characterized by a unidirectional power flow from centralized generation to consumers. Power system restructuring gave more emphasis to a modern view by introducing DG into distribution systems, bringing about a bi-directional power flow. In fact, the issue of implanting distributed generation into the distribution system was suggested supposing DGs will always be operating in a grid-connected mode. Anyway, few years later, it has been discerned that several operational issues are associated with distributed generation when operating in an island mode where [4].

Power quality in reverse consists of frequency deviation, voltage fluctuation, harmonics and reliability of the power system. Moreover, one of the specialized notions developed by DG interconnection is accidental islanding [5]-[6]. Islanding condition causes abnormal operation in the power system and negative effects on protection, operation, and handling of distribution systems. Therefore, it is immediately to effectively detect the islanding conditions and promptly disconnect DG from the network.

Within this condition, a supposed island is developed, resulting in unexpected results that may consist of an increased intricacy of orderly reconstruction (out of phase switching of re-closers leading to damage of the DG, nearby loads, and utility equipment), a reduced stability of system voltage and worst of all, an increased risk to related maintenance personnel. In other words, under the scenario of islanding, line crewmembers may err in evaluation the load-side of the line as passive where distributed generations are in fact feeding power to loads; therefore endangering the life of operators and in the intervening time providing explanations on the importance of a credible alert behavior to such events. Therefore, during the interruptions of utility power, the related DG must detect the loss of utility power and disconnect itself from the power grid quickly [7].

There are many suggested procedures for detection of an island [8]-[17]. The NDZ can be described and elaborated on as the domain (in terms of the power difference between the DG inverter and the load or load parameters) in which an islanding detection scheme under examination fails to detect this situation [10]. The second characteristic is related to the type of loads (potential loads inside island), which can be modeled as a parallel RLC circuit. This circuit is mainly utilized because it causes more difficulties for islanding detection techniques than others. [11]. Most islanding detection procedures suffer from large NDZs [18]-[19] and/or have a run-on time between half a second to two seconds [20], and therefore cannot be utilized for uninterruptible independent operation of an island. These procedures can be mainly categorized into far and local procedures. Local procedure can be divided into active and passive ones. It is clear that these procedures may have better reliability than local procedures; so they are expensive to performance and hence uneconomical. These schemes consist of power line signaling and transfer trip [21]-[22]. Local procedures count on the information and data at the DG site. Passive procedures depend on measuring specific system parameters and do not interfere with the DG operation. One of the simplest passive methods used in islanding detection is over/under voltage and frequency. In any event, if the load and the generation on the island are precisely compatible, the change in voltage and frequency may be very trivial and within the thresholds, therefore causing an undetected islanding situation. The other inactive procedures have been suggested on the basis of controlling rate of change of frequency (ROCOF), phase angle displacement, rate of change of generator power output, impedance monitoring, the THD producer and the wavelet transform function [23]. These offer superior sensitivity as their settings allow detection to take place within statutory limits, but their settings must be attentively chosen to prevent mal-operation during network faults.

In this paper, a passive Neuro-wavelet based islanding detection procedure decreasing the NDZ to as close as possible and to hold the output power quality fixed has been extended. The suggested strategy utilizes and combines wavelet analysis and artificial neural network to detect islanding. The suggested strategy is based on the transient voltage signals generated during the islanding event. Discrete wavelet transform is powerful of decomposing the signals into different frequency bands. It can be used in finding discriminative characteristics from the achieved voltage signals. Then, the characteristics are fed to a trained artificial neural network model which is well trained puissant of differentiating among islanding event and any other transient events such as switching or temporary fault. The trained compositor was tested with the use of novel voltage waveforms. The obtained results demonstrate that this approach can detect islanding events with high degree of accuracy.

The problem model is suited for usages of Interactive Honey Bee Matting optimization (IHBMO) to find the optimal weight parameters of neural network. The HBMO is a partly recently developed procedure that has been empirically demonstrated to perform well on many of these optimization problems [24]-[27]. This optimization method has been widely utilized to solve different problems particularly the locating problems in power system. This work shows the combining procedure by regarding the universal gravitation between queen and drone bees for the standard HBMO algorithm to recover condition called the IHBMO.

This paper is organized as follows. Section II presents the problem statement. The impact of the interface control on the NDZ of over/under voltage and frequency protection (OVP/UVP) and OFP/UFP is discussed in Section III. Section IV presents the suggested procedure structure. In section V the mentioned IHBMO is demonstrated. Section VI provides the simulation results to verify the effectiveness of the suggested procedure and the conclusion will be presented in the last section of this paper

2. Problem Statement

The islanding detection schemes suggested in literature can be categorized into two classes: far and local as demonstrated in Figure 1. Islanding detection schemes are effectively evaluated based on the NDZ. The NDZ corresponds to the range of active and reactive load-generation mismatches within the island in which the islanding detection method fails to identify the islanding state.

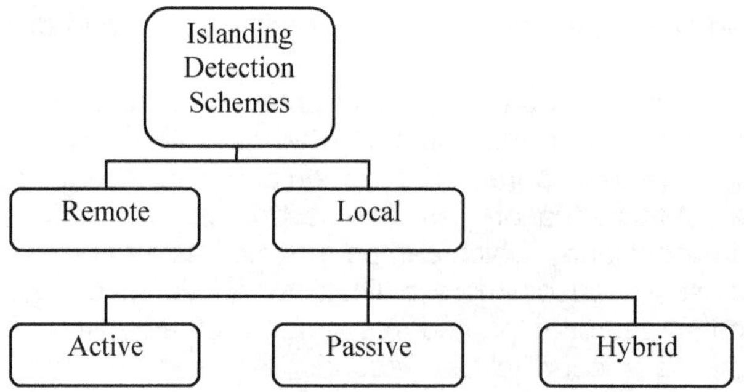

Figure 1. Classification of islanding detection schemes

This section explains a state-space mathematical model for the islanded system. It is supposed that the DG unit and the local load are stable three-phase subsystems within the island. The state space equations of the potential island in the standard state space form are.

$$\dot{X}(t) = AX(t) + Bu(t)$$
$$y(t) = CX(t)$$
$$u(t) = v_{td} \tag{1}$$

Where

$$A = \begin{bmatrix} -\frac{R_t}{L_t} & \omega_0 & 0 & -\frac{1}{L_t} \\ \omega_0 & -\frac{R_1}{L} & -2\omega_0 & (\frac{R_1 C\omega_0}{L} - \frac{\omega_0}{R}) \\ 0 & \omega_0 & -\frac{R_1}{L} & (\frac{1}{L} - \omega_0^2 C) \\ \frac{1}{C} & 0 & -\frac{1}{C} & -\frac{1}{RC} \end{bmatrix}$$

$$B^T = \begin{bmatrix} \frac{1}{L_t} & 0 & 0 & 0 \end{bmatrix}$$

$$C = \begin{bmatrix} 0 & 0 & 0 & 1 \end{bmatrix}$$

$$D = \begin{bmatrix} 0 \end{bmatrix}$$

$$X^T = [i_{td} \quad i_{tq} \quad i_{Ld} \quad v_d] \tag{2}$$

Figure 2 demonstrates the step response of system in the islanding mode. There sponse time constant of the island system is chosen as the analyzing time of ANFIS system output.

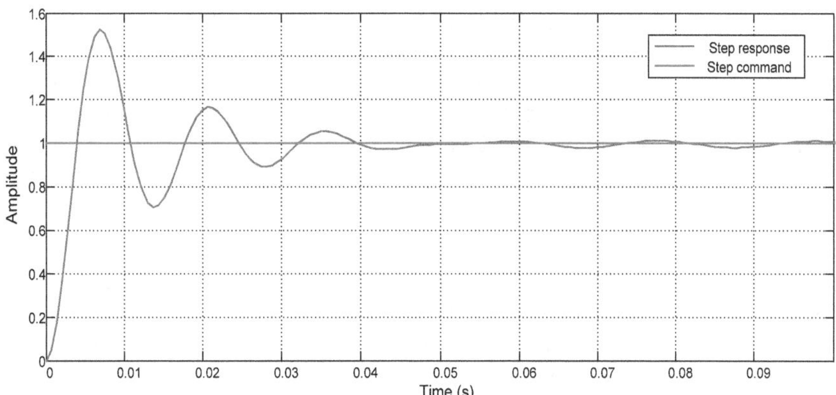

Figure 2. The step response of system in the islanding mode

The system under the research includes one 80kW inverter based DG connected to an RLC load having a quality factor of 1.8 and a grid. The system, controller, and load parameters are given in [6]. The performance of the DG under normal and islanded operating conditions was studied and simulated on MATLAB/SIMULINK. The inverter performs two major functions:

A) Controlling the active power output of the DG and, in some times, injecting a suitable amount of reactive power to decrease a power quality problem.
B) On the basis of the IEEE Standard 1547, the DG should be provided with an anti-islanding detection algorithm, which could be performed utilizing the inverter interface control.

The DG interface control is designed to supply constant current output as demonstrated in [6]. For this interface control, both I_d and I_q components of the DG output current are controlled to be equal to a predetermined value (I_{dref} and I_{qref}). The DG was operated at unity power factor by determining I_{qref} to zero. Specifically, parallel RLC loads with a high Q factor often present problems for island detection. The quality factor Q is explained by

$$Q_f = R\sqrt{\frac{C}{L}} \tag{3}$$

And is the ratio of the quantity of energy stored in the load's reactive components to the quantity of energy dissipated in the load's resistance. (For example, for $Q = 2$, there is twice as much energy stored in the L and C of the load as is being dissipated in R) Loads that are near resonance at ω_0 and have a high Q–factor are the ones that cause difficulty in islanding detection. Unhappily, the level of real or reactive power mismatch is not exceptionally determined by load parameters. Particularly, the reactive power consumption of the load is given by

$$Q_{Load} = V_{rms}^2 [(\omega L)^{-1} - (\omega C)] = \Delta Q \tag{4}$$

Equation (3) clearly demonstrating that there are infinitely many combinations of L and C that will produce the same ΔQ .

3. Non Detection Zone (NDZ)

A procedure causing a smaller NDZ will be able to detect the islanding more reliably. It can either be demonstrated in terms of power mismatch or in terms of the R, L, and C of the load. In [6],[28], an approaching representation of the NDZ for OVP/UVP was extracted. An exact and precise representation of the NDZ is presented in this part of paper. The paper studies the NDZ of an OVP/UVP and OFP/UVP islanding scheme when implemented for constant current controlled inverters. In order to determine the quantity of mismatch for which the OVP/UVP and OFP/UFP will fail to detect islanding, the quantity of active power mismatch in terms of load resistance can be demonstrated as follows:

$$\Delta P = 3V \times I - 3(V + \Delta V) \times I = -3V \times \Delta V \times .I \tag{5}$$

Which V and I represent the rated current and voltage, respectively. In distribution network, voltage values between 0.88 pu and 1.1 pu are in acceptable range for voltage relays. These voltage levels are equivalent to $\Delta V = -0.12$ and $\Delta V = 0.1$, respectively. The estimated inequality quantity for our test network (the inverter rated output power is 80kW), are 9.6 kW and -8kW, respectively. Frequency and voltage of an RLC load has the active and reactive power as follows:

$$P_L = \frac{V_L^2}{R_L}$$

$$Q_L = V_L^2\left(\frac{1}{\omega L} - \omega C\right) \tag{6}$$

Where V, ω, P and Q are the load voltage, frequency, active power and reactive power, respectively. In usual operating situations, a common coupling point voltage is caused by the power grid, and distributed generation system has no control overvoltage and until it is connected to the network the voltage is unchanged at nominal value of 1 pu. Once the island is took place, distribution system cannot control the voltage and the amount of active power imbalance determines the voltage deviation from the nominal values. Since the output power of the inverter is in unity power factor, before islanding reactive power of load is supplied just by network and after islanding the quantity of reactive power imbalance is equal to the consumed load before islanding, hence we have:

$$\Delta Q = 3\frac{V^2}{\omega_n L}\left(1 - \omega^2 LC\right) = 3\frac{V^2}{\omega_n L}\left(1 - \frac{\omega_n^2}{\omega_r^2}\right) \tag{7}$$

Where ω_n and ω_r are system frequency and resonance frequency of load, respectively. Reactive power imbalance results in the resonance frequency, then the frequency alters after the islanding happening equals the difference between network frequency and load resonance frequency.

$$\omega_r = \omega_n \pm \Delta\omega \;, \omega_r = \frac{1}{\sqrt{LC}} \tag{8}$$

Thus, the reactive power imbalance needed for certain changes in frequency can be achieved by,

$$\Delta Q = 3\frac{V^2}{\omega_n L}\left(1 - \frac{f_n^2}{(f_n \pm \Delta f)^2}\right) \tag{9}$$

In distribution network of Iran, the acceptable frequency range is between 49.7 and 50.3Hz which are equal to $\Delta f = -0.3$ and $\Delta f = 0.3$ Hz. In this paper test system, as shown in Figure 3 the quantities of reactive power imbalances are 3.05 and -5.16 kVA, respectively.

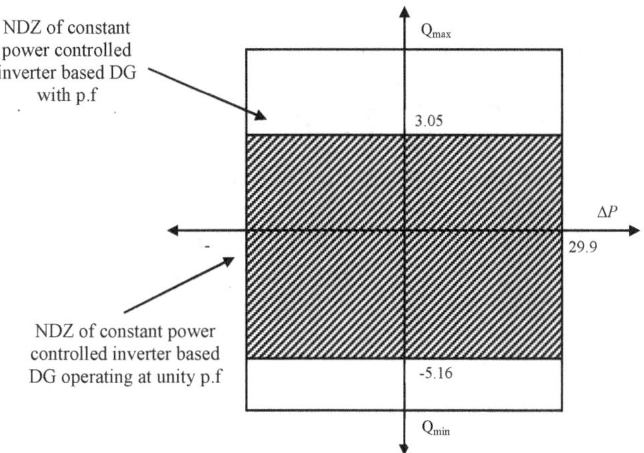

Figure 3. NDZ for the constant current interface controls for distributed generation

4. Interactive Honey Bee Mating optimization

In this part the standard HBMO closely presented. For better illustration see ref [24]. At the beginning of the flight, the queen is initialized with some energy content and returns to her nest when her energy is within some threshold from zero or when her sperm theca is full. In extending the algorithm, the usefulness of workers is limited to brood care, and therefore, each worker may be demonstrated as a heuristic which acts to improve and/or take care of a set of broods. A drone mates with a queen probabilistically using an annealing function as:

$$prob(Q,D)=e^{\frac{-\Delta(f)}{S(t)}} \tag{10}$$

Where Prob (Q, D) is the probability of adding the sperm of drone D to the spermatheca of queen Q (that is, the probability of a successful mating); Δ(f) is the absolute difference between the fitness of D (i.e., f (D)) and the fitness of Q (i.e., f (Q)); and S(t) is the speed of the queen at time t. After each transition in space, the queen's speed, S(t), and energy, E(t), decay utilizing the following equations:

$$S(t+1)=\alpha_{HBMO}\times S(t)$$

$$E(t+1)=E(t)-\gamma_{HBMO} \tag{11}$$

Where αHBMO(t) is speed deceasing factor and γHBMO is the quantity of energy deceasing after each transition (α, γϵ [0,1]).

to overcome the drawbacks of classic HBMO, the Interactive Honey Bee Mating Optimization (IHBMO) algorithm is suggested based on the structure of original HBMO algorithm. The gravitational force between two particles (F12) with the mass of the first and second particles (m1 and m2), and the distance between them (r12) can be shown as:

$$F_{12}=G\frac{m_1m_2}{r_{21}^2}\hat{r}_{21} \tag{12}$$

Where, G is gravitational constant. ˆr21 denotes the unit vector in:

$$\hat{r}_{21}=\frac{r_2-r_1}{|r_2-r_1|} \tag{13}$$

In the optimization method with IHBMO algorithm, the mass m1 is replaced by the symbol, F(parenti), which is the fitness value of the queen bee that picked by applying the mating wheel selection. The mass, m2, is replaced by the fitness value of the randomly chosen drone bee and is denoted by the parameter, F(parentk). The universal gravitation is formed in the vector format. Thus, the quantities of it on different dimensions can be considered separately [27]. Therefore, r21 is calculated by taking the difference between the objects only on the concerned dimension currently and the universal gravitation on each dimension is calculated separately. In other words, the intensity of the gravitation on different dimensions is calculated one by one. Therefore, the gravitation on the jth dimensions between parenti and parentk can be shown in the below equation. Finally, the below equation can be modified in iteration interval t:

$$F_{ik_j} = G\frac{F(parent_i)\times F(parent_k)}{(parent_{kj} - parent_{ij})^2}\cdot\frac{parent_{kj} - parent_{ij}}{|parent_{kj} - parent_{ij}|} \tag{14}$$

$$child_{ij}(t+1) = parent_{ij}(t) + F_{ik_j}\cdot[parent_{ij}(t) - parent_{kj}(t)]$$

Developing the consideration of the universal gravitation between the drone bees, which is picked by the queen bee, and more than one drone bees is obtainable by adding different Fik.[arent–parenti]. Therefore, the gravitation Fik plays the role of a weight factor controlling the specific weight of [arent–parenti]. The normalization process is taken in order to ensure that Fik ~U (0, 1). Through the normalization of Fik the gravitational constant (G) can be deleted.

5. Neuro-Wavelet Based Islanding Detection

Recently, wavelet transform has been successfully implemented in solving many power system problems such as fault detection, power quality event localization and load disaggregation. The capability of wavelet in handling non-stationary signals while preserving both time and frequency information makes it a suitable candidate for islanding detection problem.

5.1. Artificial Neural Networks

ANN includes of simple processing units, called neurons, operating in parallel to solve specific problems. Figure 4 demonstrates a simple neuron with input vector P of dimension $R \times 1$. The input P is multiplied by a weight W of dimension $1 \times R$. Therefore, a bias b is added to the product WP. f is a transfer function (called also the activation function) that takes the argument n and produces the net output a.

The idea of ANN is that these parameters (w and p) are adjusted so that the network demonstrates some desired behavior. Therefore, the network can be trained to do a specific job by adjusting the weight or bias parameters [29].

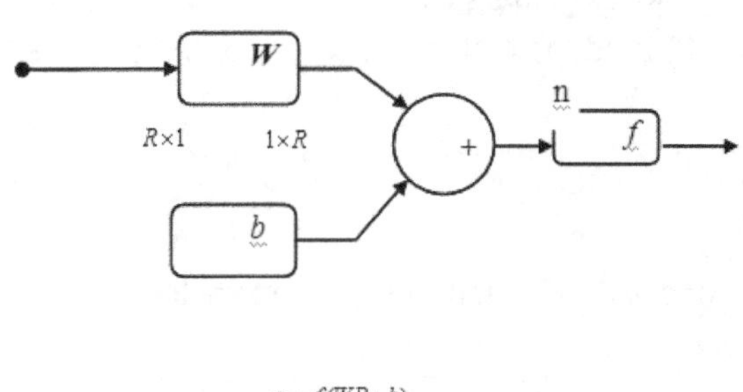

$a = f(WP+b)$

Figure 4. Neuron

A number of neurons can be combined together to form a layer of neurons. A one layer of R input elements and S neuron is represented in Figure 5.

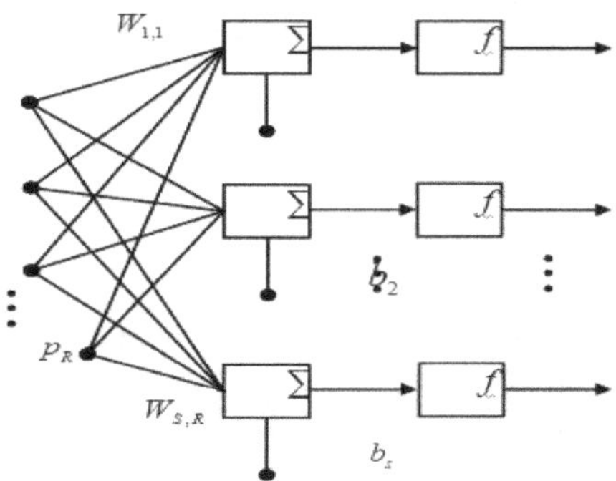

Figure 5. One layer network of R input elements and S neurons

A network can have lots of layers of neurons to form multiple layers of neurons. In Figure 6 three layers of R input elements and S neurons is demonstrated by using the IHBMO. The layer that produces the network output is called an output layer. All other layers are called hidden layers. The three layer network is described in Figure 6 has one output layer and two hidden layers.

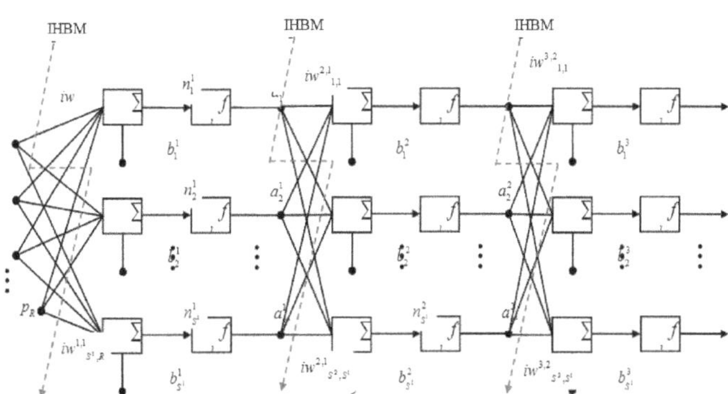

Figure 6. Three layer network of R input elements and S neurons with IHBMO

5.2. Training the ANN

A training algorithm is described as a method of updating the weights and biases of a network so the network will be able to perform the specific design task. The training algorithm is divided into two main categorizes: supervised learning, and unsupervised learning. ANN is categorized under supervised learning. In the training stage, the training data set and the corresponding targets are added into the model. Once the network weights and biases are initialized, the network is ready for training. The weights and biases are then adjusted in order to minimize the Mean Square Error (MSE). This can be achieved using the gradient of the MSE.

Wavelet Transform

WTs can be mainly divided into two classifications: Continuous Wavelet Transform (CWT) and Discrete Wavelet Transform (DWT). The continuous wavelet transform W (a,b) of signal f(x) with respect to a wavelet ф(x) is given by [30]-[31]:

$$W_{(a,b)} = \frac{1}{\sqrt{a}} \int_{-\infty}^{+\infty} f_{(x)} \phi_{\left(\frac{x-b}{a}\right)} dx \tag{15}$$

Where scale parameter a controls the spread of the wavelet and translation factor b determines its central position. ф(x) is named mother wavelet. A W(a,b) coefficient, shows how well the original signal f(x) and the scaled/translated mother wavelet match. Thus, the set of all wavelet coefficients W (a,b), associated to a specific signal, is the wavelet showing of the signal with respect to the mother wavelet. It is known as the Discrete Wavelet Transform (DWT):

$$W_{(m,n)} = 2^{-(m/2)} \sum_{t=0}^{T-1} f_{(t)} \phi_{\left(\frac{t-n.2^m}{2^m}\right)} \tag{16}$$

Where T is the length of the signal f(t). The scaling and translation parameters are functions of the integer variables m and n (a=2m, and b=n.2m); t is the discrete time index. Daubechies wavelet group is one of the most suitable wavelet families in analyzing power system transients as investigated in [32]. In the present study, the db1 wavelet represented in Figure 7 (with two filter coefficients) has been used as the mother wavelet for analyzing the transients associated with islanding. db1 is a short wavelet and therefore it can efficiently detect transients.

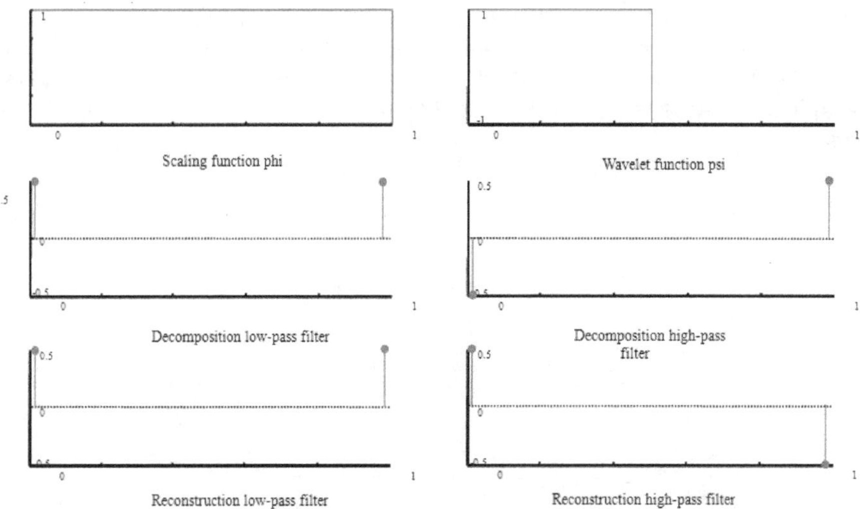

Figure 7. db1 mother wavelet

WT will be carried out on the achieved voltage signals to extract the characrtistics. The main target of characrtistic extraction is to identify particular signature of the voltage waveforms that can detect islanding and differentiate between islanding and any other transient condition. A transient signal can be completely separated into smoothed signals and detailed signals for L wavelet levels. In wavelets applications, Daubechies wavelet family is one of the most suitable wavelet families in analyzing power wavelet has been utilized as the mother wavelet for extracting the energy content of the detail coefficient of voltage waveforms. db1 is a short wavelet and thus it can efficiently detect transients. The frequency band information of the wavelet analysis is demonstrated in Table 1. The sampling frequency is 10 kHz.

Table 1. Frequency band information for the different levels of wavelet analysis

Wavelet Level	Frequency Band (Hz)
1-D1	2500-5000
2-D2	1250-2500
3-D3	625-1250
4-D4	312.5-625
5-D5	156.25-312.5
6-D6	78.125-156.25
7-D7	39.0625-78.0625
A7	19.5-39.025

The energy content in the details of each decomposition level for all voltage signals was calculated using the detail coefficients in the corresponding level. The energy content in the other decomposition levels can be estimated as:

$$\|ED_{1a}\| = \left[\sum_k d_k^2 \right]^{1/2} \tag{17}$$

Where ED1a is the energy content of D1 for voltage signal of phase (a) and dk is the kth coefficient in the first decomposition level. After collecting the characteristics for the simulated different cases, the characteristics will be fed to a trained ANN in order to identify whether the event took place is islanding or non-islanding event. The flow chart of the suggested procedure is demonstrated in Figure 8.

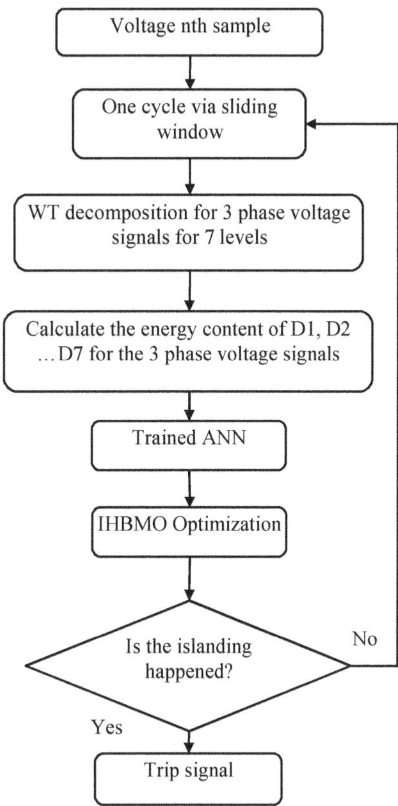

Figure 8. The flow chart of proposed technique for islanding detection

6. Simulation Results

In this part, the test system in Figure 9 has been simulated by MATLAB/Simulink. The system, DG, and load parameters are listed in [33]. The Effective voltage waveform of the common coupling point for islanding mode is shown in Figure 10.

Figure 11 to 13 show an example of the wavelet details (D3, D4 and D6) of 2 cycles voltage waveform acquired from phase (a) at DG when four different events have took place. These events have taken place at time = 1.5 sec. In the suggested figures the islanding when power match (a), islanding when power mismatch (b), switching of load (c), and three-phase-to-ground fault (d), are shown. It is noticed from the graphs that during the fault the details dropped to zero since the DG is equipped with UOF/UOV protective relay which isolated the fault. Also, the Performance matrix of DG is demonstrated in Table 2.

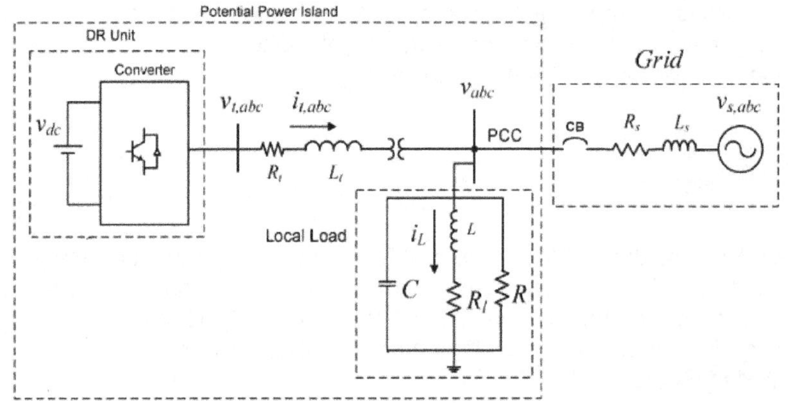

Figure 9. Schematic diagram of a grid-interfaced DG unit

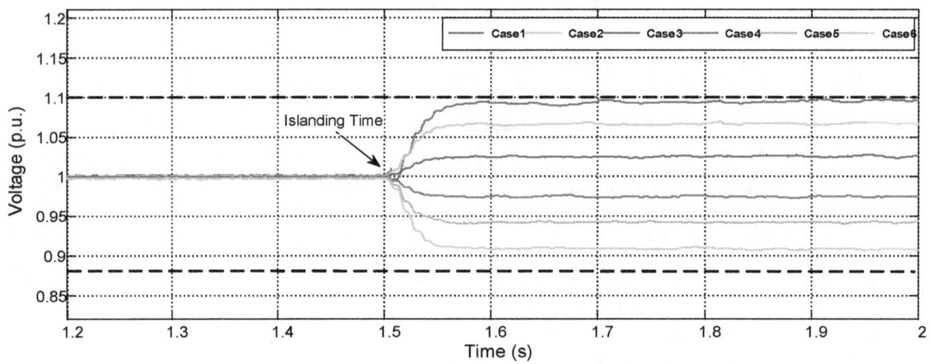

Figure 10. Effective voltage waveform of the common coupling point for islanding mode

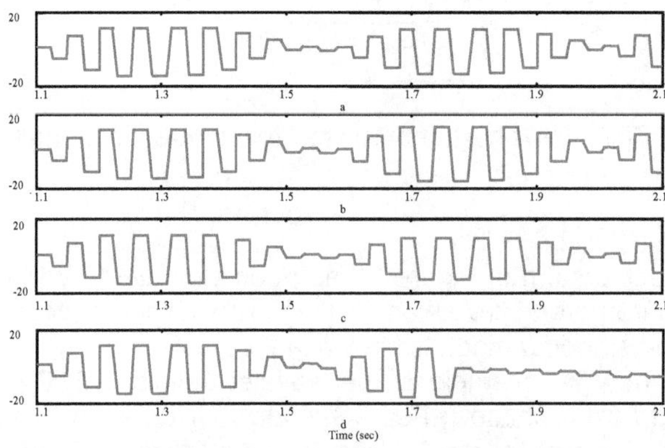

Figure 11. Wavelet detail D-3 of voltage waveform in case of (a) islanding when power match, (b) islanding when power mismatch, (c) switching of load, and (d) three-phase-to-ground fault.

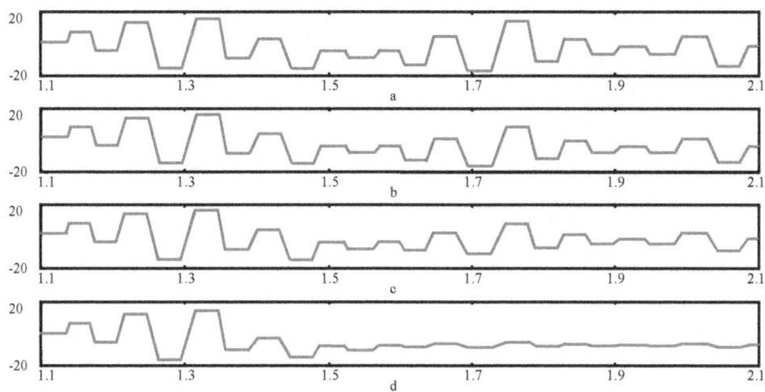

Figure 12. Wavelet detail D-4 of voltage waveform in case of (a) islanding when power match, (b) islanding when power mismatch, (c) switching of load, and (d) three-phase-to-ground fault.

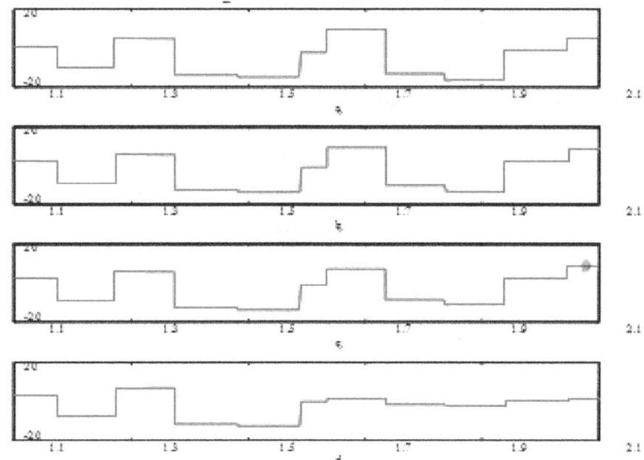

Figure 13. Wavelet detail D-6 of voltage waveform in case of (a) islanding when power match, (b) islanding when power mismatch, (c) switching of load, and (d) three-phase-to-ground fault.

Table 2. The Performance matrix of DG

		Islanding	Non-islanding			
		Matched	Switching of load	Switching of Capacitor bank	Faults	Normal operation
Islanding	Matched	20	0	0	0	0
Non-islanding	Switching of load	1	9	0	0	0
	Switching of Capacitor bank	0	0	10	0	0
	Faults	0	0	0	10	0
	Normal operation	0	0	0	0	10
Accuracy		98.70%				

6.1. Voltage deviation (Voltage Swell, Voltage Sag) in islanding detection method

Islanding detection procedure should be kept safe from voltage changes. By adding an adaptive system with a delay system, one can keep safe the detection procedure from the voltage changes. Therefore, the inverter output current is monitored continuously and once the difference between this current and threated current is observed the comparator detects automatically these abnormal conditions. These abnormal conditions can be sign of either: a) an electrical island, b) voltage deviation. Table.3 shows Voltage relay responses when an abnormal condition is observed in the standard distribution networkIEEEStd.1547. Simple voltage relays should detect the voltage changes at the appropriate time and then eliminates the distributed generation from the grid.

Table 3. Voltage relay responses

Voltage range (% of base voltage)	Clearing time (s)
V<50	0.16
50%<V<88	2
110<V<120	1
V>120%	0.16

Performance of the suggested procedure is analyzed in this mode for the various conditions which are given in the Table 4.Three phase voltage waveform of the common coupling point for each case are represented in Figure 14. Also rate of change of active power for all cases studied are shown in Figure 15. Immediately following these change at the time t = 1 sec, rate of change of active power for each condition are increased or decreased. Finally, in Fig.16 the output of detection method for all studied cases are shown. It is obvious from Figure 16 that after all studied cases value of the proposed algorithm has not changed and the output of proposed method is remained 0. Thus, the suggested procedure does not send a trip signal to distributed generation and works in a reliable mode.

Table 4. The various condition for voltage deviation test

	Case1	Case2	Case3	Case4	Case5	Case6
	One-Phase Fault	Two-Phase Fault	Two-Phase Fault	Three-Phase Fault	Three-Phase Fault	Three-Phase Fault
Fault Resistance (Ω)	1	0.1	1	0.05	0.1	1
Ground Resistance (Ω)	0.1	0.1	0.1	0.1	0.1	0.1

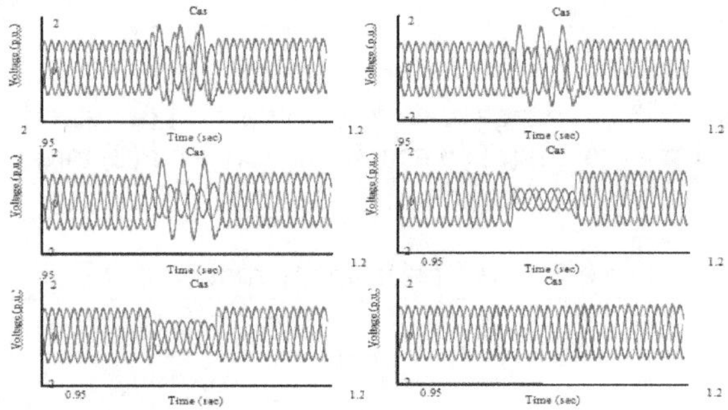

Figure 14. Three phase voltage waveform of the common coupling point for voltage deviation mode

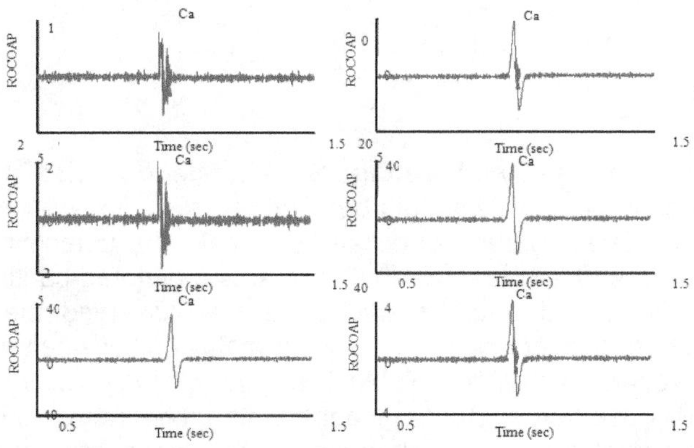

Figure 15. rate of change of active power for voltage deviation mode

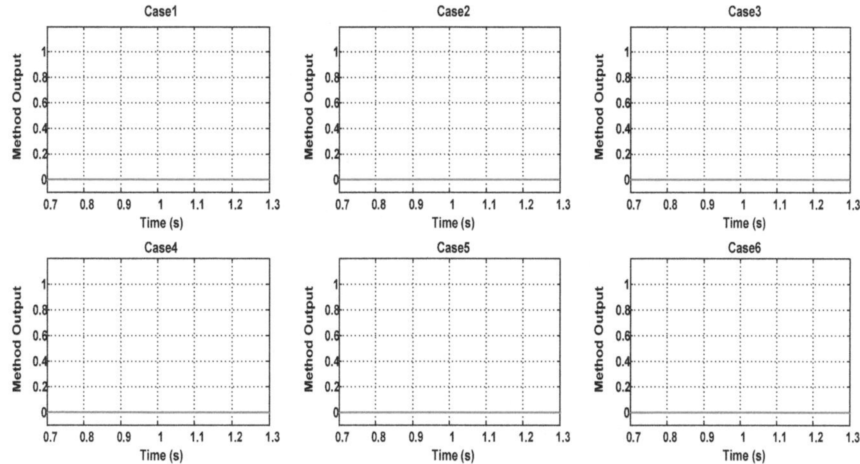

Figure 16. the output of detection method for voltage deviation mod

7. Conclusion

A new technique for islanding detection of distributed generation is suggested on the bases of Hybrid Wavelet Neural Network with Interactive Honey Bee Matting Optimization (IHBMO). Many schemes have been suggested to detect islanding such as passive, active and communication based procedures. Passive techniques work well when there is power imbalance between the power generated from the DG and the power consumed form the load. From the other point of view, active techniques affect the power quality and do not perform well in the presence of multiple DGs. The energy content of wavelet details are then calculated and fed to a trained ANN which is capable to differentiating between islanding and non-islanding events. Following the increased number and enlarged size of distributed generating units installed in a modern power system, the protection against islanding has become extremely challenging nowadays. The main emphasis of the proposed scheme is to decrease the NDZ to as close as possible and this procedure can also overcome the problem of setting the detection thresholds inherent in the existing techniques. Simulation results verified that the islanding cases were successfully detected in case of adding untrained identical DGs since the voltage transients generated is identical.

References

[1] EF. El-Saadany, HH. Zeineldin, AH. Al-Badi. Distributed generation: benefits and challenges. International Conference on Communication, Computer & Power. 2007: 115-119.

[2] A. Algarni. Operational and planning aspects of distribution systems in deregulated electricity markets. *Ph.D. dissertation.* University of Waterloo, Waterloo, ON, Canada. 2009.

[3] HH. Zein El Din, EF. El-Saadany, MMA. Salama. Islanding Detection of Inverter Based Distributed Generation. in IEE Proceedings in Generation, Transmission and Distribution. 2006; 153(6): 644–652.

[4] MA. Eltawil, Z. Zhao. Grid-connected photovoltaic power systems: Technical and potential problems—A review. *Renewable and Sustainable Energy Reviews.* 2009; 14: 112–129.

[5] United States of America. *Congress of the U.S., Congressional Budget Office.* Prospects for distributed electricity generation. September. 2003.

[6] Zeineldin HH, El-Saadany Ehab F, Salama MMA. Impact of DG interface control on islanding detection and non-detection zones. *IEEE Trans Power Delivery.* 2006; 21(3): 1515–23.

[7] IEEE Standard for Interconnecting Distributed Resources into Electric Power Systems. IEEE Standard 1547TM. June 2003.

[8] Hernandez-Gonzalez G, Iravani R. Current injection for active islanding detection of electronically-interfaced distributed resources. *IEEE Trans Power Delivery.* 2006; 21(3): 1698–705.

[9] Karimi H, Yazdani A, Iravani R. Negative-sequence current injection for fast islanding detection of a distributed resource unit. *IEEE Trans Power Electr.* 2008; 23(1): 298–307.

[10] Ropp ME, Begovic M, Rohatgi A. Analysis and performance assessment of the active frequency drift method of islanding prevention. *IEEE Trans Energy Convers.* 1999; 14(3): 810–6.

[11] Hung GK, Chang CC, Chen CL. Automatic phase-shift method for islanding detection of grid-connected photovoltaic inverters. *IEEE Trans Energy Convers.* 2003; 18(1): 169–73.
[12] Song kim. Islanding Detection Technique using Grid-Harmonic Parameters in the Photovoltaic System. *Energy Procedia.* 2012; 14: 137-141.
[13] Wen-Jung Chiang, Hurng-LiahngJou, Jinn-Chang Wu, Kuen-Der Wu, Ya-Tsung Feng. Active islanding detection method for the grid-connected photovoltaic generation system. *Electric Power Systems Research.* 2010; 80(4): 372-379.
[14] F. Hashemi, A. Kazemi, S. Soleymani. A New Algorithm to Detection of Anti-Islanding Based on dqo Transform. *Energy Procedia.* 2012; 14: 81-86.
[15] Mohammad A. Choudhry, Hasham Khan. Power loss reduction in radial distribution system with multiple distributed energy resources through efficient islanding detection. *Energy.* 2010; 35(12): 4843-4861.
[16] Jang S, Kim K. An islanding detection method for distributed generation algorithm using voltage unbalance and total harmonic distortion of current. *IEEE Trans Power Delivery.* 2004; 19(2): 745–52.
[17] Lopes LAC, Zhang Y. Islanding detection assessment of multi-inverter systems with active frequency drifting methods. *IEEE Trans Power Delivery.* 2008; 23(1): 480–6.
[18] ME. Ropp, M. Begovic, A. Rohatgi, GA. Kern, RH. Bonn, S. Gonzalez. Determining the relative effectiveness of islanding methods using phase criteria and nondetection zones. *IEEE Trans. Energy Conv.* 2000; 15(3): 290–296.
[19] HH. Zeineldin, T. Abdel-Galil, EF. El-Saadany, MMA. Salama. Islanding detection of grid connected distributed generators using TLS-esprit. *Electric Power Syst. Res., Elsevier.* 2007; 77(2): 155–162.
[20] SJ. Huang, FS. Pai. A new approach to islanding detection of dispersed generators with self-commutated static power converters. *IEEE Trans. Power Del.* 2000; 15(2): 500–507.
[21] ST. Mak. A new method of generating TWACS type outbound signals for communication on power distribution networks. *IEEE Trans. Power App. Syst.* 1984; PAS-103(8): 2134–2140.
[22] W. Xu, G. Zhang, C. Li, W. Wang, G. Wang, J. Kliber. A power line signaling based technique for anti-islanding protection of distributed generators—Part I: scheme and analysis a companion paper submitted for review.
[23] Cheng-Tao Hsieh, Jeu-Min Lin, Shyh-Jier Huang. Enhancement of islanding-detection of distributed generation systems via wavelet transform-based approaches. *International Journal of Electrical Power & Energy Systems.* 2008; 30(10): 575-580.
[24] Javidan J, Ghasemi A. Environmental/Economic Power Dispatch Using Multi-objective Honey Bee Mating Optimization. *International Review of Electrical Engineering (I.R.E.E.).* 2012; 7(1): 3667-75.
[25] Niknam T, Mojarrad HD, Meymand HZ. A new honey bee mating optimization algorithm for non-smooth economic dispatch. *Energy.* 2011; 36: 896-908.
[26] Niknam T. An efficient hybrid evolutionary algorithm based on PSO and HBMO algorithms for multi-objective Distribution Feeder Reconfiguration. *Energy Conversion and Management.* 2009; 50(8): 2074-82.
[27] Ghasemi A, Shayanfar HA, Mohammad SN, Abedinia O. Optimal Placement and Tuning of Robust Multimachine PSS via HBMO. in Proceedings of the International Conference on Artificial Intelligence, Las Vegas, U.S.A. 2011: 201-8.
[28] Z. Ye, A. Kolwalkar, Y. Zhang, P. Du, R. Walling. Evaluation of anti-islanding schemes based on nondetection zone concept. *IEEE Trans. Power Electron.* 2004; 19(5): 1171–1176.
[29] C. Parameswariah, M. Cox. Frequency Characteristics of Wavelets. IEEE Power Engineering Review. 2002; 22.
[30] Rocha Reis AJ, Alves da Silva AP. Feature Extraction via Multiresolution Analysis for Short-Term Load Forecasting. *IEEE Trans. on Power Systems.* 2005; 20: 189-198.
[31] Conejo AJ, Plazas MA, Espinola R, Molina AB. Day-ahead electricity price forecasting using the wavelet transform and ARIMA models. *IEEE Trans. on Power Systems.* 2005; 20: 1035-1042.
[32] S. Chen. Feature selection for identification and classification of power quality disturbances. Power Engineering Society General Meeting. 2005; 3: 2301-2306.
[33] Farid Hashemi, Noradin Ghadimi, Behrooz Sobhani. Islanding detection for inverter-based DG coupled with using an adaptive neuro-fuzzy inference system. *Electrical Power and Energy Systems.* 2013; 45(1): 443–455.

Permissions

All chapters in this book were first published in IJEEI, by IAES Indonesian Section; hereby published with permission under the Creative Commons Attribution License or equivalent. Every chapter published in this book has been scrutinized by our experts. Their significance has been extensively debated. The topics covered herein carry significant findings which will fuel the growth of the discipline. They may even be implemented as practical applications or may be referred to as a beginning point for another development.

The contributors of this book come from diverse backgrounds, making this book a truly international effort. This book will bring forth new frontiers with its revolutionizing research information and detailed analysis of the nascent developments around the world.

We would like to thank all the contributing authors for lending their expertise to make the book truly unique. They have played a crucial role in the development of this book. Without their invaluable contributions this book wouldn't have been possible. They have made vital efforts to compile up to date information on the varied aspects of this subject to make this book a valuable addition to the collection of many professionals and students.

This book was conceptualized with the vision of imparting up-to-date information and advanced data in this field. To ensure the same, a matchless editorial board was set up. Every individual on the board went through rigorous rounds of assessment to prove their worth. After which they invested a large part of their time researching and compiling the most relevant data for our readers.

The editorial board has been involved in producing this book since its inception. They have spent rigorous hours researching and exploring the diverse topics which have resulted in the successful publishing of this book. They have passed on their knowledge of decades through this book. To expedite this challenging task, the publisher supported the team at every step. A small team of assistant editors was also appointed to further simplify the editing procedure and attain best results for the readers.

Apart from the editorial board, the designing team has also invested a significant amount of their time in understanding the subject and creating the most relevant covers. They scrutinized every image to scout for the most suitable representation of the subject and create an appropriate cover for the book.

The publishing team has been an ardent support to the editorial, designing and production team. Their endless efforts to recruit the best for this project, has resulted in the accomplishment of this book. They are a veteran in the field of academics and their pool of knowledge is as vast as their experience in printing. Their expertise and guidance has proved useful at every step. Their uncompromising quality standards have made this book an exceptional effort. Their encouragement from time to time has been an inspiration for everyone.

The publisher and the editorial board hope that this book will prove to be a valuable piece of knowledge for researchers, students, practitioners and scholars across the globe.

List of Contributors

M. Satyanarayana and P. Satish Kumar
Department of Electrical Engineering, University college of Engineering, Osmania University (Autonomous), Hyderabad, Telangana, India

DVN. Ananth
Department of EEE, Viswanadha Institute of Technology and Management, Visakhapatnam, 531173, India

GV. Nagesh Kumar
Department of EEE, GITAM University, Visakhapatnam, 530045, Andhra Pradesh, India

Raju Basak and Arabinda Das
Electrical Engineering Department, Jadavpur University, Kolkata -700032, West Bengal, India

Amarnath Sanyal
Calcutta Institute of Engineering and Management, Kolkata -700040, West Bengal, India

Saifullah Khalid
Department of Electrical Engineering, IET Lucknow, India

Mounir Taha Hamood
Department of Electrical Engineering Collage of Engineering, Tikrit University, Tikrit Iraq

K Karthik, T Gunasekhar and M Anusha
Dept of Computer Science and Engineering, K L University, Vijayawada, Vaddeswaram 522502, India

D Meenu
Dept of Computer Science and Engineering, Ideal College of Engineering, Kolkata, India

M Ferni Ukrit
Department of CSE, Sathyabama University, Chennai, Tamilnadu, India

GR Suresh
Department of ECE, Easwari Engineering College, Chennai, Tamilnadu, India

Asmita Dhokrat, Sunil Khillare and C Namrata Mahender
Dept. of Computer Science and IT, Dr. Babasaheb Ambedkar Marathwada University Aurangabad, Maharashtra-431001, India

Vartika Keshri and Prity Gupta
Department of Electrical and Electronics Engineering, Oriental college of Technology, Bhopal

Naresh Vurukonda and B Thirumala Rao
Department of CSE, KLUniversity, Vijayawada, A.P, India

Debajyoti Bose and Amarnath Bose
Department of Electrical, Power and Energy, University of Petroleum and Energy Studies, Dehradun 248007, India

Ravindrakumar Yadav
Babaria Institute of Technology, Varnama, Vadodara, Gujarat, India

Ashok Jhala
RKDF College of Engineering, Bhopal, India

Madhumita Kathuria and Sapna Gambhir
Department of Computer Engineering, YMCA University of Science and Technology Faridabad, India

Jide Julius Popoola and Damian E. Okhueleigbe
Department of Electrical and Electronics Engineering, Federal University of Technology, Akure, Nigeria

Isiaka A. Alimi
Department of Electrical and Electronics Engineering, Federal University of Technology, Akure, Nigeria
Instituto de Telecomunicações, DETI, Universidade de Aveiro, Aveiro, Portugal

SA Gawish, SM Sharaf and MS El-Harony
Department of Electrical Power and Machines Engineering, Faculty of Engineering of Helwan, University of Helwan

Abayomi Isiaka O. Yussuff and Nana Hamzat
Lagos State University, Lagos, Nigeria

Nor Hisham Haji Khamis
Universiti Teknologi Malaysia, Skudai, Malaysia

Ika Puspita and A. M. Hatta
Photonics Engineering Laboratory, Department of Engineering Physics, Institut Teknologi Sepuluh Nopember

Sachin Sharma, Alok Pandey and Nitin Kumar Saxena
MIT Moradabad, UP India

Osman Abd Allah, Mohammed Abdalla and Alaa Awad Allah
Sudan Academy of Science, Khartoum, Sudan

Suliman Abdalla
Sudan Atomic Energy Commission, Khartoum, Sudan

Amin Babiker
Faculty of Engineering, Alneelain University, Khartoum, Sudan

K. Lenin
Department of EEE, Prasad V.Potluri Siddhartha Institute of Technology, Kanuru, Vijayawada, Andhra Pradesh-520007

Kamal Sehairi and Chouireb Fatima
Laboratoire LTSS, Departement of Electrical Engineering, Université Amar Telidji Laghouat, Route de Ghardaia, Laghouat 03000, Algeria

Cherrad Benbouchama and Kobzili El Houari
Laboratoire LMR, École Militaire Polytechnique, Bordj El Bahri, Algeria

Nasser Yousefi
Young Researchers and Elite club, Ardabil Branch, Islamic Azad University, Ardabil, Iran

Index

CPSIA information can be obtained
at www.ICGtesting.com
Printed in the USA
LVHW061944261219
641636LV00003BA/9/P